KU-022-426

Contents

Chapter 1 The manufacturing industry

Summary

When you have read this chapter you should be able to:

- Understand what is meant by the term 'manufacturing'
- Appreciate the role of our manufacturing industries as a means of national wealth creation
- Appreciate the range of products produced by UK manufacturers and their importance in terms of national wealth creation and employment opportunities
- Understand the meaning of 'scale of production'
- Understand the meaning of such terms as 'gross domestic product' (GDP) and 'gross national product' (GNP)
- Understand the meaning and importance of the 'balance of trade'
- Understand the factors affecting the location of manufacturing industry
- Recognize the key stages of production
- Recognize the scales of production
- Recognize a typical production system.

1.1 What is manufacturing?

Manufacturing can be defined as *the conversion of raw materials into useful articles by means of physical labour or the use of power driven machinery.*

In prehistoric times, cave dwellers found that if a piece of flint stone was struck with another stone the flint could be turned into a sharp spear head or arrow head. They took the flint as their raw material and, by means of physical labour, converted it into something useful. They were manufacturing. By converting a useless piece of flint into a useful tool, they made it easier to defend themselves and to hunt for food in order to provide the tribe with its next meal.

Manufacturing has come a long way since then. Today, although there are still many skilled craftsmen who make things by hand, the manufacturing industries that are of significant economic importance to the nation are usually large concerns employing hundreds and in some cases thousands of people. They need large buildings, complex machinery and major investments of capital. They increasingly market and sell their products not just in one country but around the world.

Manufacturing as we understand it today began in what is called the *industrial revolution.* The UK was the first nation to undergo the change from a largely agricultural economy to full-scale industrialization. The foundations for this industrial revolution were laid in the seventeenth century by the expansion of trade, the accumulation of wealth and social and political change. This was followed in the eighteenth century by a period of great discoveries and inventions in the fields of materials, transportation (better roads, canals and railways), power sources (steam), and in the mechanization of production.

The momentum for change originated with the mechanization of the textile industries (spinning and weaving) which, in turn, resulted in these cottage industries being replaced by the factory system. This increased the need for machines and the power units that would drive these machines more reliably than water wheels that came to a halt during periods of drought. The mechanical engineering industry was born out of the demand for machines and motive power sources of ever increasing size and sophistication. Fortunately, at this time, a series of inventors and engineers in the UK developed the early steam pumping engines used in mines into 'rotative' engines suitable for driving machines. The transportation of raw materials and finished goods was transformed by the development of the railways and by steam powered boats capable of crossing the oceans of the world. At last, the manufacturing industries and the transportation of their goods were free from the vagaries of the weather and the limitations of horse drawn vehicles.

Mechanization and reliable power sources enabled UK manufacturers to turn out more goods than the UK alone needed. These surplus goods were sold abroad, pointing the way to the worldwide trade in all sorts of manufactured goods that we take for granted today. The ships taking the products manufactured in the UK to other countries brought raw materials from those countries back to the UK to sustain further manufacture.

Manufacturing is a commercial activity and exists for two purposes:

- *To create wealth.* There is no point in investing your money or other people's money in a manufacturing plant unless the return on the investment is substantially better than the interest that your money could earn in a savings account.
- *To satisfy a demand.* There is no point in manufacturing a product for which there is no market. Even if there is a market (*a demand*), there is no point in manufacturing a product to satisfy that demand unless that product can be sold at a profit.

We have already defined manufacturing as the conversion of raw materials into useful articles by means of physical labour or the use of power driven machinery. When this conversion takes place there is *value added* to the raw materials. This increase in value represents the creation of wealth both for the owners of the manufacturing enterprise itself and the nation as a whole. Manufacturing makes a vital contribution to the local and national economies of the UK. Prior to the Second World War the UK was still a major manufacturing country exporting a wide range of products all over the world. This predominant position in manufacturing and global trade has diminished rapidly since the Second World War with the rapid development of industrial economies in Asia and the Far East where labour is plentiful and cheap, also manufacturing and the marketing of manufactured goods has become global. Throughout the world, national companies have merged into multinational companies with the financial strength and resources to buy their raw materials in bulk and establish factories wherever suitable labour is most plentiful and cheapest.

An example of industrial globalization is shown in Fig. 1.1. In this example a leather footwear product is designed, developed and marketed in Western Europe. The hides for the leather are sourced in Eastern Europe. The hides are tanned, cut and sewn to make the uppers in Indonesia and then sent to Thailand to be assembled to the soles and finished. The finished shoes are sent in bulk to a distribution warehouse in the UK before being

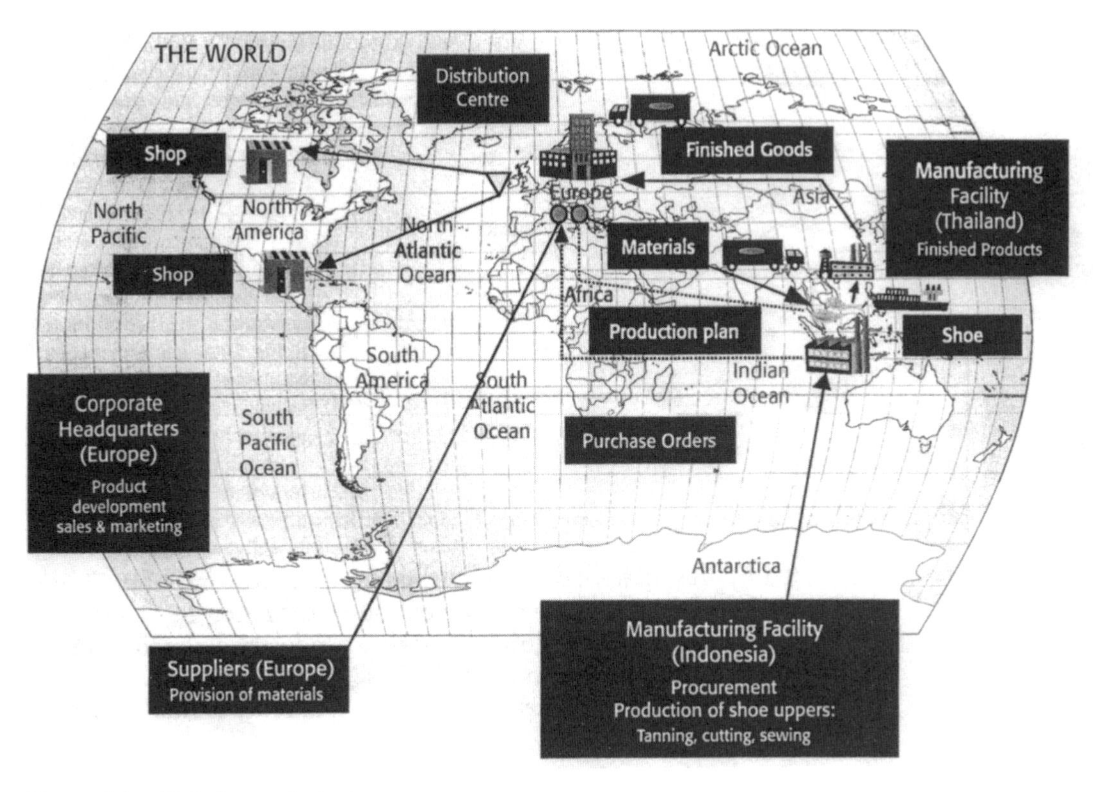

Figure 1.1 Global manufacture.

individually packed and despatched to retail shops throughout the world. Globalization has largely come about through the ease of modern communications via e-mail and the Internet. Also, the use of computer numerically controlled (CNC) machines enables high quality manufacturing to be achieved with relatively low cost, low skilled labour.

Test your knowledge 1.1

1. Describe briefly what is meant by 'manufacture'.
2. State the main purposes of manufacture.
3. Explain what is meant by 'global manufacture'.

1.2 The main UK manufacturing sectors

Despite its now diminished role, manufacturing industry is still vital to the British economy. The income generated by manufacturing and the hundreds of thousands of jobs which the manufacturing industry creates is still essential if the UK is to sustain an acceptable level of employment and pay its way in the world. Increasingly, however, we are becoming dependent on the income

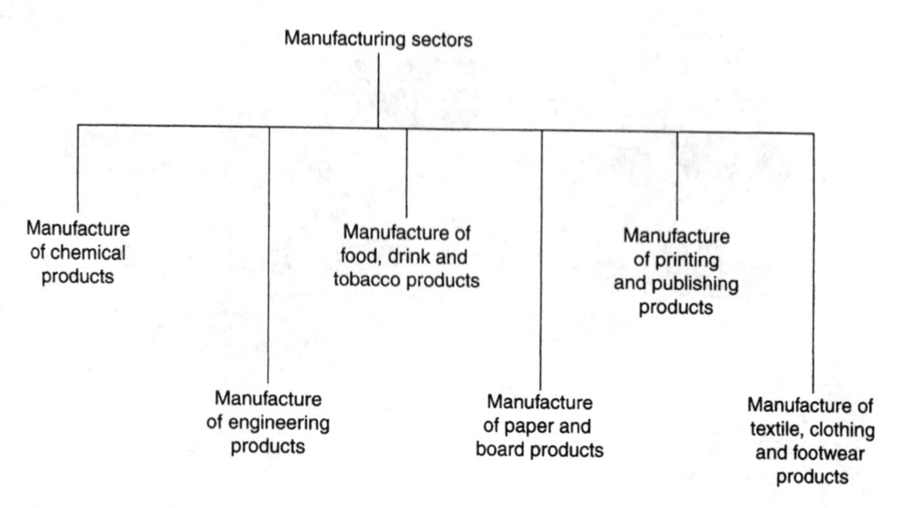

Figure 1.2 Main UK manufacturing sectors.

generated by the *service industries* of banking, insurance and in-tourism to make good the shortfall in the balance of trade (see Section 1.8.3). Manufacturing takes many different forms. To make the industry easier to study, we will break it down into six main *sectors*. These sectors are shown in Fig. 1.2.

Let's now look at each of these sectors in turn in order to find out what it is that they actually produce.

1.3 The chemical industry

Chemical manufacturing covers a range of different activities, starting with the manufacturing of the basic chemicals (feed stocks) from raw materials. These feed stocks are then used by other parts (subsectors) of the industry in more complicated processes. Chemicals are defined as being either *organic* or *inorganic*.

Organic chemicals consist of complex *carbon compounds*. Carbon forms more compounds than any other chemical. It forms the complex long chain molecules found in animal and plant life (hence the name *organic* chemicals). Carbon also forms the long chain molecules found in the byproducts of the fossil fuel industries and are widely used in the manufacture of plastic materials. We ourselves are built from carbon chains. Some organic chemical substances upon which we depend are shown in Fig. 1.3. As you will see, many organic chemicals are substances that occur naturally. Other organic chemicals are synthetic and are manufactured from a variety of raw materials.

Inorganic chemicals are all the substances that are *not* based upon complex carbon chain compounds. For example chemicals made from mineral deposits such as limestone, rock salt, sulphur, etc., as shown in Fig. 1.4.

Both organic and inorganic chemicals are manufactured and used in very large quantities, and this is reflected in the size of the plants in which they are made. In addition, many chemicals, if they are not properly handled, can be extremely dangerous. Computerized and automated processes are widely used in chemical engineering to avoid the possibility of catastrophic accidents

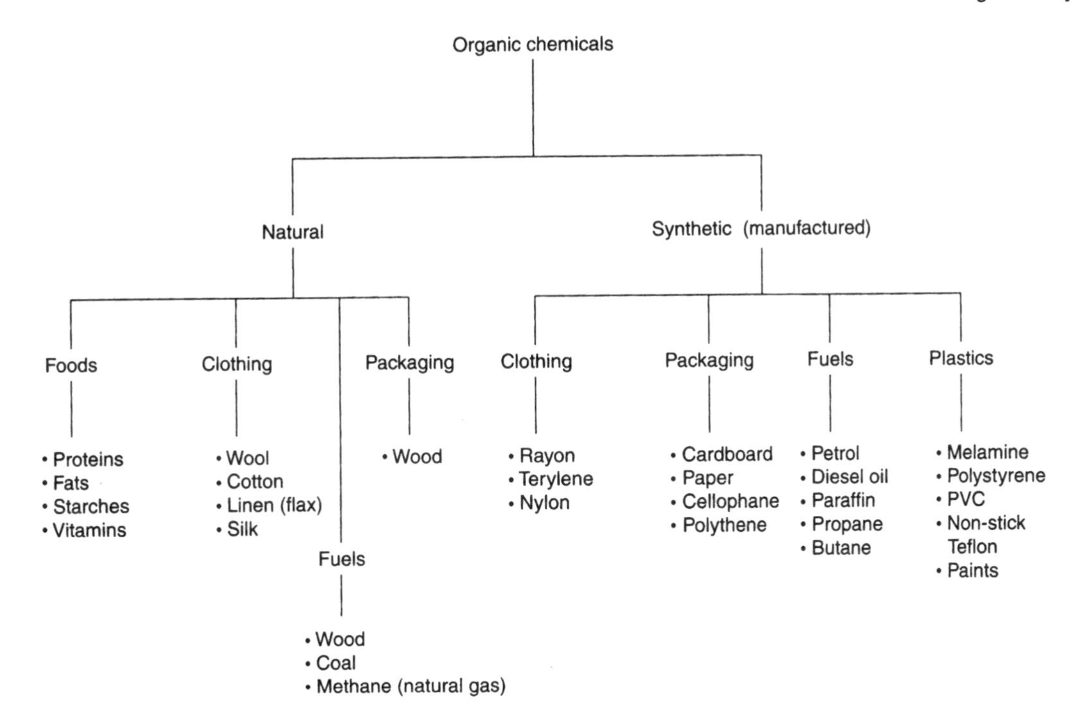

Figure 1.3 Some organic chemicals.

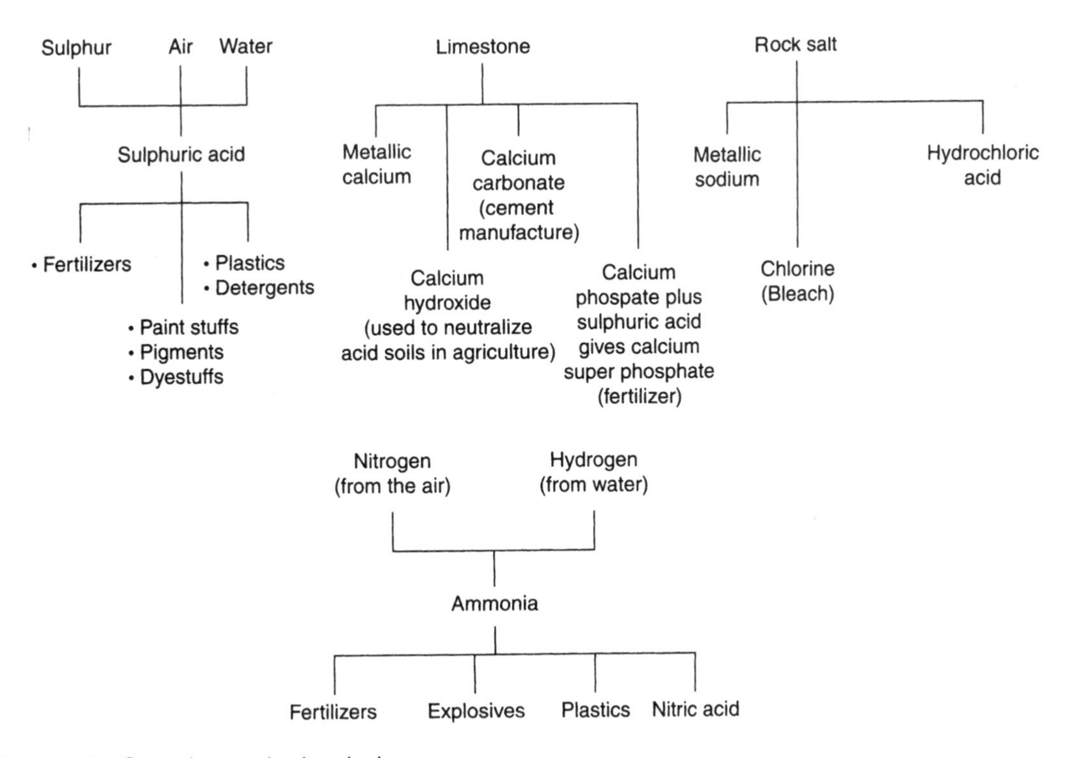

Figure 1.4 Some inorganic chemicals.

caused by human error. The size of the factories, the complexity of the manufacturing processes and the need for stringent safety precautions means that the whole business is very expensive, with the result that there are only a few, very large companies that dominate the sector.

In the UK, the chemical industry makes products such as plastics, fertilizers, paints, adhesives, explosives, synthetic fibres, and provide gases such as oxygen, acetylene, nitrogen, argon, and carbon dioxide for industry and gases such as oxygen, and anaesthetics for hospitals. Some chemical manufacturing companies specialize in domestic and health care products. These range from medicines and disinfectants to cleaning materials and polishes. The companies may be UK owned or they may be the UK branches of international (global) companies.

Another factor which ensures that the chemicals manufacturing sector is dominated by a handful of large operators is that they have to put a lot of money every year into research and development (R&D). Sometimes that research will come to nothing and the money invested will be lost. Sometimes the research will be successful and will be developed into a product that will generate a lot of income. Only very large global companies can afford to take this sort of gamble. The profits from the successful research not only has to pay for the money lost on unsuccessful research, it also funds the next generation of research.

Many manufactured chemical products play an important part in everybody's daily life. You can use a detergent to wash up after cooking a meal or you can use it to clean your car. Dyes give colour to your clothes, your wallpaper and your carpets. Fertilizers ensure that there is enough food in the shops at a price you can afford.

Without paints all the woodwork in your home would soon decay and rot, while your car or your motorbike would quickly become rusty. Pharmaceuticals – medicines – can keep us healthy and, in extreme cases, save our lives.

Polymers (so-called plastic materials) are used for clothes, ropes, electrical insulators, sports gear and a host of other products upon which modern society depends. Without plastics there would be no CDs, no computers, not even the old-fashioned records made from vinyl pressings, and there would be no tape from which audio and video recording media are made. Without printing ink you wouldn't be able to read the sports pages in your newspaper because there wouldn't be a newspaper.

You may think that, as a law abiding citizen, explosives play no part in your life. However, explosives are widely used in mining and quarrying for the raw materials which are processed to produce things which we use everyday – for example, bauxite is mined to be processed into aluminium and there are few households which do not own at least one aluminium saucepan.

As you can see, the chemicals industry plays a part in nearly every aspect of our daily lives, from the clothes we wear to the games we play and the cars we drive. Life would be very different without them.

Test your knowledge 1.2

1. Use the Internet to find the names of the main chemical manufacturing companies operating in the UK.
2. List the main product groups for each of the companies listed in question 1.

3. The BOC Group plc manufactures and supplies industrial gases. Find out which of these gases are extracted from atmospheric air and briefly explain what they may be used for.
4. Name the products associated with the following companies:
   (a) Glaxo Smith Kline plc;
   (b) Fisons plc;
   (c) ZENECA Group plc;
   (d) Procter and Gamble plc.

1.4 The engineering industry

The engineering industry is very diverse and we need to divide it up into a number of subsectors in order to simplify our study of it. The main subsectors are shown in Fig. 1.5. All these subsectors use metal and plastic raw materials. Some use naturally occurring substances such as wood for pattern-making for the metal casting industry which also uses sand for the moulds into which the molten metal may be poured. Natural rubber obtained from the sap of the *Hevea brasiliensis* tree is used for making road vehicle and aircraft tyres. Glass is used in the manufacture of optical instruments and as fibres for reinforcing plastics. The main raw materials for making glass are silica sand, soda ash (crude sodium carbonate) and quick lime (obtained from limestone). Various metal oxides may be added depending upon the properties required. These various substances are mixed together and melted in furnaces to make glass.

Before we start to look at the various manufacturing companies, we need to consider the sourcing of the raw materials. We have already seen that plastic materials are produced by the chemical industry for manipulation by spinning, weaving, moulding and fabrication into finished products by other industries. As stated above, glass is also produced by chemical reactions between various substances in the glass-making 'tank' furnace. Metals are also extracted from mineral ores by chemical reactions. Other physical/chemical processes are used to refine the raw material into useful materials for the engineering industry. The demarcation between chemical engineering and materials engineering quickly becomes blurred.

At the time of the industrial revolution, the UK was largely self-sufficient in metals. In the 'Black Country' – so-called because the green pastures of

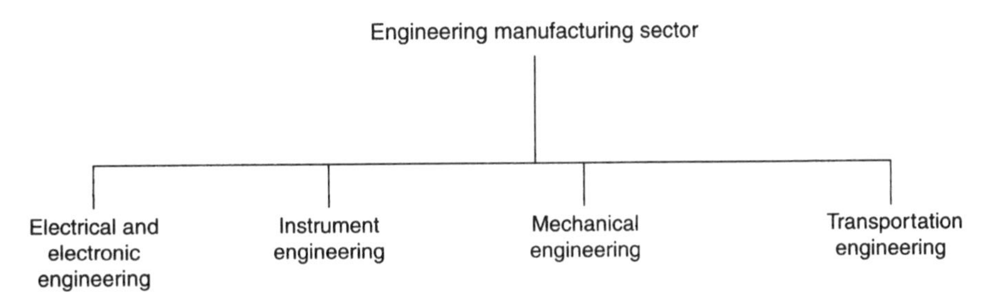

Figure 1.5 Engineering manufacturing sectors.

the area soon became blackened by the pollution from the iron foundries in the early days of the industrial revolution – which is that area surrounding the town of Dudley in the Midlands – seams of iron ore and coal lay side by side, whilst limestone lay under Dudley Castle. These are the ingredients needed for the extraction of iron ore by the chemical reactions occurring in blast furnaces. Tin, copper and zinc ores were to be found in Cornwall. Unfortunately the demand was great and these resources were small and soon worked out. Iron ore deposits were found elsewhere in the UK, but since the end of the Second World War these have also been worked out.

Metals are still extracted and refined in the UK but almost entirely from imported metallic ores. Some metals such as aluminium and copper have to be refined by electrolytic processes using vast amounts of electrical energy. These are mostly refined where hydro-electric plants can generate electricity cheaply. Except to a limited extent in Scotland, there are no hydro-electric schemes in the UK. The continental countries of Europe are more fortunate in this respect.

Iron and steel is still manufactured in the UK but on a much-reduced scale. Because the basic raw material (iron ore) is imported, most iron and steel works are sited on the coasts of Yorkshire, Humberside, and South Wales near to large ports. Increasingly it is cheaper to import bulk steel and castings from countries in the Far East and South America where labour and other costs are far lower. The manufacture of non-ferrous metals and precious metals (gold, silver and platinum) is, surprisingly, mostly centred in the Midlands despite the need to import the raw materials. In addition, non-ferrous metals are also manufactured on Tyneside and in London, Avonmouth and South Wales.

Now let's look at those subsectors of the engineering industry that consume the materials we have just considered. They use these materials to manufacture a vast range of products for an equally vast range of markets at home and overseas. These subsectors will be considered in greater detail starting with electrical and electronic engineering.

1.4.1 Electrical and electronic engineering

First, let's consider the difference between electrical and electronic engineering.

- *Electrical engineering* is concerned with heavy current devices used in electricity generation and transmission, and in consuming devices such as motors, heaters and lighting.
- *Electronic engineering* is concerned with relatively small current devices such as computers, telecommunications, navigational aids, and control systems. It is also concerned with the manufacture of components such as transistors, diodes, capacitors, resistors and integrated circuits for use in such devices.

The electrical engineering subsector not only makes equipment such as generators, transformers, and the cables without which no electricity would reach homes and workplaces, it also produces a huge range of products which are complete in themselves and which you are likely to find in the home. Some of these are set out in Fig. 1.6. Domestic appliances are also referred to as 'white goods' and as 'consumer durables'. First because they often come in white enamelled cabinets and second because they are used by domestic consumers and the products themselves are durable.

The electronic subsector manufactures products that are used in industry, commerce, service industries and the home. There is often a considerable level of overlap. Desktop computers are equally likely to be found in the home as in offices, banks, hospitals, and in manufacturing and design companies. The software will differ, but the basic machines are the same. Dedicated computers may be used for machine tool control, in the engine management systems of road vehicles, and navigational aids in aircraft and ships. Navigational aids linked with satellites are now small enough for use in cars. Similar hand-held devices are now available for hikers. Electronic control systems incorporating dedicated computers are used to control domestic devices such as washing machines, toasters and refrigerators. Programmable logic controllers operate the traffic lights at crossroads and can be reprogrammed easily as traffic patterns change. Figure 1.7 shows a broad breakdown of the products manufactured in this subsector.

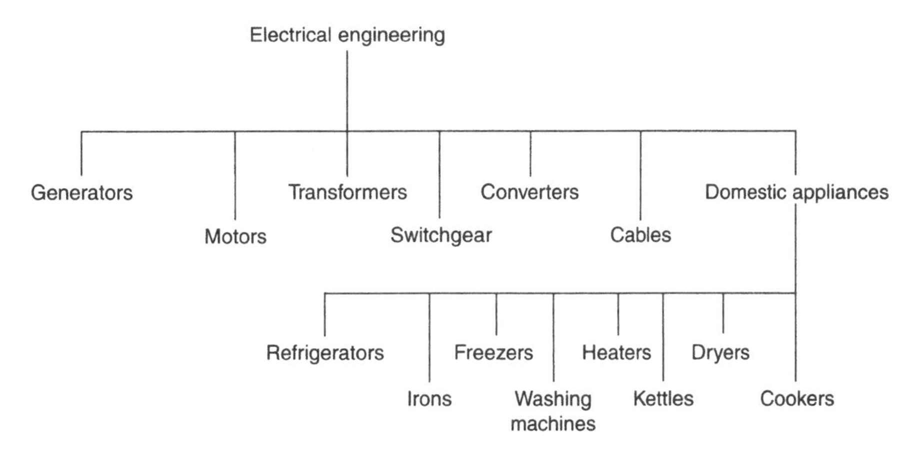

Figure 1.6 Products of the electrical engineering sector.

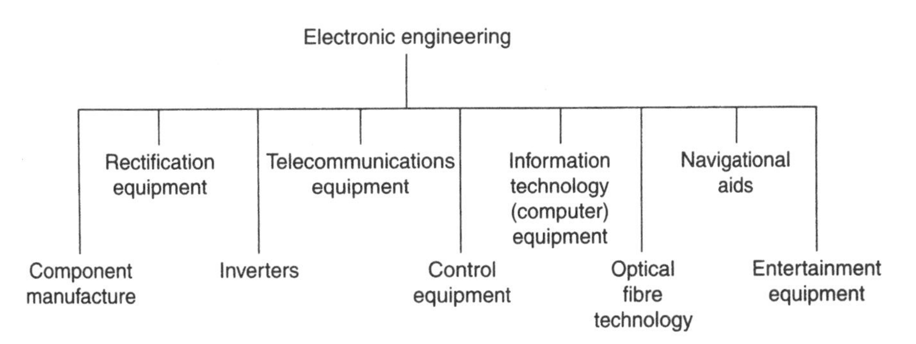

Figure 1.7 Products of the electronic engineering sector.

1.4.2 Instrument engineering

This subsector is concerned with the manufacture of scientific and industrial instruments for measurement, control and diagnosis. Most UK companies are relatively small specializing in satisfying niche markets, the larger instrument manufacturing companies operating in the UK are multi-nationals. Figure 1.8 shows the general breakdown of the products associated with this subsector.

1.4.3 Mechanical engineering

Although classified here as a subsector, mechanical engineering covers such a wide range of engineering activities that it needs to be broken down still further and this is done in Fig. 1.9. Let's now consider some typical products from each of these groups.

We will consider the groups in alphabetical order starting with agricultural machinery. (Tractors will be considered under *transportation*.) Typical products in each group are shown in Figs 1.10 to 1.17 inclusive.

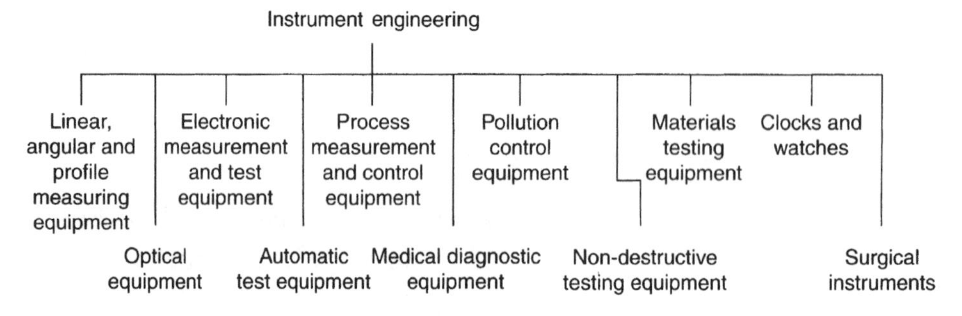

Figure 1.8 Products of the instrument engineering sector.

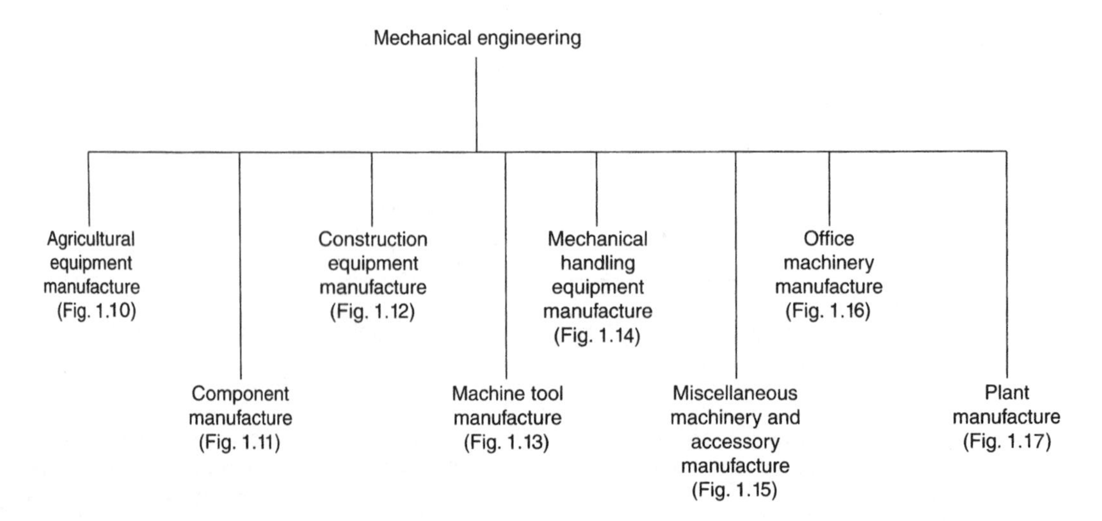

Figure 1.9 The mechanical engineering sector.

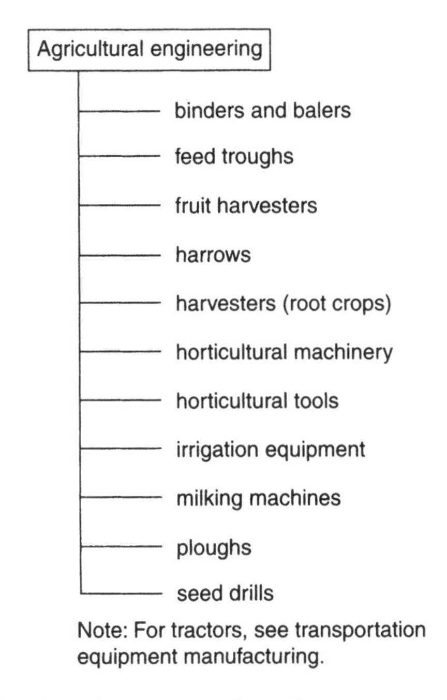

Figure 1.10 Agricultural equipment manufacturing.

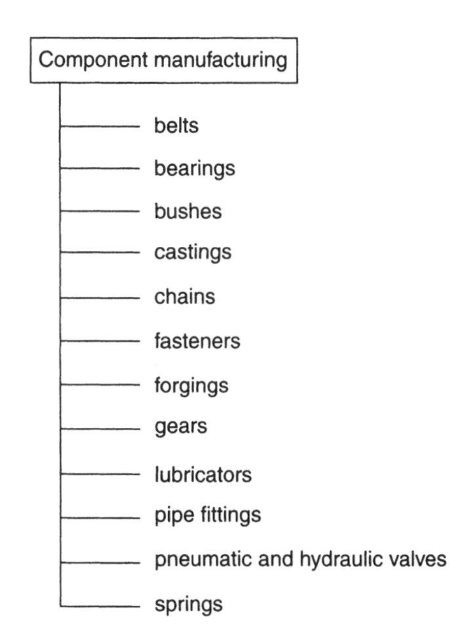

Figure 1.11 Component manufacturing.

1.4.4 Transportation equipment manufacturing

Again, this subsector of the engineering industry covers such a wide range of engineering activities that it needs to be broken down still further as shown in Fig. 1.18.

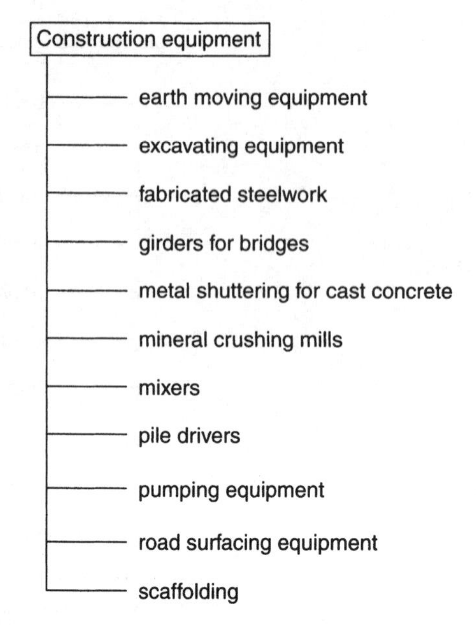

Figure 1.12 Construction equipment manufacturing.

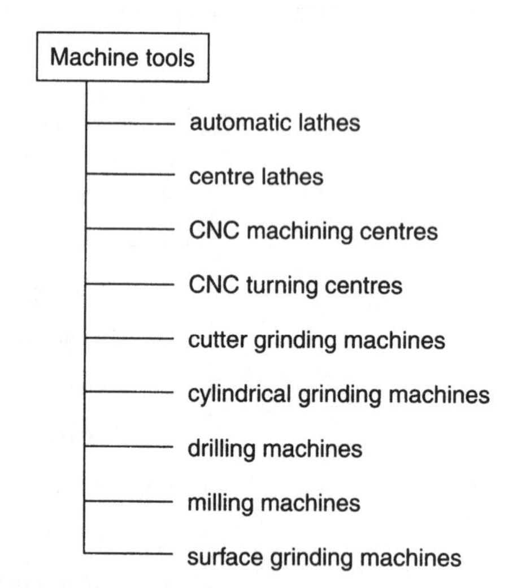

Note: CNC = Computer numerically controlled.

Figure 1.13 Machine tool manufacturing.

At one time the UK led the world in the manufacture of all types of transportation equipment. Again, however, this lead was lost following the Second World War. From being the largest manufacturer of steam ships we are now the smallest among the maritime nations of Europe. Most of our shipbuilding capacity is kept for strategic purposes to manufacture and maintain vessels for the Royal Navy. Very little merchant shipping is now made and maintained in UK yards. We were the first nation to develop the steam locomotive and develop a railway system. We went on to build and

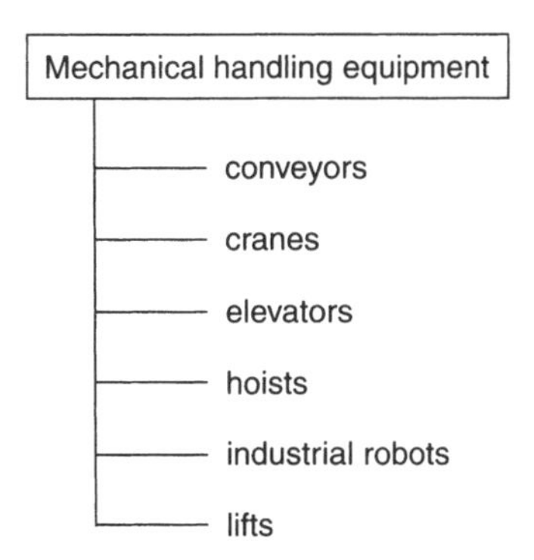

Figure 1.14 Mechanical handling equipment manufacturing.

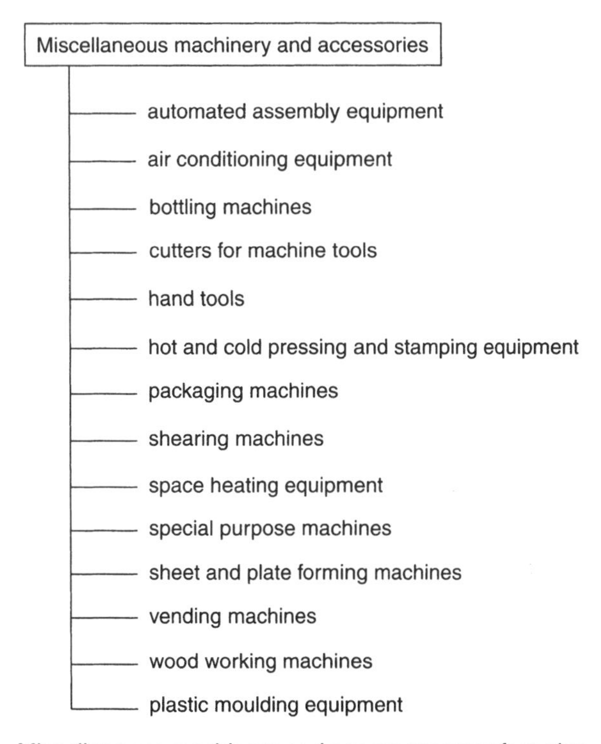

Figure 1.15 Miscellaneous machinery and accessory manufacturing.

equip railways for other countries throughout the world. However, nowadays diesel and electric railway locomotives are no longer produced in the UK, and there is only limited production of rolling stock and permanent way equipment. Also road transport has largely superseded rail transport as it can provide door to door delivery with minimum handling. Passenger traffic is still an important part of the rail system since it is 'self-loading'. Railways and shipbuilding were intensive users of coal, iron and steel. The run-down of these industries has had a disastrous 'knock-on' effect on iron and steel manufacture and the heavier sectors of the UK manufacturing industry, for example marine engines.

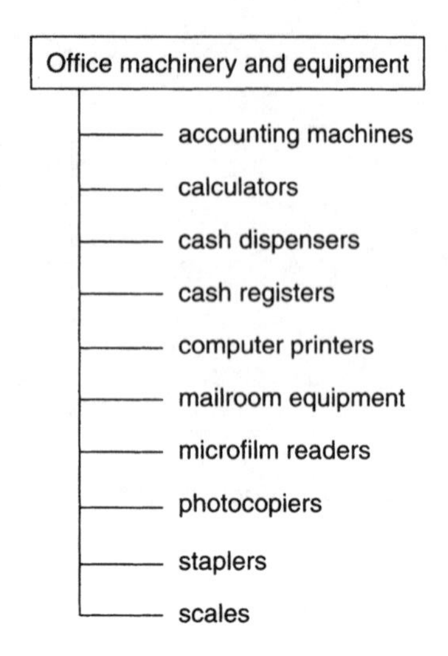

Figure 1.16 Office machinery and equipment manufacturing.

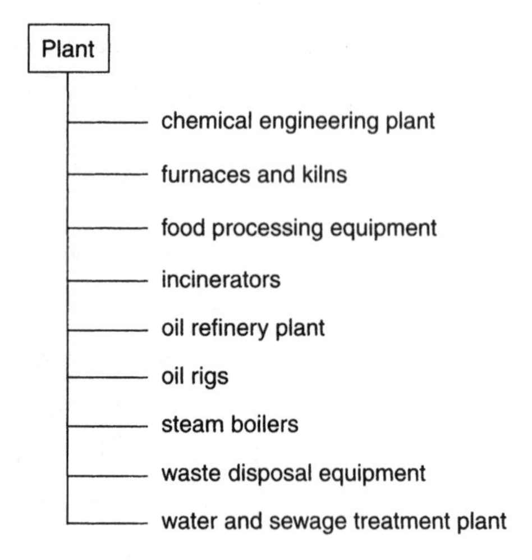

Figure 1.17 Plant manufacturing.

The UK owned car industry has fared little better and many well-known names prior to the Second World War have disappeared. With the sale of Rolls-Royce, the only major UK owned car-maker left is the MG Rover Group. All the others have either closed or been taken over by multinational companies based overseas. However these overseas companies manufacturing in the UK – such as Toyota, Ford (who now own Jaguar Cars and Aston Martin), Nissan, Honda and Peugeot – ensure that the UK is still a major exporter of road vehicles. Further, the UK still makes specialist sports cars on a small scale such as the Morgan. Also, the UK is still a world leader for the design, development and manufacture of racing cars. Small- and medium-sized commercial vehicles are still manufactured in the UK for the home market and for export.

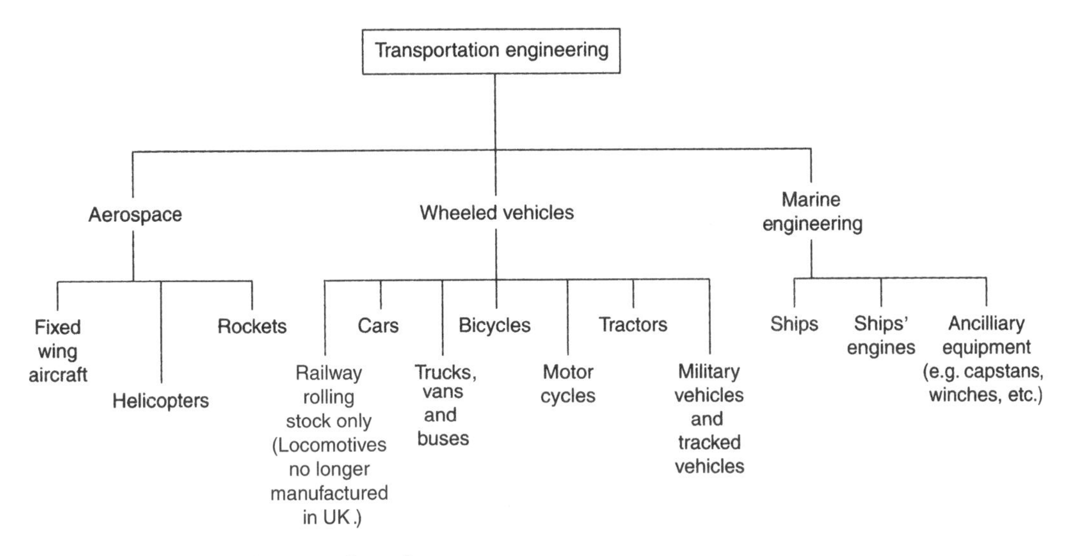

Figure 1.18 Transportation manufacturing sector.

The lead in motor cycle manufacture passed to the Japanese after the Second World War. Only Triumph remains as a large-scale UK manufacturer of motor cycles. No pedal cycles are now manufactured entirely in the UK, although some specialist cycles using imported frames are assembled here using mainly imported components and accessories. Most cycles are imported from the Far East and Asia where labour and other costs are much lower than in the UK. The aerospace industry is still a major contributor to the transport subsector of the engineering industry. This is largely due to its importance as a manufacturer of defence equipment. The cost of the design and development of modern aircraft means that, outside the USA, new projects demand international co-operation as, for example, the European Air Bus and the new Euro-fighter. The UK aerospace industry and aviation equipment manufacturing industries, including Rolls-Royce aircraft engines, are major employers and exporters and are of major importance to the UK economy.

Test your knowledge 1.3

1. List the main subsectors of the engineering industry.
2. Use the Internet to find four major UK companies in each of the sub-sectors listed in question 1.
3. Use the Internet to find out what the companies named in question 2 manufacture and where they are located. Suggest reasons for their location.
4. There are many engineered products in your home. Find out:

   (a) How many are produced in the UK.
   (b) The names of the companies responsible for their manufacture.
   (c) The sub-sector of the engineering industry to which these firms belong.

1.5 The food, drink and tobacco products industries

The food, drink and tobacco industries are also extremely diverse. Again we need to divide them up into a number of subsectors in order to simplify our study of them. The main subsectors are shown in Fig. 1.19.

1.5.1 Food industries

Let's commence by considering some of the many aspects of the food manufacturing industries, both fresh foods and processed foods.

Bakery products

A few large companies such as Rank, Hovis McDougal and Allied Bakeries dominate the bakery industry and also manufacture flour both for the wholesale and retail markets. The range of products manufactured by the bakery industry is indicated in Fig. 1.20. In addition, there are a number of smaller,

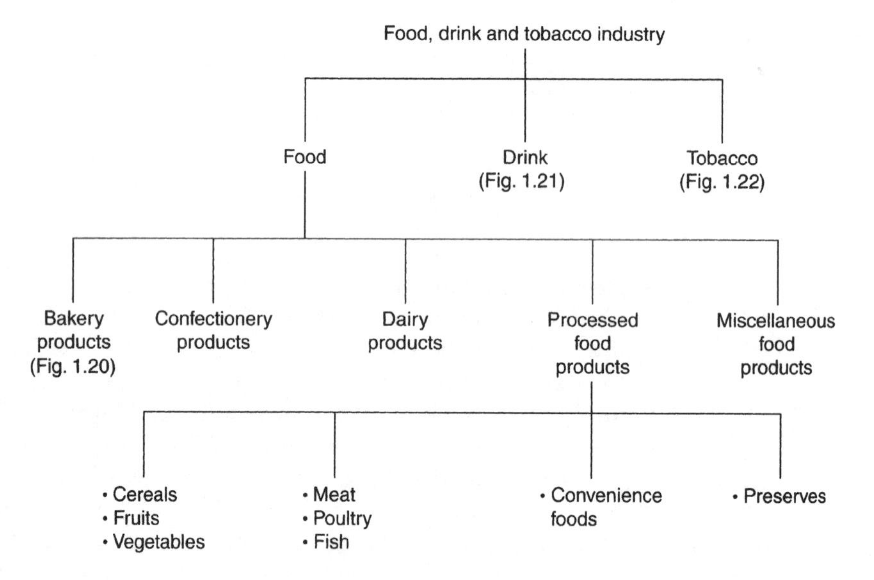

Figure 1.19 The food, drink and tobacco products manufacturing sector.

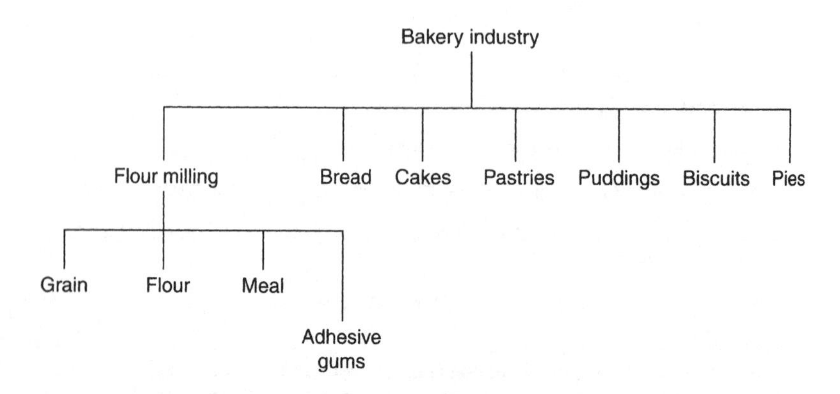

Figure 1.20 The bakery products manufacturing industry.

independent manufacturers making specialist products such as cakes, pastries and such things as pies and puddings – the last two obviously overlap with those parts of the food industry which provide the fillings. Just as in baking, the majority of biscuits are manufactured by a few large companies.

Confectionery products

The largest manufacturers in this subsector of the food sector are: Cadbury Schweppes, Mars and Rowntree Mackintosh, this latter company being part of the international Nestlé Group. They manufacture chocolates, sweets, drinking chocolate, cocoa and chocolate biscuits. Confectionery products also include 'fancy' cakes. A large number of smaller firms manufacture sweets and fancy cakes. Although some, trading under their own brand names, are nevertheless owned by the larger groups.

Dairy products

Milk is produced by cows owned by dairy farmers. It is a natural process. A cow only produces milk after calving. So you have to produce calves as well. Some female calves are retained to replace older milking cows when their milk yields fall. These older animals are unsuitable for human consumption but, if free from disease, can form the feed stock for the pet food industry. The majority of calves – male and female – are an important by-product of the dairy industry and are sold on for fattening for the fresh and processed meat industry. This helps to keep the cost of milk and dairy products at an affordable level. The production of milk and meat is inexorably linked by the laws of nature.

Milk is the bulk raw material of the dairy industry that this industry then processes. The milk that comes from the farm is not the same milk that is delivered to your doorstep – it is pasteurised and purified before it reaches you. Pasteurising is a heat treatment process that kills harmful bacteria and also assists the milk in staying fresh. Some milk is further processed into butter or cheese, yoghurt and cream. Milk from which the cream (fatty content) has been removed is sold as low fat, skimmed or semi-skimmed milks. Canned milk is usually either condensed or evaporated to drive off the water content and concentrate the nutrients.

Processed foods (cereals, fruit and vegetables)

Many foods are processed in some way before being made available to the professional chef, cook or the domestic consumer. From the earliest days cereal crops have had to be harvested, threshed to remove the seeds and these seeds or 'ears' have had to be ground to produce flour for cooking. Nowadays we expect our potatoes to be cleaned and prepacked for sale in the supermarkets. In view of the quantities involved the cleaning, grading and packing is a highly mechanized industry. Potatoes are also processed into chips and crisps. The chips are often partly cooked and 'oven ready'. Both these processes are highly mechanized in large and expensive plants that only major companies can afford.

Although many fruits and vegetables are sold fresh when they are in season, soft fruits and vegetables are often frozen so that they are available all year round. The speed of modern transport and the fact that the seasons alternate between the northern and southern hemispheres of the world, means that fresh produce is much more widely available than it used to be. Automated sorting, grading and packaging plants coupled with flash freezing facilities require a large investment and access to mass markets on a global scale so again food processing is 'big business' involving the largest companies. The same produce is also processed into sauces. Some fruits are canned in syrups to preserve them. This again requires large and expensive automated plants. The engineering industry provides the cans and machinery. *You cannot dissociate the various manufacturing industries; they all depend on each other.*

Processed foods (meat, poultry and fish)

Meat may be sold raw through butchers' shops or it may be bought processed in some form to reduce the time spent in the kitchen in its preparation. For example, ready cooked and canned meats, sausages and frozen poultry. Meats preserved by smoking and curing in salt has long been used for the production of ham and bacon. Although fish is still available fresh both on the bone and filleted, it is now available frozen, smoked, canned and also reconstituted as into fish fingers. A byproduct of the fish processing industry is fish blood and bone fertilizer for the horticultural industry.

Processed foods (convenience)

These consist of boxed and frozen complete, precooked meals containing meat, vegetables and a sauce. They merely have to be reheated in a microwave cooker ready for serving. Vegetarian meals are available as well. Potato crisps and similar prepacked snacks can be included in this category.

Processed foods (preserves)

This sector of the industry includes tinned fruit, vegetables, soups, bottled pickles and sauces, together with preserves such as jams and marmalades.

Miscellaneous food products

Under this heading can be included the production of a wide range of products such as:

Horlicks	Ovaltine	cocoa	drinking chocolate
sugar	table salt	vegetable oils	white fats
margarine	coffee	tea	yeast products
mustard	spices	starch products	pet and farm animal foods

The above are but a few examples of an ever growing range of such products.

1.5.2 The drinks industry

The drinks industry produces all those drinks that – apart from the addition of water in some cases – are ready for immediate consumption. Again, it eases our study of this subsector of the food, drink and tobacco industries to divide it up further as shown in Fig. 1.21.

Soft drinks

These can be grouped as shown above in Fig. 1.21. Let's now look at the individual categories in more detail.

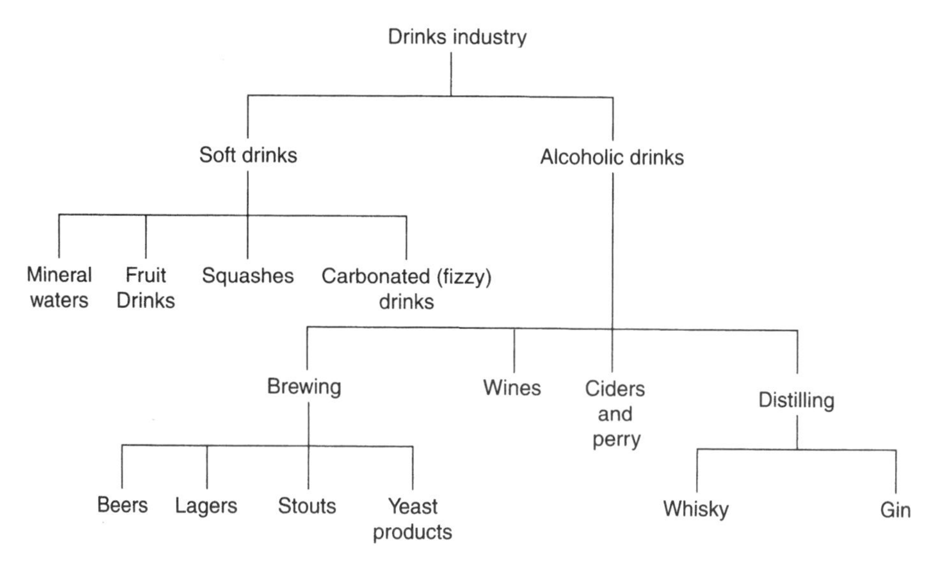

Figure 1.21 The drinks manufacturing industry.

Mineral waters are enjoying an expanding market both for sparkling (carbonated) and still waters. These are filtered and bottled at springs that are free from contamination. The naturally dissolved mineral salts in these waters improve their flavour. Many people find them more palatable than tap water.

Fruit drinks made by crushing fruits and extracting the juice may be produced as frozen concentrates requiring thawing out and the addition of water, or bottled ready for use. When bottled ready for use they are often referred to as *cordials*. They may be pure or they may contain preservatives and added sweeteners. The fruits used are typically oranges, lemons, limes, grapefruits, blackcurrants and unfermented apple juice.

Squashes are manufactured concentrates that require diluting with water to taste. They may contain some fruit concentrates but they contain mainly flavourings, sweeteners, preservatives, added vitamins and colorants. They cannot be considered as 'whole fruit' drinks. The so-called 'slimline' and 'diet' drinks contain artificial sweeteners in place of sugar.

Carbonated (fizzy) drinks include such products as cola, lemonade, tonic water, ginger beer, ginger ale, soda water and others. These may be drunk by themselves or used as *mixers* with wines and spirits. The matter is slightly confused because fruit drinks such as tomato juice and grapefruit juice may also be used as mixers. The 'fizz' is produced by adding carbon dioxide gas under high pressure so that it is absorbed into the liquid during processing.

Brewing

By volume, brewing is the biggest sector of the industry producing alcoholic drinks, providing *beers* (made from grain), *ciders* (made from apples) and, to a smaller degree, *perry* which is made from pears. Beer accounts for the overwhelming bulk of this market, producing draught, bottled and canned beers including bitters, lagers and stouts. The bulk of this production is in the hands of a small number of multinational companies. However, there is a niche market for small breweries producing specialist products.

Brewing can be traced back to the ancient Egyptians. Prior to the 15th century most brewing in the British Isles was carried out in the monasteries where the monks produced *ales*. Early in the 15th century hops with their preservative and flavouring properties were added and ale became known as 'biere'. This continental spelling was quickly corrupted to 'beer' and this name continues today.

Beer is made from crushed, malted barley steeped in boiling water. The liquid (*wort*) is run off for further processing and the residue known as 'brewers' grains' becomes a byproduct for livestock feedstuffs. Again notice how the waste from one industry is an essential raw material for another. Hops and, in some cases sugar, are added to the liquid which is again boiled. The liquid is again run off and the residual hops become a byproduct used in market gardening as a soil conditioner. It is widely used for mushroom growing. Finally yeast is added to the *wort* to cause fermentation. This is the stage where some of the natural sugars break down and become alcohol and carbon dioxide gas is released.

The carbon dioxide gas is collected and dissolved under pressure in carbonated drinks. So carbonated soft drinks depend on brewing for their 'fizz'. The yeast reproduces itself to form a frothy mass, most of which is skimmed off. Some is kept for future production and the surplus is a byproduct used for the production of bread and yeast foods (such as Marmite). The yeast is rich in proteins and vitamins of the B complex. The liquid now stands for some days to 'mature' and is then filtered and put into bottles or barrels. Lagers are matured for very much longer than ordinary beers. Stouts are made from roasted barley grains, sometimes with sugar. The roasting caramelizes the barley to give the stout its distinctive flavour and dark colour.

This topic area has deliberately been dealt with in some detail to reinforce how some processes can often generate byproducts that are essential feed stocks for other unrelated industries. It is important to realize that the manipulation of markets, however well intentioned, in one product area can often have serious economic consequences by affecting the steady flow of byproducts that are the essential raw materials for other industries. For example, the ash from coal-fired power stations is the raw material for the manufacture of insulation blocks used in building. Sulphur products in the flue gases of power stations burning coal are removed to prevent 'acid rain'. To remove the sulphur products, the flue gases are passed over lime beds where they react with the lime to form calcium sulphate that is the raw material for making plaster and plaster boards for the building industry.

Spirit distilling

Drinks with the highest alcohol content are spirits. These are manufactured by a distillation process. The main distilled spirits manufactured in this country are whisky (distilled in Scotland and Ireland), gin, and vodka. Other spirits which are popular, but which are produced abroad, are brandy and rum. Scotch whisky is distilled from a malted barley mash and may be sold as pure malt whisky or a number of whiskies may be mixed and sold as blended whisky. Irish whiskey is distilled from a rye mash. After distillation the spirit is left to mature in oak casks for a number of years. Much of the flavour and the colouring comes from the 'tannin' in the oak. Gin is also distilled from a

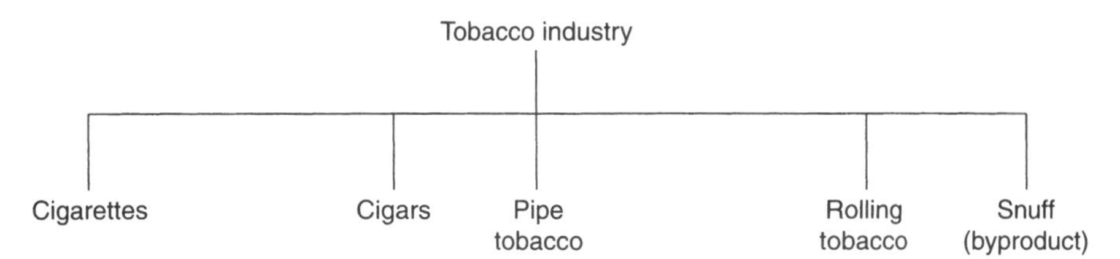

Figure 1.22 The tobacco products manufacturing industry.

grain mash but, in this case, it is flavoured with juniper berries. It may be dry (unsweetened) or slightly sweetened. The alcoholic drinks industry is very heavily taxed and contributes huge sums in revenue to the government.

Wines

The UK also has a small wine industry based mainly on vineyards in the south of England and the Midlands (grapes do not flourish in the colder climates of the more northern parts of Britain). Almost all the British wines are white. Wines are also manufactured on a small scale from imported grapes. As with brewing most of the small independent manufacturers have been absorbed by the larger companies who now control most of the industry. This has come about because of the level of demand and the economics of large-scale production.

1.5.3 The tobacco industry

The tobacco industry is capital intensive because of the high level of automation involved, particularly in the bulk manufacture of cigarettes. The capital-intensive nature of the industry leads to it being controlled by four major companies together with a number of smaller companies that make specialist products for the wealthier smoker. The main groups of tobacco products are shown in Fig. 1.22. Like alcoholic drinks, the tobacco industry is heavily taxed and is a major source of revenue for the government.

The combined food, drinks and tobacco sector of UK industry is very important to the economy. As well as providing for the home market it also exports its products abroad and earns a significant amount of foreign currency.

Test your knowledge 1.4

1. Name the main sub-sectors of the UK food, drink and tobacco industry.
2. Name typical examples of manufactured products from each group that you can find in your home and, where applicable, name the companies.
3. Use the Internet to find out where the companies named in question 2 are located and possible reasons for their location.
4. Write a brief report on the importance to the UK economy of the food, drink and tobacco industries.

1.6 Paper and board manufacture

This sector of the manufacturing industry produces the paper for newspapers and books, writing paper, wrapping and packing papers, toilet rolls, kitchen paper and tissues. It also makes boards, boxes, bags and all manner of packaging, as well as such things as wallpaper. The industry divides into two sub-sectors, one sub-sector making paper and the other involved in the manufacture of boards (cardboards and mill boards), as shown in Fig. 1.23. An expanding sub-sector of the paper and card manufacturing industry is involved in the recycling of waste paper to reduce the demand on traditional sources of raw materials, such as softwood forests.

1.6.1 Printing and publishing

Obviously there are areas of common interest between this sector and the paper and board industry, since the latter produces the materials used by the printing and publishing industry. The printing and publishing sector produces not only newspapers, magazines and books but also stationery, official documents, greetings cards, banknotes and postage stamps. The printing industry has seen tremendous changes over recent years, with computerization doing away with many of the skilled printing trades. Mergers in the newspaper, magazine and book publishing sectors have resulted in the industry being controlled by a small number of large companies. Only newspapers are produced 'in house', magazines, books and other products rely upon the support of a large number of smaller companies specializing in such skilled trades as computerized typesetting, printing and binding. There are a number of general printers who produce stationery and advertising matter for industry and commerce. No detailed figures are included for this sector of manufacturing which is relatively small, compared with those previously discussed, and has little influence on the national economy.

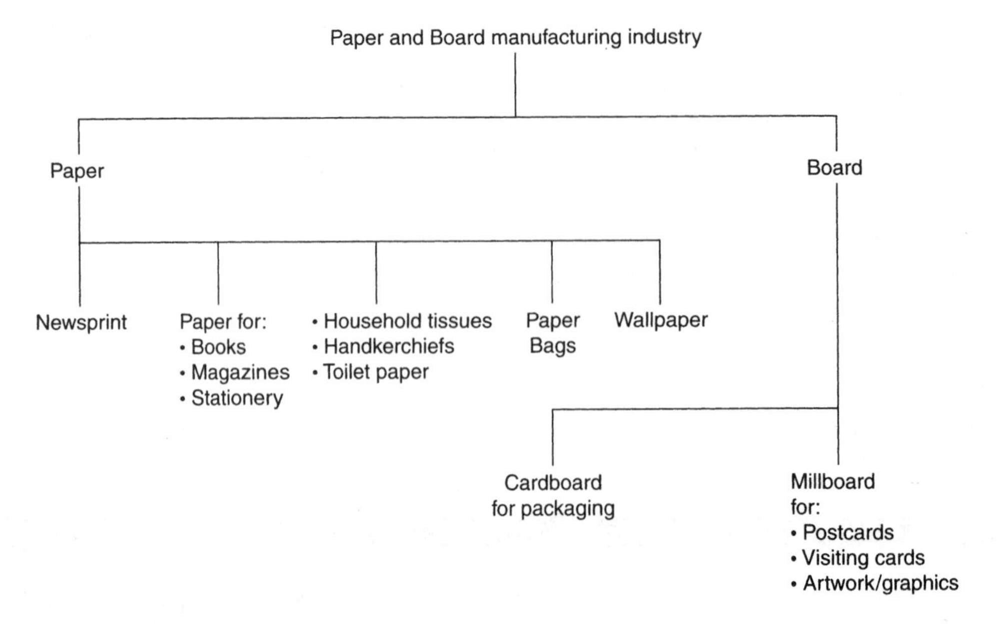

Figure 1.23 The paper and board manufacturing industry.

Test your knowledge 1.5

1. Use the Internet to name the main manufacturers of bulk paper and board in the UK, state where they are located, and suggest possible reasons that could have influenced their location.
2. Find out and explain the essential differences between publishing and printing.
3. Use the Internet to name the main publishing groups concerned with the production of:
   (a) newspapers;
   (b) magazines;
   (c) books.

1.7 Textiles, clothing and footwear

Like so many other sectors of the UK industry, the manufacture of many textiles and articles of clothing and footwear has increasingly moved abroad to countries where costs are lower. As on previous occasions it is necessary to break down this sector into a number of sub-sectors as shown in Fig. 1.24. The manufacture of textiles and fabrics requires expensive and large machines, therefore it is concentrated in a relatively small number of major companies. There are some smaller companies making specialized fabrics that are relatively costly. However making up the raw materials into garments is still largely labour intensive and, despite the decline in the clothing industries of the UK, the number of employees remains surprisingly high. Imported fabrics from the Far East and Eastern Europe have seriously challenged the manufacture of textiles in the UK. This has resulted in a substantial reduction of this sector of the manufacturing industry. Fashion design is an important

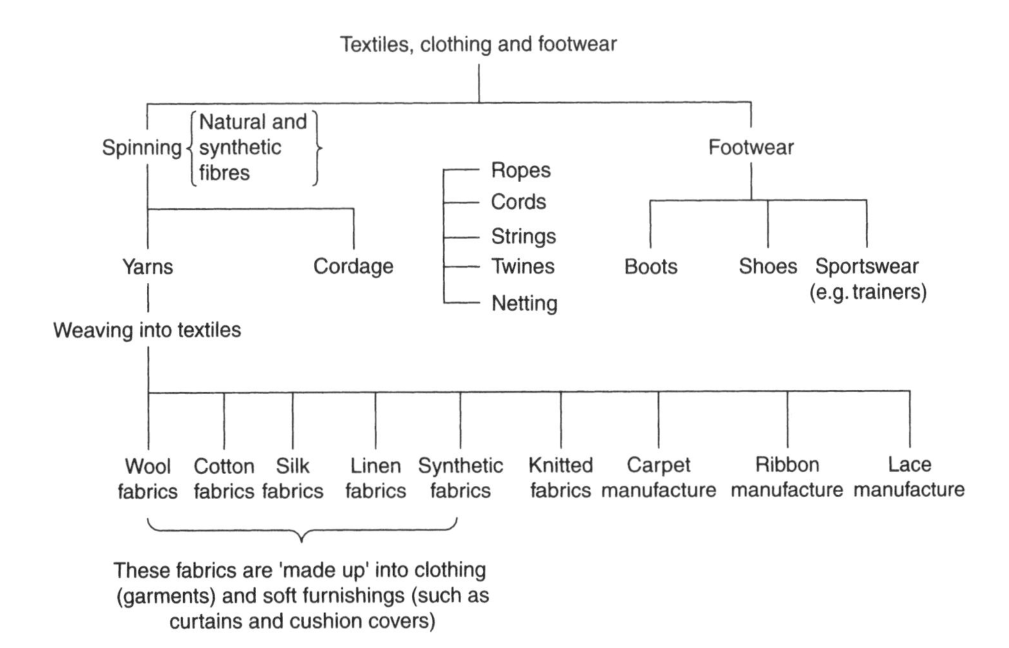

Figure 1.24 The textiles, clothing and footwear manufacturing industry.

sector of the clothing industry in the UK even if the designs themselves are made up in the Far East and shipped back to the West for marketing and distribution.

1.7.1 The textile industry

Textiles are made by spinning the wool and fur of animals, or fibres obtained from plants, into yarns that are then dyed and woven into textiles. Some textiles have the pattern printed on to them after weaving, whilst in others yarns of different colours are used. Some textiles are made from synthetic fibres or blends of natural and synthetic fibres. Let's now look at the main groupings of these products.

Wool

The UK is still one of the world's largest manufacturers of woollen textiles with the industry based mainly in Yorkshire and Scotland. The raw material (fleece) is obtained from the annual shearing of sheep. Sheep grow a heavy coat in the winter for protection against the cold and this is sheared off in the spring so that they are not affected by the heat of the summer. Some of this raw material is obtained from sheep bred in the UK but much is imported, coming largely from Australia and New Zealand. The fleece is spun into yarn and the yarn is either woven or knitted.

Linen

This is woven from fibres from a plant called *flax* which is grown mainly in Southern Ireland. However it is also grown in Northern Ireland and, to a lesser extent, in Scotland. Fine linen products such as tablecloths, handkerchiefs and other household textiles are manufactured in Northern Ireland, whilst heavy duty linens and canvases as used for tents, tarpaulins and sails are mainly manufactured in Scotland.

Cotton

This does not grow in the UK and raw cotton was imported mainly from America, India and Egypt. The raw cotton was spun and woven into textiles mainly in Lancashire, originally because the damp atmosphere of Lancashire aided the spinning process and helped to prevent the fibres from becoming brittle and breaking. At first, cotton was imported from America and the port of Liverpool grew up to service this industry as the major west-coast port servicing this transatlantic trade. The cotton industry has suffered in recent years from the growth of textile manufacturing in the Far East and most cotton fabrics are now imported ready woven from abroad. As previously stated, cotton is widely grown in the Middle East and the Far East where there is abundant cheap labour to process it into textiles that are more profitable to export than the raw cotton.

Synthetic fibres

The manufacture of synthetic fibres is a branch of the plastics industry. The plant is capital intensive and the industry of synthetic fibre manufacture is dominated by a few large companies. The fibres are manufactured from rayon (cellulose base), nylon and polyester (terylene). These fibres are spun into yarns that are then woven into textiles. One of the problems with synthetic fibres is that they are largely non-absorbent and, therefore, difficult to dye. Further, although strong, they tend to fray when subject to abrasive wear. Frequently textiles are made from mixtures of cotton or wool with synthetic fibres as these are more pleasant when worn next to the skin.

Carpet manufacture

Britain also has a substantial carpet-making industry, including famous names such as Wilton and Axminster that have an international appeal and reputation. These high quality carpets are woven and have a cut pile. Cheaper carpets use synthetic fibres for the pile or mixtures of synthetic and natural fibres. Kidderminster in the Midlands was also a major carpet manufacturing area but it has suffered from cheap foreign imports in recent years. The foundation or backing of carpets is largely made from the jute fibre.

Jute manufacture

Unlike cotton fibres which are obtained from the seed pod of the plant, jute is made from fibres found in the stems of the *Corchorus* plant grown mainly in Pakistan. It is spun and woven into coarse cloth in both Pakistan and India. Raw jute is imported into the UK where it is also manufactured into yarns and hessian fabric. As previously stated, jute is widely used as the foundation or backing for carpets and rugs. A coarse cloth called *hessian* is made from jute and this is used for making sacks. Hessian is also used in the manufacture of upholstery.

Miscellaneous

Not all textile yarns are made up into cloths and fabrics. Some are converted into lace, ribbons, and cordage used in the manufacture of strings, twines, nets and ropes. The UK has one of the largest cordage manufacturing industries in Europe.

1.7.2 The clothing industry

Most of the textiles produced in this country are used in the making of clothes. The UK has one of the largest clothing manufacturing industries in Europe. Unlike many other sectors of British industry previously discussed, the clothing sector is spread over a very large number of small concerns. British clothing and fashion designers have an international reputation for excellence. The industry is labour intensive so it is being affected by cheap imports from low cost areas abroad.

- The clothes are cut out by manually controlled or computer controlled, power driven shearing machines.
- They are then made up on power driven sewing machines operating very much faster than their domestic counterparts.
- Finally the garments are pressed and packaged.

Computerization is starting to creep into the larger concerns, especially for tailoring where CNC-controlled cutting-out machines are being used. This enables an individual customer's measurements to be entered into the computer and the suit is cut out exactly to size ready for making up.

Knitwear is largely manufactured in Scotland and in the East Midlands. Hand-made knitwear is largely made in the Scottish islands as a 'cottage industry'. Mass produced knitwear is made in and around the cities of Leicester and Nottingham. Unlike woven cloths, the yarn in knitwear is made directly into the finished garment ready for wear. There is no cutting out and no waste. Knitting machines are frequently computer-controlled nowadays and the patterns are also designed on computers and downloaded directly on to the knitting machines. This allows the process to respond quickly to changes in customers' requirements. Typical products are underwear, socks, stockings, tights, gloves, jumpers, pullovers and cardigans. Hats, gloves and fur goods (natural and imitation) are also made in the UK, but in relatively small quantities by specialist firms.

1.7.3 Footwear

The footwear industry in the UK has suffered in recent years from the importation of cheap products. Leather has largely given way to synthetic matersials and composites. Fashion products tend to be made in Italy and sportswear is largely imported from low cost regions. Nevertheless quality footwear is still made in the UK and there are also small firms who make boots and shoes to individual order.

Test your knowledge 1.6

1. Name the main types of fabrics and textiles manufactured in the UK and name the major companies producing them.
2. Examine your own clothes.
   (a) From what sort of textiles are they made?
   (b) Try and find out the names of the companies that made up the clothes.
   (c) What size firms are they and where are they located?
3. Use the Internet to name the main boot and shoe manufacturers in the UK and find out the range of products they make and how many people they employ.
4. Use the Internet to name the main manufacturing regions in the UK for textiles, clothes and footwear and suggest reasons that could have influenced their location.

Finally, this review only covers the larger manufacturing sectors in the UK. In addition there is the ceramics industry making goods ranging from bricks, roofing tiles and drain pipes to ceramic tiles, sanitaryware and tableware. There are also the woodworking industries producing goods ranging from joinery items like window frames, door frames, doors and staircases to the finest furniture making. No book could cover every branch of manufacturing still carried on in the UK.

1.8 The importance of the main manufacturing sectors to the UK economy

So far we have looked at the main manufacturing sectors of the UK. Let's now look at the economics of manufacture. Suppose you manufacture bird tables in your garage as a paying hobby. At the end of the year you can sit back in pride for you have employed and paid several members of your family for their help, you have produced a thousand bird tables selling at £12.00 each and you can show a turnover of £12 000.00. Unfortunately each bird table cost you £15 to make, so you made an overall loss of £3000.00. Not much wealth creation here; in fact you are worse off than when you started despite that impressive turnover. To be successful in business, companies have to make a profit and create wealth for themselves and for the nation.

You may wonder how the manufacturing sectors can contribute to the national wealth when so many of them are multi-nationals based outside the UK and when so many UK companies have branches abroad. Let's look at this more closely. Two financial terms that you may have seen used in the newspapers and on television are:

- Gross domestic product (GDP)
- Gross national product (GNP).

1.8.1 Gross domestic product (GDP)

This is the total value of all the output produced within the borders of the UK. So GDP is defined as the value of the output of all the resources situated within the physical boundaries of the UK no matter where the owners of those resources live. So the UK branch of a multi-national company based abroad contributes to the GDP. On the other hand the wealth created by the output of an overseas branch of a UK company is not included in the GDP. Further, outputs in this context not only include manufactured products (*visible earnings*), but services such as banking and insurance (*invisible earnings*) as well. In many ways, the GDP is more important in assessing the economic health of the nation than the GNP to be considered next.

1.8.2 Gross national product (GNP)

This is the total value of all the goods and services produced by companies owned by UK residents and UK financial institutions. These UK-owned companies can lie not only within the physical borders of the UK, but overseas as well. For example, the wealth created by the overseas branch of a UK-based company would count towards the GNP. On the other hand, unlike the GDP, wealth created within the UK by companies owned by foreign interests do not count towards the GNP. Therefore the GNP is defined as the value of the total outputs of all the resources owned by citizens of the UK wherever the resources themselves may be situated. Again, outputs include the invisible earning of the service industries as well as manufactured goods.

1.8.3 Balance of trade

The UK is not self-sufficient in food, raw materials or manufactured goods. In recent years the manufacturing base of the UK has shrunk as a result of firms being forced out of business by cheaper foreign imports. This is serious because we have to be able to pay our way in the world. To remain financially sound we must ensure that our income exceeds our expenditure. This applies both to our personal domestic lives and to the nation as a whole.

On a national basis the total income we earn from abroad for our exports must 'balance' or exceed the total payments we must make abroad to pay for our imports. This 'balance' is called the *balance of trade*.

With the shrinkage of its manufacturing base, the UK balance of *visible* exports (e.g. manufactured goods, foodstuffs and fossil fuels) is insufficient to pay for our imports. Fortunately we have a relatively healthy sale of services abroad (e.g. banking, insurance, and consultancy) and these *invisible* exports usually give us a trading balance, as does *inward investment* when foreign companies build factories in the UK. When you go abroad on a holiday you take money out of the country and the cost of your holiday counts as an import. When tourists come to this country they bring money into the country. Therefore in-tourism counts as an export. The tourist industry is a vitally important invisible earner for the nation.

It would appear that the manufacturing industries of the UK have much to answer for since, overall, we import more manufactured goods than we export. In fact you may wonder why we persist in maintaining a manufacturing industry at all. However, care must always be exercised when interpreting statistics.

- Not all the products imported in any one sector are raw materials for manufacturing. Many are finished products sold directly to the consumer. When the cost of these is stripped out of the equation the balance in favour of manufacturing becomes much more acceptable.
- If we reduced or closed down the manufacturing industries of this country, then many more imported goods would be required to satisfy the national domestic market. The more of these goods we can manufacture in the UK the fewer we have to import. Further, the cost of imported raw materials is always lower than the cost of finished products.
- Finally, we must remember that manufacturing employs a great many people in the UK. Any reduction in manufacturing would increase unemployment. Only persons who are employed can create wealth for the nation. Remember that it is estimated that the 4.3 million people representing over 20 per cent of all employed persons are engaged in the manufacturing industries, and that manufactured goods account for roughly 80 per cent of all UK exports and 20 per cent of the gross domestic product (GDP).
- Therefore it should be the aim of the nation to expand rather than contract its manufacturing base so that more citizens of the UK can be employed in wealth creation and contribute to a sound national economy and balance of payments.

Test your knowledge 1.7

1. Write a brief report on the importance of the manufacturing industries to the UK economy in terms of:

   (a) Wealth creation;
   (b) Balance of trade;
   (c) The beneficial effects of manufacturing on society.

1.9 Industry and its location

Many factors have influenced the location of industry in the UK over the centuries and from time to time these factors have changed as a result of the changing needs of the population, as a result of technological advance, and as a result of political expediency.

1.9.1 Raw materials

Even with present-day rapid and easy transport systems, many industries still need to be located near to sources of raw materials. Makers of iron and steel, for example, will tend to locate their factories near to the supplies of iron ore. This is because iron ore is bulky and the further it has to be transported, the higher the cost. But the industry also has to take into account the location of fuel supplies, such as coal, which is also bulky and costly to move over long distances. Almost all the iron ore used today together with the coal used in its extraction is imported. Therefore most iron and steel manufacture is carried on mainly in coastal locations. Similarly, sugar beet refineries are located near to the beet growing areas of East Anglia and Lincolnshire, and paper-making is usually sited near to conifer forests.

1.9.2 Human resources

The traditional industries of the UK were highly regionalized for reasons of access to raw materials, skilled labour, and sources of energy such as water power and, later, coal for steam engines. The bulk transport of raw materials is no longer a problem and the electricity grid provides power wherever it is required. However, the traditional industries of cotton and woollen cloth manufacture, pottery manufacture, automobile manufacture and shipbuilding are still carried on, albeit on a much reduced scale, mainly in their original locations because of the availability of a skilled and specialist workforce. The introduction of automated manufacturing processes has reduced the need for large numbers of traditionally skilled workers and manufacturing concerns are often re-located elsewhere for different reasons such as government financial incentives to attempt to alleviate pockets of unemployment.

1.9.3 Incentives

When the older industries start to close down (for example deep mined coal) whole communities are often left without employment. To offset the deprivation that this causes, local, regional and national government together with the European Union (EU Development Fund) steps in with grants and loans to set up *Enterprise zones* to attract new industries to such areas. The incentives include low interest loans, low rents and rates, estates of ready built 'start-up' factory premises, business parks, and grants. These new industries will usually be quite unrelated to the industries they are replacing and much retraining of the local labour force will be required. Government funding will also be provided for this as well. Therefore the traditional geographical location of industries is becoming blurred and distorted for social and political reasons unrelated to the basic logistics of manufacture.

1.9.4 Energy sources

The replacement of water power and steam with electricity, coupled with modern transport systems means that manufacturing industries no longer have to be sited by rivers or coal fields. However, this does not apply to the energy providers themselves. Electricity generating stations still have to be located either close to our remaining coal fields or close to special coal handling ports for cheap, imported coal. They also have to be adjacent to an adequate water supply such as a large river. Although the boiler water is recirculated, a constant supply of fresh water is required to make up the inevitable losses that occur from the cooling towers due to evaporation.

Oil terminals also have to be sited near to deep-water harbours that can handle the giant oil tanker ships. Oil refineries are inevitably built adjacent to such oil terminals to ease the bulk handling of crude oil that can do so much environmental damage if a spillage occurs. Natural gas is usually piped ashore and coastal sites are chosen for the purification of gas and the extraction of any useful byproducts before it is compressed and fed through the network of pipes, forming the 'gas-grid', to those areas where it is required.

Many smaller electricity power stations now use alternators driven by gas turbines similar to the jet engines of aircraft. The exhaust gases from these gas turbines, still very hot, are then used to produce steam in boilers. This steam is then used in conventional steam turbine-alternator sets to generate electricity. Such composite plants can be started up quickly to satisfy peak demands. Since they run on natural gas, they are also much cleaner and more efficient than traditional coal-fired power stations. Unfortunately natural gas supplies are very limited compared with oil and coal.

1.9.5 Transport and markets

Transport and markets have always been interlinked. Transport has been and still is a necessary evil. It in no way enhances the value or marketability of the finished goods. If a manufacturer could have every supplier and customer next door to the factory, then that manufacturer would be delighted. Unfortunately such an ideal is not possible, so transport facilities are a very important factor in locating a manufacturing plant. The need to transport raw materials and finished goods is an added cost that needs to be kept to a minimum to keep a manufacturer's products competitive. Today we trade all over the world and refer to 'global markets'.

The cost of the transport of raw materials and finished goods has to be added to the cost of the product and paid for by the consumer. If the raw materials are bulkier and heavier than the finished goods, then the manufacturing plant is more likely to be sited near to the source of raw materials or the docks at which it is imported. On the other hand if the finished goods are bulky and heavy compared with the materials and equipment from which they are made then the manufacturing plant will be sited nearer to the market. For example it is easier to transport the steel and engines to a shipyard on the coast, than to transport the finished ship to the sea.

Transport has to be matched to the product. It is uneconomical to use air-freight to transport heavy and bulky building supplies and machinery and these are better transported by road, rail, or sea. On the other hand some perishable goods that have to arrive fresh for the market can be economically transported by air-freight. Letters and parcels can also be transported by air-freight as speed is essential, as it is in the transportation of emergency medical supplies.

All manufacturing plants must be located near to appropriate transport systems. It is no use setting up a manufacturing plant just because a green-field site has been made available in a centre of abundant cheap labour, with a government grant to build and equip it. The site must be accessible for incoming raw materials and have adequate transport facilities for its finished products.

Therefore communications and ease of access to the market are vital considerations. A company which imports its raw materials from Scandinavia or which exports its products to Scandinavia, for example, will locate its factories in areas with good communications to ports on the east coast of England and Scotland. Companies trading with Europe will tend to site their factories in the south east of the UK to take advantage of the cross channel ferries and the cross channel rail link.

Test your knowledge 1.8

1. List the main factors that influence the siting of a manufacturing company.
2. Name a major iron and steel making plant in the UK, state where it is located and discuss the reasons that dominated the choice of location.
3. Suggest reasons why a major tobacco products firm is located at Bristol.
4. Suggest reasons for the siting of the Imperial Chemical Industries chemical plant at Billingham.

Key notes 1.1

- Manufacturing is the conversion of raw materials into useful articles by means of physical labour or the use of power driven machinery.
- The purpose of manufacturing goods is to satisfy a demand and to create wealth.
- The main manufacturing sectors in the UK are the chemical industry, the engineering industries, food and drink processing industry, tobacco products industry, paper and board making industry, printing and publishing industry, textiles, clothing and footwear industry.
- The engineering industry may be broken down into the following sectors: electrical engineering, electronic engineering, instrument engineering, transportation engineering and mechanical engineering.
- The mechanical engineering sector may, itself, be broken down into the following sub-sectors: agricultural equipment engineering, component manufacturing, constructional equipment manufacturing, machine tool manufacturing, mechanical handling equipment manufacturing, miscellaneous machinery and accessories manufacturing, office equipment manufacturing, plant manufacturing.
- Other industries may be similarly broken down into various sectors and sub-sectors.

1.10 Key stages of production (example)

Before a product can be manufactured there are two important functions that must be considered. One is *marketing* and the other is *design*. It has already been stated that the purpose of manufacturing is to satisfy a market requirement and to create wealth. There would be no point in manufacturing a product that no one wanted to buy or which could not be made and sold at a profit. Marketing and selling are often confused. Marketing is concerned with investigating the need for a product, testing and analysing the requirements of prospective customers and determining the price they may be willing to pay. Marketing also investigates the products of a company's competitors and investigates how those products may be improved upon. Prior to the launch of a new product the market must be prepared by a suitable advertising campaign. Only then does the sales team move into action and follow up the marketing campaign with the procurement of orders. Therefore selling is the procurement of orders.

Once the marketing team has established the requirement for a new or improved product a detailed design specification must be drawn up in consultation with the production engineering department to ensure that the design can be manufactured on a profitable basis. Prototypes can then be made and tested and samples sent to selected customers to evaluate. Once the teething problems have been corrected, bulk manufacture can commence. Let's now consider the key stages of production.

We have already seen how the finished product of one industry is often the raw material of another industry. Also that the waste product of one industry can be an essential raw material for yet another industry. In this way many quite different manufacturing industries can be closely linked as suppliers and users of each other's products and of each other's waste. Therefore it is often difficult to decide exactly where a *production system* starts and finishes.

For the purposes of this book we will try to simplify the *key stages of production* by limiting ourselves to the processes that occur *within a single plant*. For example let's consider a loaf of bread. For simplicity we will ignore the growing of the corn and the milling of the grain to make the flour. Let's only consider what happens in the bakery and see how the key stages of production shown in Fig. 1.25 can be applied to our loaf of bread. Remember that the grain from the farmer is the raw material for the flour miller and the flour from the mill is one of the raw materials for the baker. Also, by-products from flour milling are the natural gums dextrin and gluten. These are the raw materials for the adhesives used in the manufacture of envelopes, stamps and food labels that may have to be licked. Being safe natural products there is no possibility of toxic contamination in their use.

1.10.1 Sourcing and procurement

Before the raw materials can be prepared for processing, they must be bought. This is a two-stage process. First, the materials must be *sourced*, then they must be *procured*.

Sourcing is finding suitable sources of raw materials for a particular manufacturing process. It can range from scanning the trade directories and the websites of suitable suppliers to global exploration for mineral resources (metal ores) and fossil fuels.

Procurement is the process of acquiring the raw materials once they have been sourced. This involves negotiating a suitable price, scheduling the delivery quantities and dates, and placing a contract with the supplier. The supplier usually arranges suitable transport.

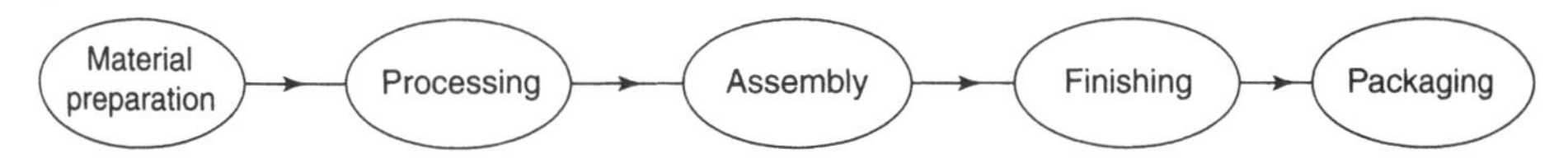

Note: 1. Not all products require all these stages.
2. Not all products require all the stages in this order.

Figure 1.25 Key stages of production.

1.10.2 Material preparation

The various ingredients for making our loaf will be checked for quality and freedom from contaminants. The ingredients will be measured out into the quantities required for the type of bread being made. Typically 250 kg of flour, 140 litres of water, 4 kg of salt and 2.5 kg of compressed yeast to make a batch of the basic dough.

1.10.3 Processing

The above ingredients are kneaded into a dough by machinery and various other ingredients such as fat, milk powder and sugar are added according to the type of bread being made. The dough is then allowed to stand and ferment for several hours, after which it is automatically divided up and machine moulded to shape before it is baked for approximately three quarters of an hour at a temperature of 230°C. The bread is then allowed to cool.

1.10.4 Finishing

In this example we can assume that we require a sliced loaf. Therefore the finishing process will be that of automatically slicing the loaf by machine.

1.10.5 Packaging

Since we are considering the manufacture of a sliced loaf, the bread will be automatically wrapped to keep it fresh. The packaging will be printed with the manufacturer's logo for advertising purposes, together with the list of ingredients and nutritional information now required by law. A flow chart for key stages in the production of our loaf of bread is shown in Fig. 1.26.

1.11 Key stages of production (general)

Material preparation refers to the inspection and preparation of the incoming materials ready for processing and is very important in achieving and maintaining the quality of the finished product. Some further examples of material preparation are:

- The washing and grading of fruit and vegetables in the food processing industries.
- The unpacking and thawing out of frozen produce such as fish and poultry prior to processing.
- The cutting of cloth to size and shape in the clothing industry and the cutting of leather to size and shape in the footwear industry.

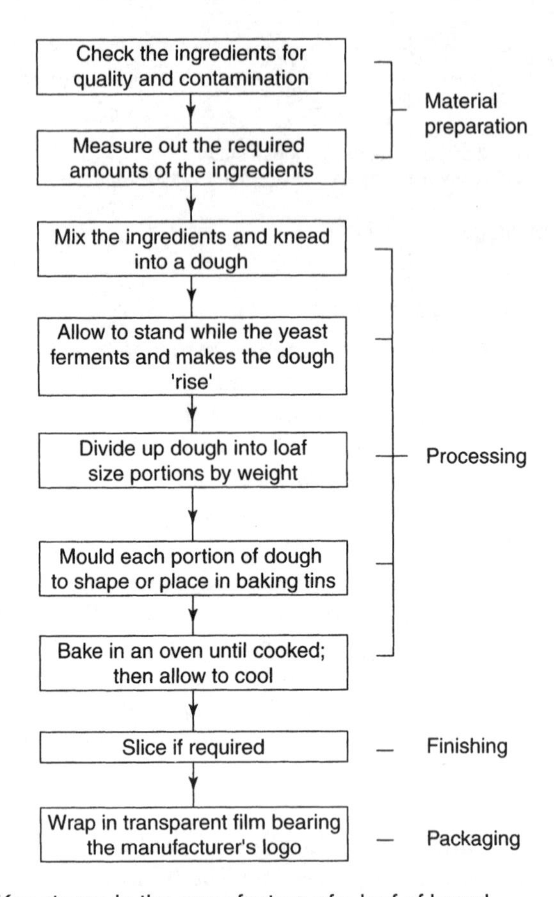

Figure 1.26 Key stages in the manufacture of a loaf of bread.

- The blending, cutting and shredding of leaf tobacco ready for the manufacture of cigarettes and cigars.
- The degreasing and surface treatment of metal products before processing.

Processing refers to the conversion of materials from one form to another. As we have already seen, a bakery converts flour, water and yeast into bread. Some further examples of processing are:

- The casting of metal ingots into useful shapes such as automobile cylinder blocks in a foundry.
- The cutting of components from steel bars in the machine tools of an engineering workshop.
- The forming of plastic moulding powders into finished products by compression moulding and injection moulding machines in the plastics industry.
- The slicing and precooking of potatoes to make 'oven-ready' chips in the food industry.
- The machining of wood to make the component parts of doors, window frames and staircases in the joinery industry.

- The weaving of yarns into fabrics and the manufacture of clothes from fabrics.

Finishing is any treatment applied to manufactured goods to give them either an attractive appearance or to protect them from deterioration or both. Some examples of finishing are:

- The painting of exterior woodwork. This improves the appearance and also preserves the wood and protects it from the onset of wet rot. The paint system should consist of a primer, an undercoat and a topcoat (gloss) containing a varnish to seal the paint system and wood against the penetration of moisture.
- The painting of steelwork to prevent rusting. This requires a different type of primer paint containing corrosion inhibitors rather than the fungicide of a wood primer. In both cases the primer provides a 'key' to help the undercoats and topcoat adhere to the object being painted.
- Galvanizing. Goods manufactured from steel are dipped into molten zinc that forms a corrosion resistant coating on the steel.
- Electroplating such as nickel plating and chromium plating. This is less corrosion resistant than galvanizing, but has a more decorative appearance.
- Anodizing aluminium where an aluminium oxide film is built up on the metal. Sometimes the chemicals used provide an oxide film of the required colour. Sometimes the film is coloured after the anodizing process.

As well as providing protection and improving the appearance of manufactured products, finishing processes can be used for other purposes. For example

- Water-proofing fabrics for manufacturing wet-weather clothing.
- Liquid repellent treatments for carpets to protect them against staining when liquids are accidentally spilt on to them.
- Fire retardant and flame resistant treatments for furnishing and clothing fabrics.
- Crease resistant and drip-dry treatments for clothing fabrics. Essential for the clothes of people who travel a lot.

Assembly is the joining together of various manufactured parts to make a complete product. Sometimes the assembly is permanent as when the body panels of a car are welded together. At other times the assembly is temporary as in the bolted-on wheels of a car. It must be possible to easily remove a wheel if a tyre becomes punctured. Some examples of assembly processes are:

- The riveting and welding together of engineering structures to make bridges, ships and off-shore oil rigs.
- The use of synthetic adhesives in the manufacture of goods ranging from wooden furniture to the stressed components of aircraft.
- The stitching together of clothing.
- The soldering of electronic components to the printed circuit boards in products such as computers and television sets.

Packaging of finished products has become an important part of the marketing process. The requirements of packaging are:

- To protect the goods from damage in transit.
- To protect the goods from contamination.
- To protect the goods from climatic changes.
- To enable the goods to be stored correctly and without damage.
- To enable the goods contained within the packaging to be easily identified without the need for unpacking.
- To be informative and attractive to the customer.
- To comply with legislation.

Very often the packaging is the first line of communication with a prospective customer. Therefore the packaging is an important means of communicating and selling. It must be pleasing in design and it will always contain the brand name. The busy shopper in a supermarket must be able to locate and identify a particular product and brand instantly on the crowded shelves. To comply with legislation it must also state the country of origin. Food products especially are subject to legislative requirements such as nutritional information and the ingredients used.

Typical packaging materials consist of paper and boards, rigid, flexible and foam plastics. Foam plastics are frequently moulded to the shape of the product so that it is fully supported. Liquids may be contained in moulded plastic or glass bottles or in metal cans. Solid foods may be contained within metal cans (for example, corned beef) or in plastic prepacked trays covered with a transparent film (for example, bacon).

Finally, as we have seen in this section, not all the key stages are used every time. You do not have to assemble a loaf of bread. Nor do the stages always follow in the same order. The various parts of a car are painted and/or electroplated (finished) after subassembly but before final assembly.

Test your knowledge 1.9

1. List the key stages of production.
2. Describe briefly how these key stages may be applied to a product of your own choice.

1.12 Scales of production

Popular makes of motor cars are mass-produced in large quantities by increasingly automated manufacturing techniques because of the large demand for such products. On the other hand large structures, such as bridges, are usually built on a 'one-off' basis mainly by hand. However some components, such as nuts and bolts, used in the construction of a bridge may well be mass-produced on automatic machines because of the quantity required. Top restaurants will generally employ chefs to cook meals to order to a very high standard for a limited number of customers who are able and willing to pay the high cost of such labour-intensive service. On the other hand, fast food outlets are more likely to use processed foods that require the minimum of preparation. This increases the volume of meals that can be served and reduces the time and cost involved.

Therefore the *scale of production* will depend upon the type of product and the demand for that product. The broad groupings are as follows:

- Continuous flow and line production;
- Repetitive batch production;
- Small batch, jobbing and prototype production.

Let's now look at some typical applications of each category.

1.12.1 Continuous flow and line production

In *continuous flow production* the plant resembles one huge machine in which materials are taken in at one end of the plant and the finished products are continuously despatched from the other end of the plant. The plant runs for 24 hours a day and never stops. Plastic and glass sheet and plasterboard is produced in such a manner.

In *line* or *mass production* plants the manufacturing system plant is laid out to produce a single product (and limited variations on that product) with the minimum of handling. The product is moved from one operation and/or assembly station to the next in a continuous, predetermined sequence by means of a conveyor system. Individual operations are frequently automated. Such plants usually manufacture consumer goods such as cars and household appliances in large quantities in anticipation of orders. The layout of a typical flow or line production plant is shown in Fig. 1.27.

The characteristics of *continuous flow* and *line production* can be summarized as:

- High capital investment costs.
- Long production runs of the same or similar product type. In some cases production never ceases until the plant requires refurbishment or the product changes.
- Highly specialized plant resulting in inflexibility and difficulty in accepting changes in product specifications.
- Rigid product design and manufacturing specifications.
- The processing equipment is laid out to suit the operation sequence.
- Stoppages occurring in any of the production units needs to be rectified immediately or the whole production line is brought quickly to a halt.
- Components, products and subassemblies move from one work-station to the next by means of pipelines in the case of fluids, gases and powders or by mechanical conveyors in the case of solid goods.

Typical products requiring this type of production are shown in Fig. 1.28.

1.12.2 Batch production

As its name implies, this involves the production of batches of the same or similar products in quantities ranging from, say, hundreds of units to several thousand units. These may be to specific order or in anticipation of future

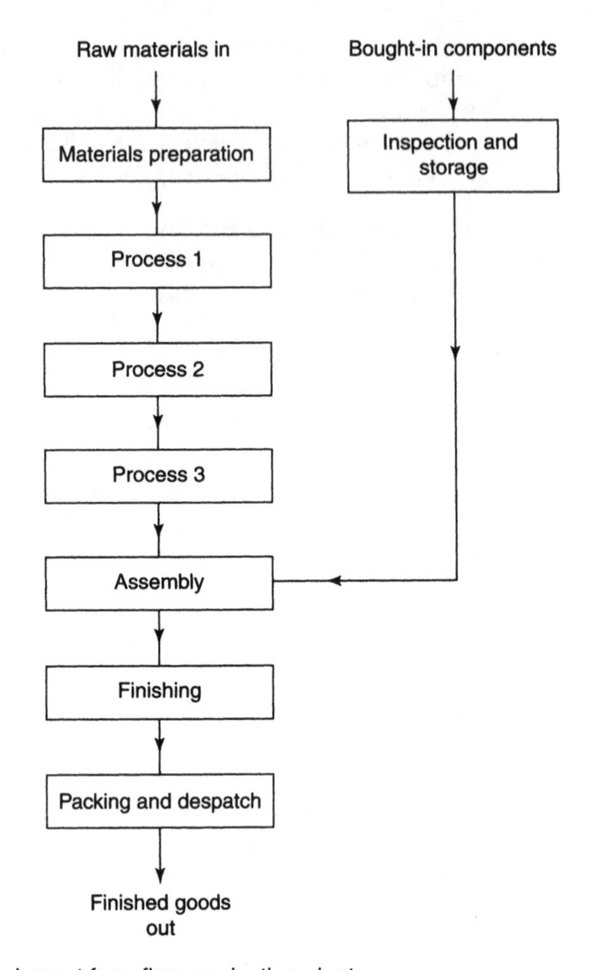

Figure 1.27 Layout for a flow production plant.

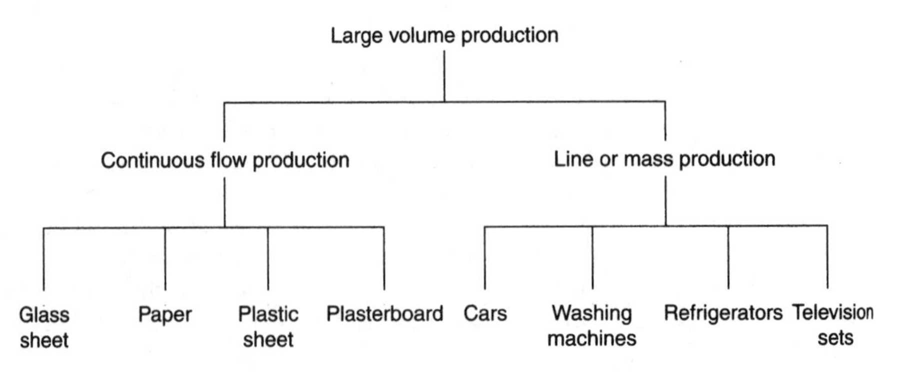

Figure 1.28 Large volume production.

orders. If the batches of components are repeated from time to time, this method of manufacture is called *repetitive batch production*. General purpose rather than special purpose machines are used and these are usually grouped according to process.

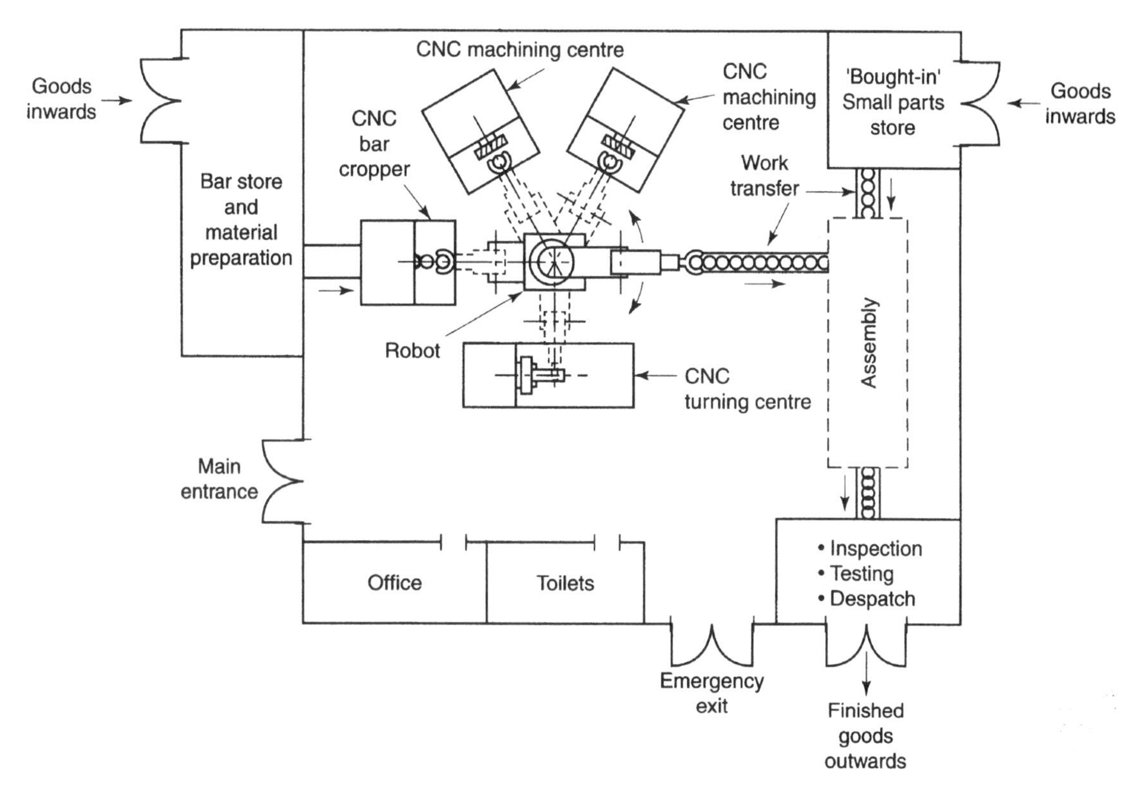

Figure 1.29 Layout for batch production (FMS cell).

Nowadays, the machines are often arranged to form *flexible manufacturing cells* in which the machines may be computer numerically controlled (CNC) and linked with a robot to load and unload them as shown in Fig. 1.29. To change the product, the computers are reprogrammed. The computer programs are kept on discs and are available for repetitive batches thus saving lead-time in setting up the cell.

The characteristics of batch production can be summarized as:

- Flexibility. A wide range of similar products can be produced with the same plant.
- Batch sizes vary widely.
- Production is relatively slow as one set of operations are usually completed before the next are commenced.
- Work in progress has to be stored between operations. This ties up space and working capital. It also has to be transported from one group of machines to the next.
- General purpose machines are used. These have a lower productivity than the special purpose machines used in line and flow production.
- It is unlikely that all the machines in any particular plant are required for every product batch. Therefore work loading will tend to be intermittent with some machines remaining idle from time to time, whilst others are overloaded resulting in production bottlenecks.

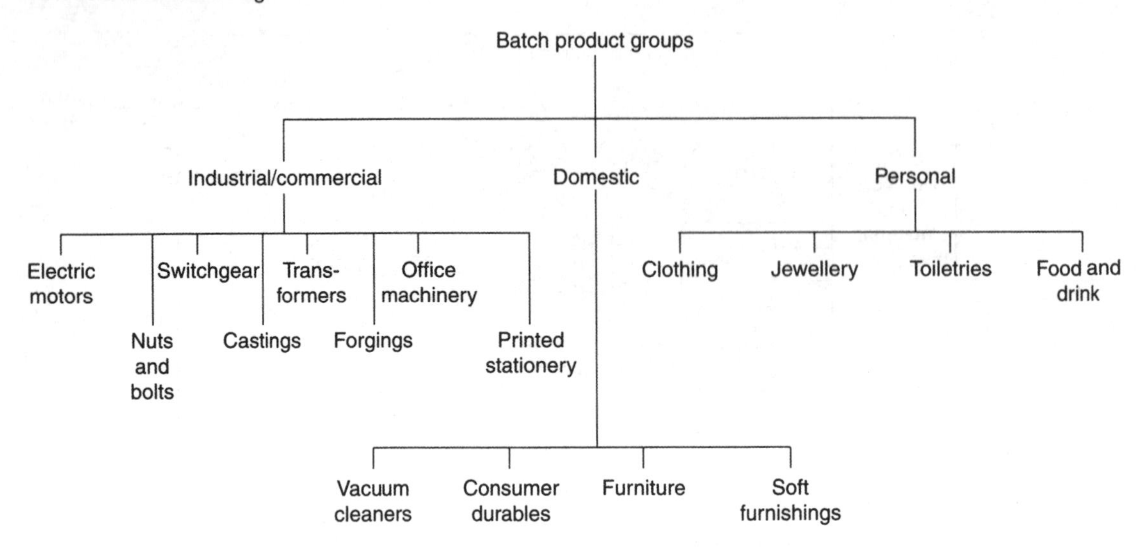

Figure 1.30 Some typical batch product groups.

- For the reasons stated above, unit production costs are very much higher than for flow and line production. Process efficiency is lower since variations will exist between the rate of production and consumer demand.
- Batch production is the most common method of manufacture in the UK.

Some typical product groups are shown in Fig. 1.30.

1.12.3 Small batch, prototype and jobbing production

This refers to the manufacture of products in small quantities or even single items. The techniques involved will depend upon the size and type of the product. For very large products such as ships, oil rigs, bridges and the steel frames of large modern buildings, the workers and equipment are brought to the job. At the other end of the scale, a small drilling jig built in the toolroom will have parts manufacture on the various specialist machining sections and will be brought to and assembled by a specialist toolroom bench-hand (fitter). Prototypes for new products are made prior to bulk manufacture to test the design specification and ensure that it functions correctly. Modifications frequently have to be made to the prototype before manufacture commences.

Workshops for small batch (100 or less) and single products such as jigs, fixtures, press tools and prototypes, are referred to as *jobbing shops*. That is, they exist to manufacture specific 'jobs' to order and do not manufacture on a speculative basis. The layout of a typical jobbing shop is shown in Fig. 1.31.

So far we have only considered an engineering example, but the same arguments apply elsewhere. For example when you order a suit from a *bespoke tailor* it will be manufactured as a 'one-off' specifically to your measurements and requirements. It will be unique and made mainly by hand in the tailor's workroom. The tailor will not manufacture on a speculative basis. However, the suit you buy from a clothing store will be one of a *batch* produced in a factory in a range of standard sizes and a range of

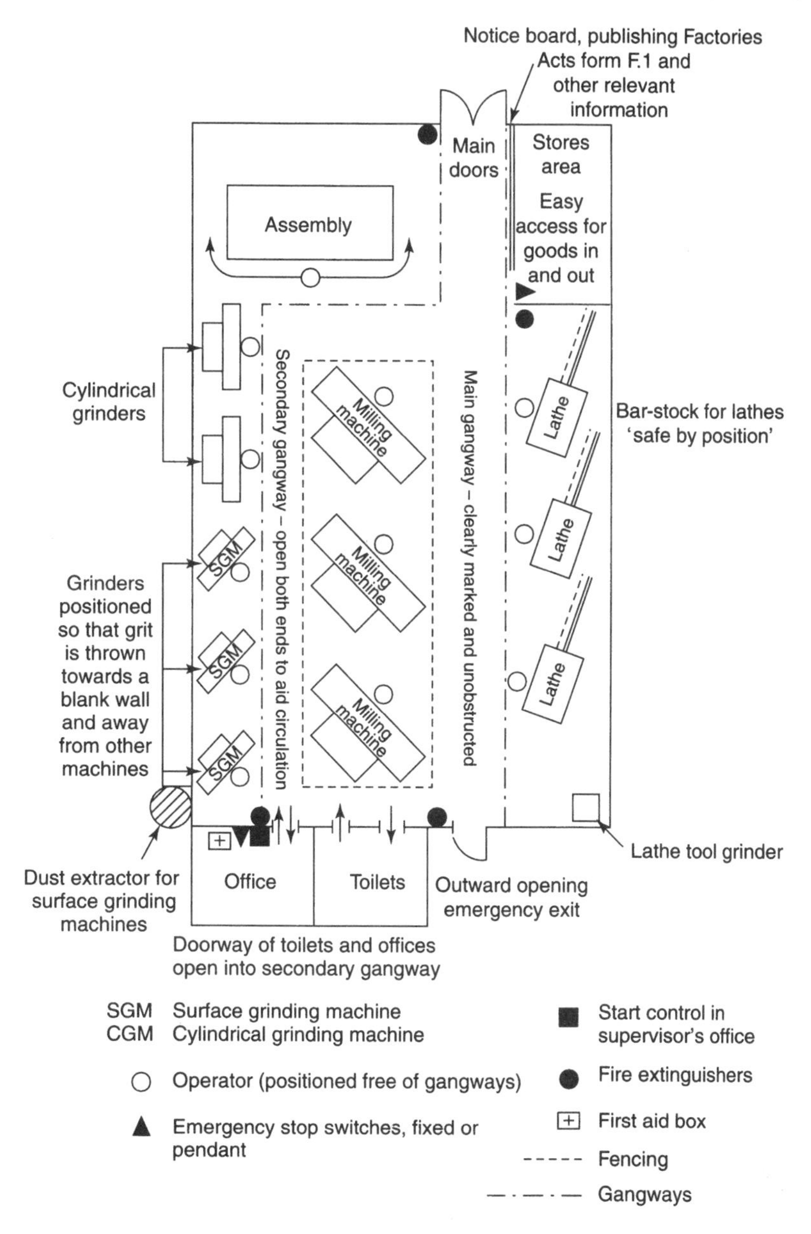

Figure 1.31 Jobbing shop layout (engineering).

standard styles. The various parts of the suit will be cut out and made up by specialist machinists. The characteristics of small batch, prototype and prototype production can be summarized as:

- Work is quoted for job by job. Nothing is manufactured for stock.
- A wide range of general-purpose machines and associated processing equipment (for example heat treatment equipment for a toolroom) is required.
- Highly skilled and versatile operators are required.

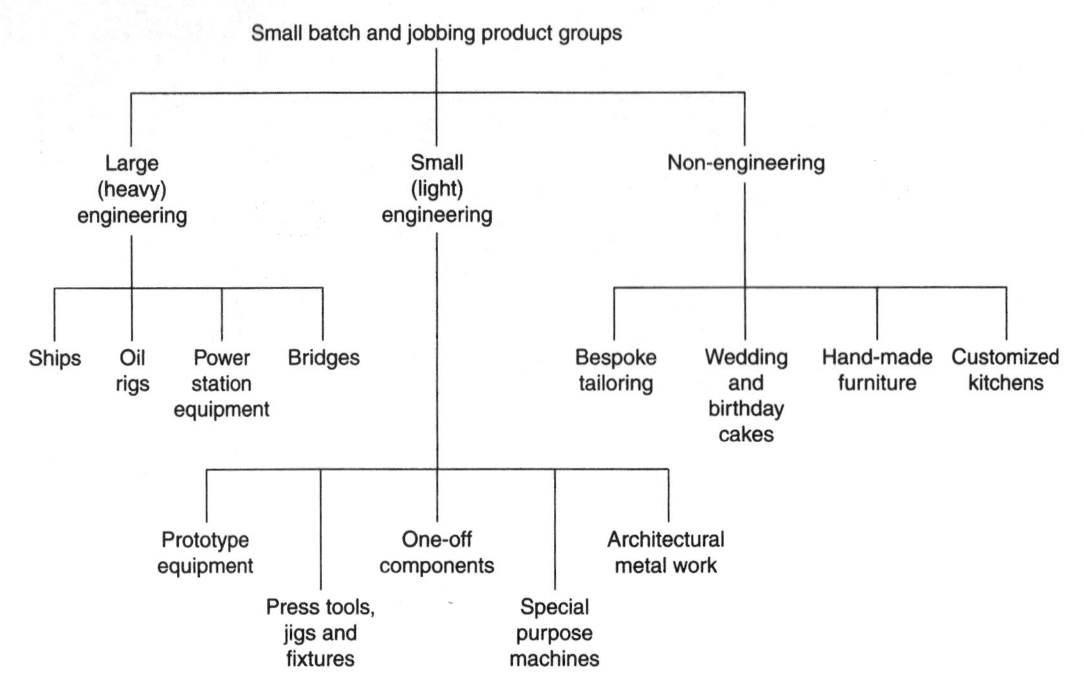

Figure 1.32 Small batch and jobbing product groups.

- Economical loading of the plant and personnel is difficult.
- Often, only outline specifications are provided, therefore design facilities are required.

Some typical products produced on a small batch or a jobbing basis are as shown in Fig. 1.32.

1.12.4 Production systems

Finally, it must be remembered that all production systems have a single aim, to convert raw materials into a finished product that can be sold at a profit. The inputs and outcomes of a typical manufacturing system are shown in Fig. 1.33. These can vary in detail depending upon the product being manufactured. However, it is no good producing goods profitably and at a price the customer is willing to pay if the quality and reliability does not provide the customer with 'value for money'. This is why, in Fig. 1.33, *quality control* embraces every stage of the production system.

Test your knowledge 1.10

1. Briefly summarize the essential differences between:

   (a) Continuous (flow) production;
   (b) Batch production;
   (c) Jobbing production.

2. Give an example where each of the production methods listed above would be appropriate and state the reasons for your choice.

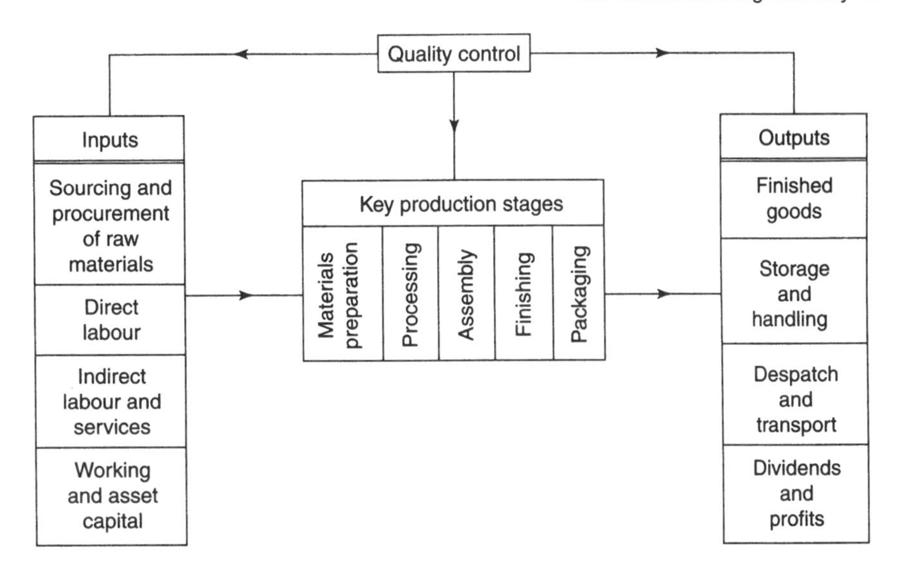

Figure 1.33 A typical production system.

Key notes 1.2

- The key stages of production are: material preparation, processing, assembly, finishing, packaging.
- The scales of production can be: continuous flow production, line (mass) production, batch production, small batch and jobbing production.
- Sourcing is finding suitable sources of raw materials.
- Procurement is negotiating the purchase and delivery of raw material.

Assessment activities

1. Describe what is meant by manufacturing and summarize the importance of the manufacturing industries for the UK in terms of:
   (a) the economy;
   (b) employment.
2. Discuss the factors that influence the location of industry both nationally and globally, with particular reference to the influence of ICT.
3. Discuss the impact of manufacturing industry on the environment and how any damaging effects can be minimized or eliminated.
4. Describe how the waste products of one industry can become the feed stock for another industry.
5. For a product of your own choosing, describe the key stages of its production and describe how the Internet could be used for the sourcing and procurement of raw materials for that product.
6. Describe the scales of production and the problems that are likely to arise when changing from one level to another. Suggest circumstances which are likely to require such changes.

Chapter 2 Design and graphical communication

Summary

When you have read this chapter you should be able to:

- Develop a design specification
- Select a suitable material from a study of material specifications and performance characteristics
- Understand how production details and constraints influence the design process
- Appreciate how quality standards affect the client design brief
- Develop design ideas
- Present a design solution
- Modify a design solution following consideration of the presentation feedback.

2.1 Product design specification

2.1.1 Introduction

Today we live in a society that is dependent upon technology. Our standard of living and the way we live relies heavily upon the use of products, designed and manufactured to make our daily lives easier. All the articles around us have been designed by someone attempting to solve a particular problem. For example, the next time you get up and prepare to go to school, college or work, think about the products around as you wash, dress and have breakfast before leaving home. The clothes you wear, the furniture and appliances in your kitchen, the cooking utensils and food you need for your breakfast, as well as the television and radio you may turn on as you get ready to go out, have been designed and manufactured in response to a particular need. Manufacture cannot exist with out design.

You should begin work on your personal design portfolio (PDP) as soon as you start the unit and then continue to collect evidence of your work as you progress through the units. Your PDP will also enable you to collect the evidence that you need to demonstrate that you have become proficient in key skills.

- *Design* is the process of preparing the detailed information that defines the product to be made.
- *Production* is the process of using the resources of materials, energy and manufacturing technology to convert that design information into saleable products.

Nowadays there is a wide range of materials and information available to us, together with the knowledge of many technologies and skills to help us when designing new products or adapting existing products. Figure 2.1 highlights the key role that the design process plays, whether the company is creating mass-produced products, chemical engineering processes, communications systems, fabric design and manufacture or food processing and drinks manufacturing.

2.1.2 Development of a design brief: the client

In Chapter 1 we defined the key purposes of manufacturing *as creating wealth* and *satisfying consumer demand*. Remember that designing is not always

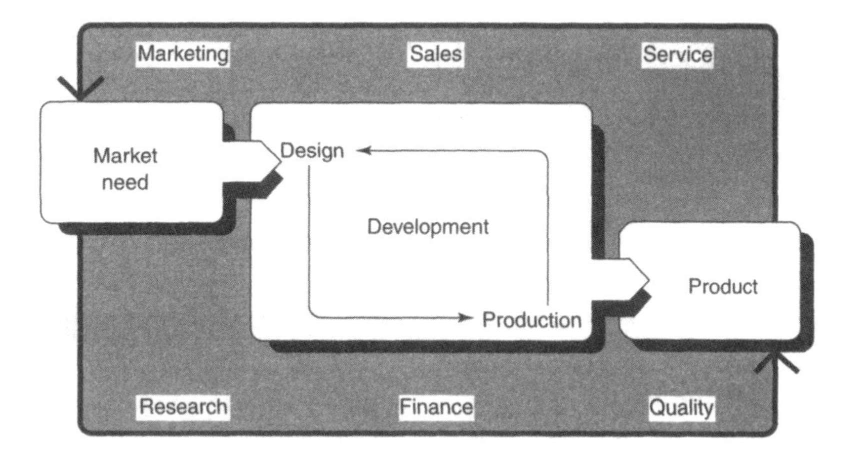

Figure 2.1 The key role of design.

about creating a brand new product or service. In most cases it's about improving or modifying an *existing* product. Just as in the natural world new species have evolved from their predecessors so, in the world of manufacturing, new products evolve from previous products by retaining the good features, eliminating the bad features and introducing technological advances where they are appropriate. The design process starts as soon as a client (customer) need is identified. The client may be:

- An individual, a company or an organization outside that for which a designer works. This is an *external client*.
- A person, a department or a division of the company for which you work. This is an *internal* client.

The relationship between a designer and his or her client is an important one. The client may only have a rough idea of what the end result may look like and how it might work. In other cases the client may be able to suggest just how the product or service will work and what it will look like. For example, in a major new product such as a new military aircraft, a ship, or a bridge, the client will have considerable technical knowledge and have a very precise specification that needs to be fulfilled. In this case the client has a *preconceived* idea of what the product or service will be. Sometimes this is a good thing because it gives the designer a starting point to work from. In many cases, however, preconceived ideas only serve to limit the range and effectiveness of the solution that the designer comes up with.

2.1.3 Identifying client need

The product must meet the needs (requirements) of the customer as closely as possible. This is often a compromise between the ideals of the customer and what can be realized within the technology available and the price the customer is prepared to pay. For a ship, an oil-rig, or a bridge for instance, the customer has expert knowledge and will produce a detailed specification to set the designer on his or her way. On the other hand, products such as cars and domestic appliances have to satisfy the requirements of very many people all of whom will have their own idea of what they want from such a product and how much they are prepared to pay. Market researchers will carry out surveys of what is available and what features need to be changed to attract customers to the new product under consideration. Many of the requirements will conflict with each other and the final design will have to be a compromise.

Client need for a new product may be triggered by any of the following:

- A specific enquiry from a customer
- The development of a new technology
- A new method of applying existing technology
- Identification of a niche in the market
- A decline in the sales revenue of existing products caused by a reduction in demand.

Next the consumer need is translated into a brief description outlining what the product has to do and what its commercial potential may be. In many industries this is called the *product concept*. If the product idea or concept is thought to have commercial potential, a *feasibility study* usually follows. The feasibility study should contain sufficient information to allow senior management to approve or reject the product idea.

2.1.4 The design brief

Before consulting a designer it can be safely assumed that the client, whether internal or external, will have carried out market research to identify the need for the proposed new product or the proposed modification to an existing product. The client will discuss the outcomes of the market research with the designer and between them they will draw up a *design brief*. This design brief will be developed into a detailed *design specification*, or alternative design specifications, for the client to choose from and to approve and from which will be made prototype products for testing and refinement before manufacture.

Therefore the design brief is a description of the client's requirements given to the designer. It should list the requirement in terms of product performance, i.e. functional details, costs and any special requirements such as times scales. It will also list quality standards and safety regulations that the product designer must observe, but must not provide design solutions.

The key requirements of a design brief can be summarized as:

- The customer's concept.
- The functional requirements.
- Market details such as affordable cost, market size, safety legislation and legal requirements.
- Quality standards.

Once the design brief has been agreed, the design process can begin.

2.1.5 The design process

The design process is the name given to the various stages that we go through when we design something. Each stage in the process follows the one that goes before it and each stage is associated with a particular phase in a design process. Remember, when designing an entirely new product or service the design process will have many more stages because it is usually

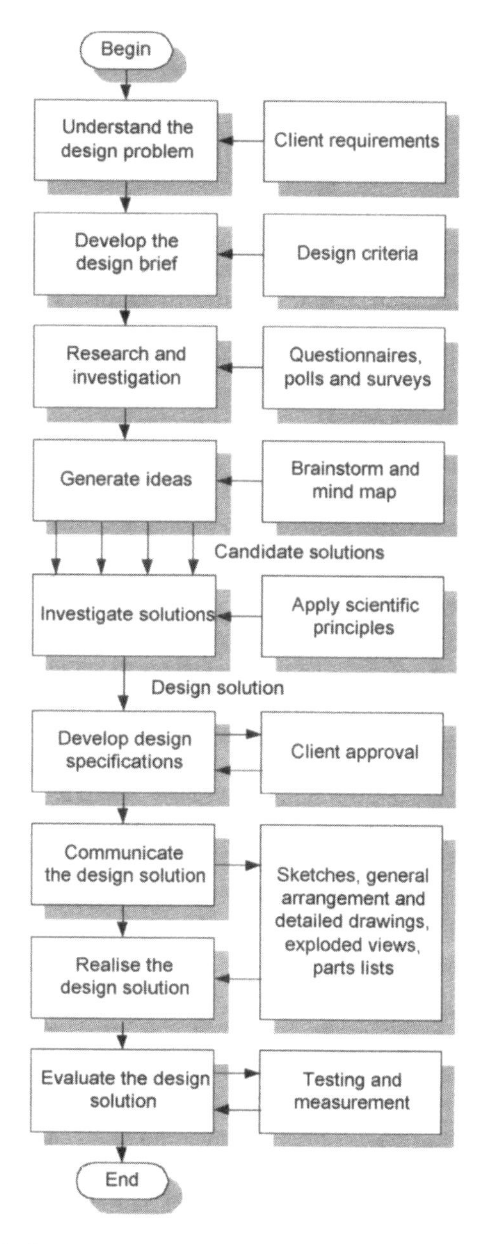

Figure 2.2 The design process.

necessary to consider a wider range of options and alternatives than when simply modifying or redesigning an existing product.

A typical design process involves the following tasks as shown in Fig. 2.2, whilst Fig. 2.3 shows key words derived from the design process.

- Understanding and describing the problem
- Developing a design brief with the client (the client may have already developed the design brief from preconceived ideas arising from previous market research)
- Carrying out research and investigation

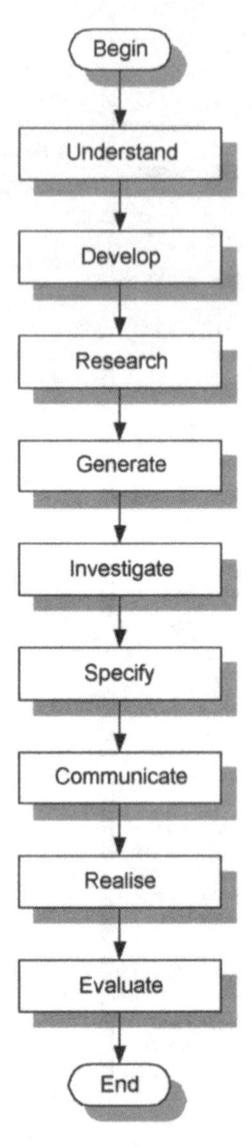

Figure 2.3 The design process described in terms of the key words for each stage of the process.

- Generating ideas and using techniques such as mind mapping
- Investigating solutions and applying scientific principles. Developing an agreed set of design specifications
- Communicating the design solution using appropriate techniques, such as technical drawings (engineering), drawings, templates and patterns (clothing and footwear), and recipes (food processing)
- Realizing the design solution
- Evaluating the design solution.

So designing a product involves a series of clearly defined tasks and each of these forms an important stage (phase) in reaching the eventual goal, *a manufactured product that can satisfy a market demand at a profit to the producer and client.*

Test your knowledge 2.1

1. Draw up a design brief for a simple, low cost product of your own choice, for example, a bird table.
2. Draw up a design brief for improving some article with which you are familiar.

2.1.6 Key features in a product design specification

Having established our design brief, the designer now has to set to work on an initial concept. Let's now look at some of the factors the designer needs to consider when developing the design brief into a working design. These factors can be summarized as:

- Aesthetic
- Contextual
- Performance
- Production parameters
- Quality standards.

Aesthetics

Aesthetic features are concerned with the appreciation of the beauty of products. In other words features that make a product look good. Ideas about what looks good will vary from person to person. But the judgement of what makes a product attractive to look at or pleasing to use or consume is based on the senses of sight, touch, taste, smell and hearing.

- Sight is influenced by the look of a product, i.e. its shape, form or colour.
- Touch is influenced by the feel of a product, i.e. its texture or surface finish. Hand-held products need to be designed to fit the shape and dimensions of the human hand.
- Taste can be sweet, sour, bitter or salty. Smells can be pleasant and exciting or unpleasant and annoying. Smell is usually related to taste.
- Hearing can be irritated by squeaks or rattles, with loud noises becoming intolerable. Excessive noises can be damaging to health. In the case of entertainment equipment, such as CD players, the *quality of the sound* is very important.

Having established that *aesthetic* features are those that make the design attractive to look at and pleasing to use, by way of an example, let's compare a Victorian mangle you may have seen in a museum with a modern wringer. The mangle would be made from wood and cast iron. It would be painted black and would be heavy, bulky and very strong. To our eyes it would be ugly and 'over-engineered'. It would have unguarded gears and would be clumsy to use. Aesthetics would not have been considered in its design. It would simply be functional.

However, the modern wringer will have been designed to satisfy a more discerning market and will be judged in competition with other modern household appliances. Aesthetics will most certainly have been considered in its design. Its lines will be smooth and pleasing to the eye. It will be easy to keep clean and light and easy to use. Modern materials will have been used and there will be no dangerously exposed moving parts. It will be light in colour and small in size to fit into a modern kitchen. Nevertheless it will still be functional. We can summarize our present aesthetic design ideas as being based upon:

- Simplicity of shape
- Smooth lines
- Balance and proportion, attractive colours
- The ability for a product blend with its environment
- Pleasant to our senses (hearing, sight, smell, taste, touch).

Contextual

Contextual design features are those features that have to be included to meet the demands of the general setting, situation or environment in which the product is to be used, that is, the relevance of a design to its working environment and the persons who will be using it. Some of the contextual factors that must be considered could be:

- Will it be aimed at a specific market (male or female) or will it have universal appeal? Will the product be used indoors or outdoors?
- If indoors, will it be used in a damp (kitchen or bathroom) or dry (living room) environment?
- Will it be used in a temperate or a tropical climate?
- What is the target age group?
- Will it be used by persons suffering from physical disabilities?

Performance

The performance of a product is measured by how well it functions when used. For example, the performance of engineering products is frequently defined by criteria such as:

- Input voltage or current
- Mechanical forces applied
- Operating life
- Operating speed range, operating temperature range
- Output power range
- Efficiency
- Performance accuracy
- Running time between services.

On the other hand, the performance of clothing products is more likely to be judged on their ability to:

- Resist wear and tear
- Resist shrinkage when cleaned or washed
- Resist fading
- Resist creasing when worn
- Remain smart in appearance over a reasonable period of time.

Therefore it becomes apparent that the performance criteria varies with the product under consideration.

Production parameters

Let's now consider the production parameters.

Size

The size of a product will influence the cost of materials used. So to keep material costs to a minimum, designers will try to keep product size as small as possible, as long as product function is not affected. If size reduction demands more precise manufacturing methods, product cost may increase. Size will also influence the space required to manufacture the product. For example, to assemble an aircraft needs much more space than that required to assemble a lap-top computer. Manufacturing space requirements will influence overhead costs and eventually total manufacturing costs.

Weight

The design weight of a product will also influence its size and the type of material used. For example, to keep weight as low as possible aluminium alloys are used in the manufacture of aircraft.

Cost

As already seen costs are an important factor, as almost any production parameter can be related to cost. The total cost of a newly designed product is the sum of such cost elements as:

- Design costs
- Development costs
- Material costs and manufacturing costs.

The designer needs to know the magnitude of these costs so that he or she can make informed choices between different materials and different production processes. The probable demand is also important since this will also influence the production process chosen.

Quantity

The scale of production used, i.e. single item, small batch, repetitive batch or continuous flow, will be influenced by the demand (product quantity). As we saw in Chapter 1, this has a significant effect on the manufacturing processes chosen and on the cost of manufacture.

Time

Time is related to the production process and influences the cost of the product. The greater the manufacturing time per unit of output, the greater will be the cost of the product. Conversely, the shorter the manufacturing time, the less will be the cost.

Quality standards

The client specifies product quantity standards. To assure that the product *conforms* to the specified quality standards, careful attention must be paid to their selection by the client. Some typical quality standards are listed below.

- Product performance standards
- Material specifications

- Manufacturing tolerances
- Surface-finish requirements.

Product quality affects cost. Generally, a low quality specification can be equated with low cost. Conversely a high quality specification can be equated with a high cost. Ideally, the product designer produces a design that strikes a balance between these extremes, so that the product has adequate quality and will conform to the requirements of the client, but at the same time is suitable for manufacture at a cost that the customer can afford.

Available labour

Any new product must take into account the availability of labour that has the appropriate skills and can be hired at an economic rate. Many companies now retain those aspects of their companies in the UK where high levels of skill are required, such as design and development, but transfer the routine production overseas where labour costs are lower.

Available plant

Any new product must be designed with the production plant in mind. If only small quantities are involved, then it may not be possible to recover the cost of new plant. In which case the new design must be capable of being made on the existing plant or, if this is not possible, it may be better to subcontract the manufacture of the parts concerned to specialist companies who have the appropriate plant and the volume of work to use it economically.

Test your knowledge 2.2

1. Choose an article of clothing or a household appliance with which you are familiar and briefly assess it in terms of:

   (a) aesthetic design;
   (b) contextual design;
   (c) production parameters.

Key notes 2.1

- Modern society depends increasingly on technology.
- The products of technology need to be designed to suit the needs of the customer.
- Design is the process of preparing the detailed information that defines the product to be made.
- Development is the process of acquiring specific information about the proposed product by constructing and testing prototypes.
- The design brief is the starting point for the product designer and summarizes the functional requirements of the customer's concept, within the constraints of the market parameters such as cost, quantity and quality standards.
- Design aesthetics are concerned with those things that make the product attractive to look at and pleasing to use.
- Contextual design features are concerned with the general setting, situation or environment in which the product is to be used and the consumer groups that will use it.
- Performance is concerned with how well the product functions when used. This will vary from product to product.
- Production parameters are concerned with such physical factors as weight, size, cost, quantity and rate of output.

- Quality standards are concerned with such factors as performance standards, material specifications, manufacturing tolerances, and surface finish requirements. This can be summed up as *fitness for purpose*.
- Production constraints refer to the availability of labour, materials, manufacturing technology, health and safety requirements and quality standards of the production equipment.
- Health and safety laws apply not only to the well-being of the persons involved in manufacture, but also to the consumer and the environment as a whole. Responsibility for health and safety at work rests equally on the employee, the employer, and the suppliers of equipment and materials.
- Design proposals are developed from the design brief.
- The design proposals are assessed for feasibility in order to select the design that will eventually be manufactured.

2.2 Materials used in manufacture

Working from the design brief, material details and constraints and production details and constraints together with quality standards are identified. A number of design proposals are then developed. Each design proposal is critically assessed and the preferred design proposal is presented to the client for approval. Let's now look at the various stages of the design proposal in more detail.

The materials specified as part of the design process will depend entirely upon the product required. The material specified must be readily available, possess the required properties and must not cost more than the customer is prepared to pay. For example, steel could be an ideal material for making a motorway or a railway bridge. It is readily available and combines high strength with relatively low cost. The fact that it is heavy is of little importance. On the other hand it would be totally unsuitable for the manufacture of an aircraft where aluminium alloys combining high strength with light weight would be more appropriate. Again, it would be inappropriate to select a heavy woollen material for summer dresses and lightweight cotton fabric for winter coats in the clothing manufacturing industry. The material must suit the product and the use to which the product is going to be put. Processed foods must not only be nutritious, they must look good, taste good and smell good to appeal to the senses, yet contain no harmful additives if they are going to attract the customer.

When comparing materials the following characteristics should be taken into account:

- Availability and forms of supply
- Properties, characteristics and performance
- Cost
- Health, safety and hygiene requirements
- Handling and storage.

Materials must also be as environmentally friendly as possible in their extraction, use and disposal and come from renewable sources wherever possible or, alternatively, the material must be capable of being recycled.

Almost all known materials are used in the course of manufacturing the enormous range of products available today, and nowhere is the range greater than in engineering, chemical and the construction industries as

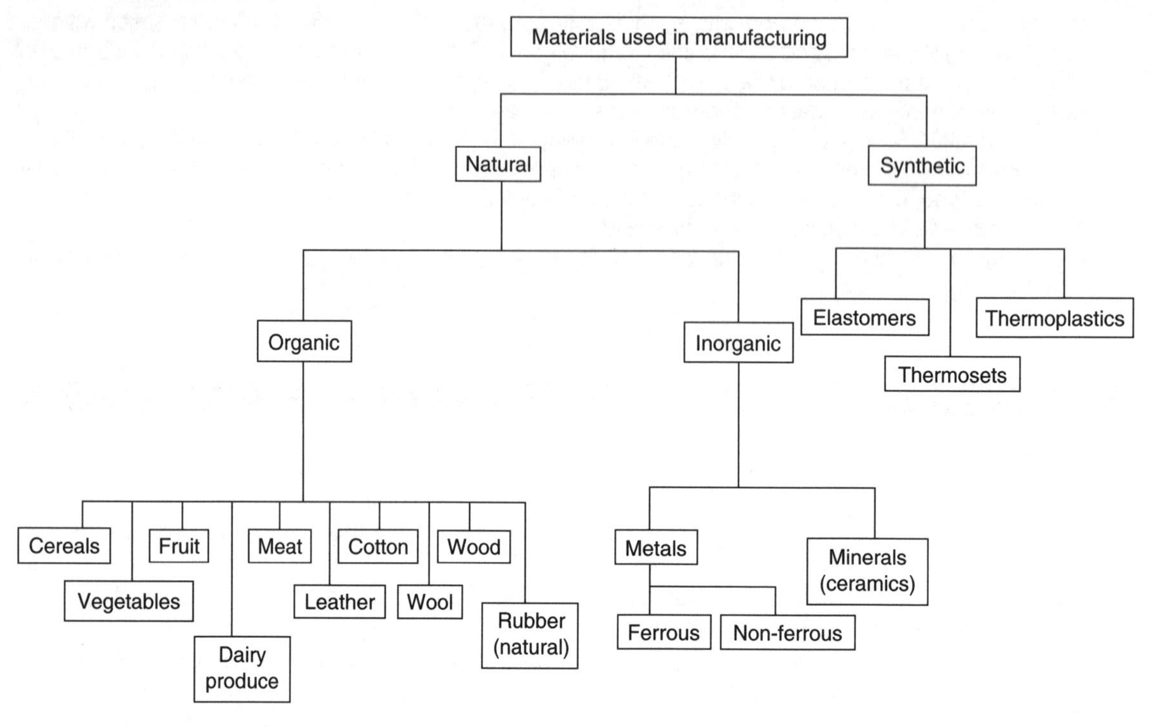

Figure 2.4 Materials used in manufacturing.

described in Chapter 1. Let's commence by looking at the various families of materials as shown in Fig. 2.4.

2.2.1 Availability and forms of supply (metals)

Let's first consider the metals. The Latin name for the metal iron is *ferrum*, so metals and alloys based on iron are referred to as ferrous metals and alloys. *Alloys* consist of two or more metals (or metals and non-metals) that have been brought together as compounds or as solid solutions to produce a metallic material with special properties. For example, an alloy of iron, carbon, nickel and chromium is *stainless steel* since it resists corrosion. Plain carbon steels are based on the metal iron together with between 0.1 per cent and 1.4 per cent carbon. The effect of the amount of carbon present on the properties of the steel is shown in Fig. 2.5. Cast irons are also ferrous metals. Grey cast irons contain between 3.2 per cent and 3.5 per cent carbon. Since the maximum amount of combined carbon that can be present is 1.8 per cent, it follows that the surplus carbon remains as flakes of graphite between the crystals of metal. It is these flakes of graphite that give cast iron its particular properties and makes it 'dirty' to machine. The compositions and typical uses of plain carbon steels and grey cast iron are shown in Table 2.1

Non-ferrous metals and alloys are the rest of the metallic materials available. For example, copper is used for electrical conductors, brass is an alloy of copper and zinc and the properties and applications of typical brass alloys are shown in Fig. 2.6. Brass alloys are difficult to cast and brass castings tend to be coarse grained and porous. Brass depends upon hot rolling from cast

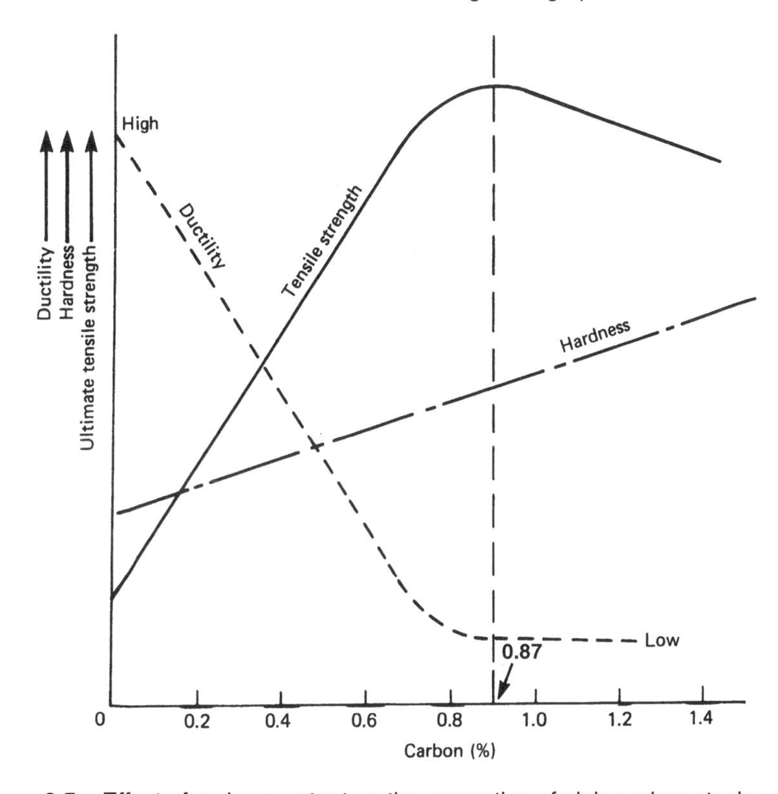

Figure 2.5 Effect of carbon content on the properties of plain carbon steels.

Table 2.1 Ferrous metals

Name	Group	Carbon content (%)	Some uses
Dead mild steel (low carbon steel)	Plain carbon steel	0.10–0.15	Sheet for pressing out components such as motor car body panels. General sheet-metal work. Thin wire, rod and drawn tubes.
Mild steel (low carbon steel)	Plain carbon steel	0.15–0.30	General purpose work-shop rod, bars and sections. Boiler plate. Rolled steel beams, joists, angles, etc.
Medium carbon steel	Plain carbon steel	0.30–0.50	Crankshafts, forgings, axles, and other stressed components.
		0.50–0.60	Leaf springs, hammer heads, cold chisels, etc.
High carbon steel	Plain carbon steel	0.8–1.0	Coil springs, wood chisels.
		1.0–1.2	Files, drills, taps and dies.
		1.2–1.4	Fine-edge tools (knives, etc.)
Grey cast iron	Cast iron	3.2–3.5	Machine castings.

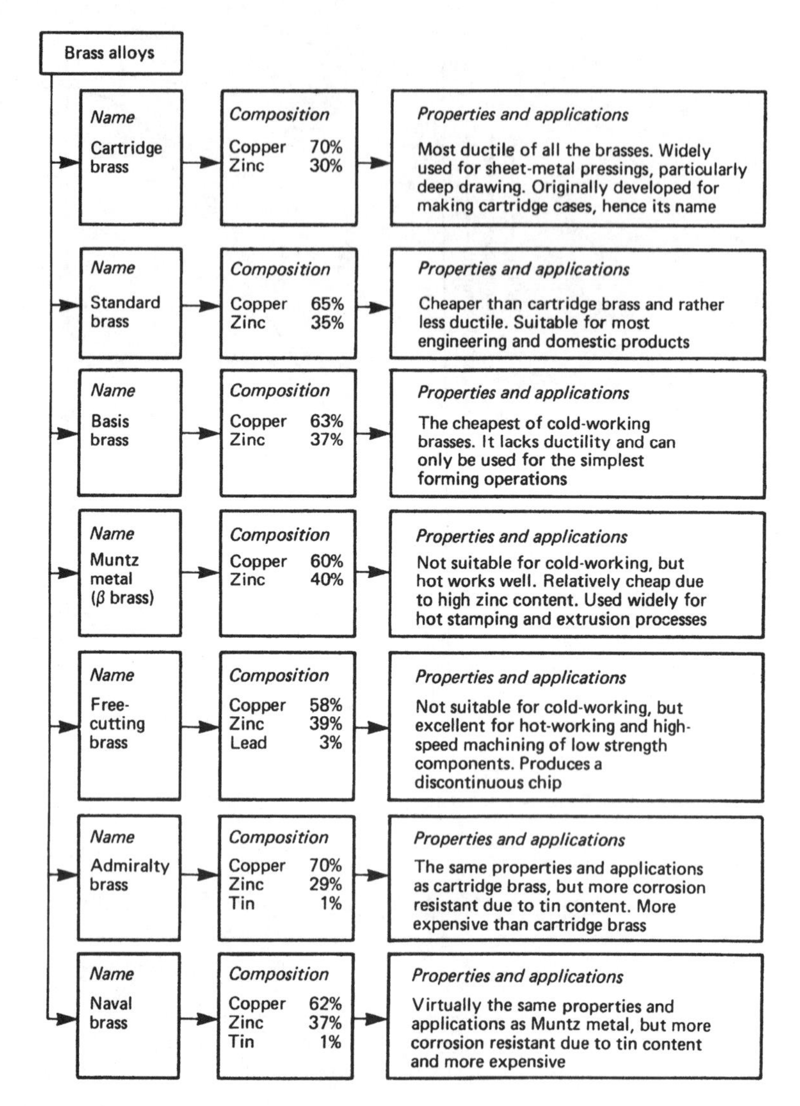

Figure 2.6 Properties and applications of brass alloys.

ingots to consolidate the metal followed by cold rolling or cold drawing to give it mechanical strength.

Tin bronze alloys consist mainly of copper and tin with a trace of zinc or phosphorus as a deoxidizing agent.

- Gunmetal bronze is an alloy of copper and tin with a small amount of zinc.
- Phosphor bronze is an alloy copper and tin with a small amount of phosphorus.

Unlike the brass alloys, bronze alloys are usually used as castings. They are extremely resistant to corrosion and wear. Gunmetal castings are used for high-pressure valve bodies and phosphor bronze castings are used for heavy-duty bearings. Some low tin content phosphor bronze alloys can be

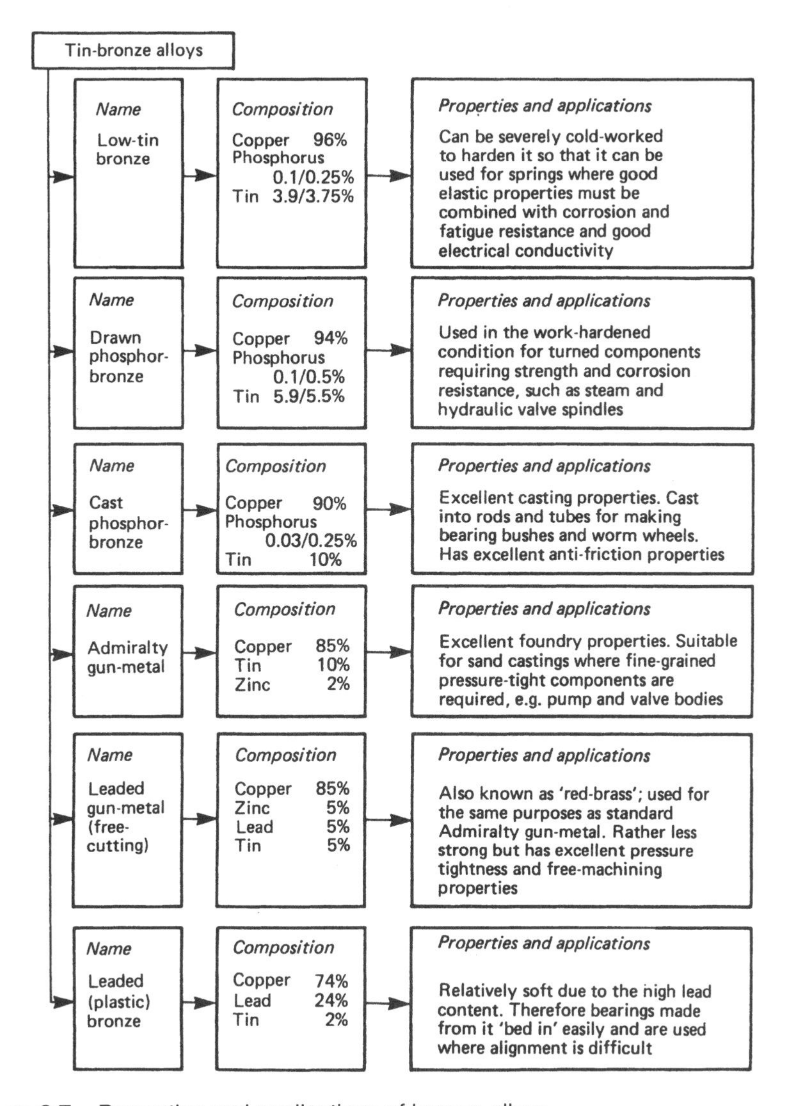

Figure 2.7 Properties and applications of bronze alloys.

extensively cold worked. Figure 2.7 lists some typical bronze alloys and their applications. Copper and its alloys are readily available, but much more expensive than plain carbon steels and are only used where their special properties are a priority.

Aluminium is not used to any great extent on its own but combined with other metals to make the light alloys used in the aircraft industry. Duralumin (one of the strongest general-purpose aluminium alloys) consists of aluminium together with copper, manganese, magnesium and silicon. Another alloy that is even stronger and used for highly stressed aircraft components and military equipment consists of aluminium together with magnesium, zinc and titanium. Although readily available, light alloys are more expensive than plain carbon steels and are mainly used where weight saving is a priority.

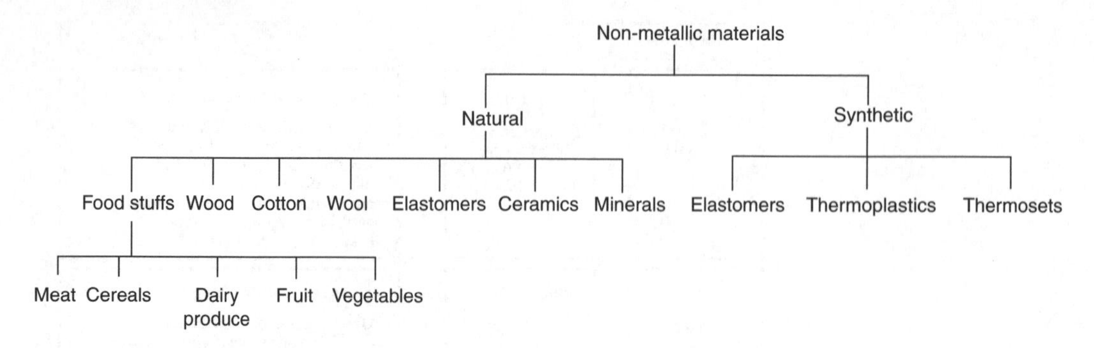

Figure 2.8 Non-metallic materials.

2.2.2 Availability and forms of supply (non-metals)

Figure 2.8 shows the main groups of non-metals. They can be natural, such as rubber, wood, cotton and wool or they can be manufactured from naturally occurring minerals as in the case of ceramics. Plastic compounds are mainly synthetic and are derived from the byproducts of oil refining and from coal derivatives. The plastic moulding materials are manufactured in bulk by the chemical engineering industries.

Ceramics

Ceramics are inorganic, non-metallic materials that are processed and/or may be used at high temperatures. The word ceramic comes from the Greek word for potter's clay, Ceramics consist mainly of silicon and/or naturally occurring metallic elements chemically combined with non-metallic elements such as oxygen, carbon and nitrogen. They are used for the manufacture of a wide range of products including:

- Glass products
- Abrasive and cutting tool materials
- Construction industry materials
- Electrical insulators
- Cements and plasters for investment moulding
- Refractory (heat resistant) linings for furnaces
- Refractory coatings for metals.

The four main groups of ceramic materials and some typical applications are summarized in Fig. 2.9.

Elastomers

Elastomers are substances that permit extreme reversible extensions to take place at normal room temperatures. Natural rubber is an obvious and important elastomer.

Natural rubber

Naturally occurring *rubber* (latex) comes from the sap of a tree called *Hevea brasiliensis.* This material is of little use in manufacturing since it becomes tacky at elevated temperatures, has a very low tensile strength at room

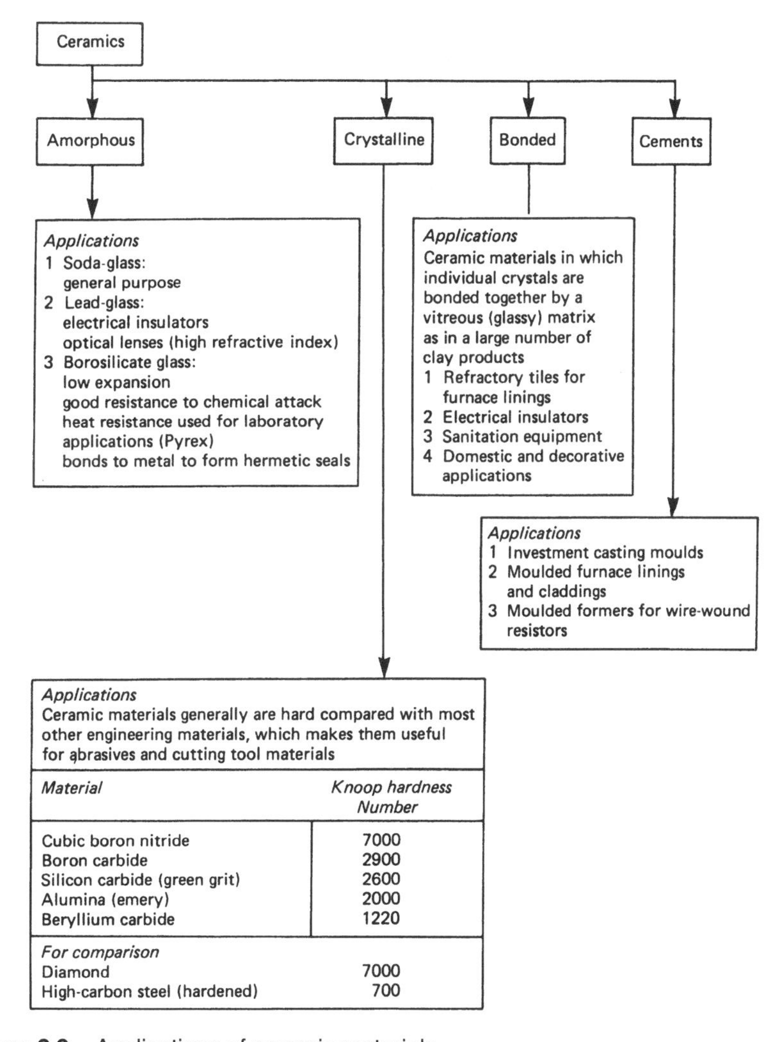

Material	*Knoop hardness Number*
Cubic boron nitride	7000
Boron carbide	2900
Silicon carbide (green grit)	2600
Alumina (emery)	2000
Beryllium carbide	1220
For comparison	
Diamond	7000
High-carbon steel (hardened)	700

Figure 2.9 Applications of ceramic materials.

temperature, and is attacked by oils and quickly perishes in the presence of the ultraviolet rays of sunlight. When compounded with other substances, such as sulphur and carbon black and then vulcanized, its working life and usefulness is extended by improving its strength and ageing characteristics. Typical products made from natural rubber include vehicle tyres (good anti-skid properties), inner tubes, footwear, respirator masks, gloves and rainwear. Although natural rubber is a naturally occurring elastomer, the term *elastomer* is usually reserved for synthetic rubbers.

Synthetic rubbers (elastomers)

The original synthetic rubber – neoprene rubber – was developed during the Second World War when access to sources of natural rubber was denied to both sides. Table 2.2 lists the properties, availability and some uses for synthetic elastomers.

Table 2.2 Elastomers

Acrylic rubbers
These are derived from the same family of polymeric materials as 'Perspex' but, in order to give them the properties associated with elastomers, they are not so heavily cross-linked. This group of rubbers has excellent resistance to oils, oxygen, ozone and ultraviolet radiation and they are used as the basis for the latex paints developed for motor vehicles.

Butyl rubber
This rubber is impervious to gases and is used as a vapour barrier and for hose linings. It is highly resistant to outdoor weathering and ultraviolet radiation and is used for construction industry sealants.

Nitrile rubber
This has excellent resistance to oils and solvents and can be readily bonded to metals. It is used for petrol and fuel oil hoses, hose linings and aircraft fuel tank linings. It is also resistant to refrigerant gases.

Polychloroprene rubber (neoprene)
This was the original synthetic rubber developed during the Second World War. It has good resistance to oxidation, ageing and weathering. It is resistant to oils, solvents, abrasion and elevated temperatures. Because of its chlorine content it is fire resistant. It is used as a flexible electrical insulator, gaskets, hoses, engine mounts, sealants, rubber cements and protective clothing.

Polysulphide rubber (thiokol)
Although this rubber has low mechanical strength, its resistance to solvents and its impermeability to gases is excellent and its weathering characteristics are outstanding. It also has good bonding properties and is widely used in the construction industries as a sealant. Thiokol and polyurethane rubbers are also used as fuels for solid-fuel rockets.

Polyurethane rubber
Polyurethane can be formulated to give either plastic or elastic properties. Although it has relatively high strength and abrasion resistance, it is of little use for vehicle tyres as it has low skid resistance. However, its outstanding service life makes it suitable for cushion tyres for warehouse trucks and forklift trucks where low speeds and dry floor conditions makes its low skid resistance more acceptable. It is also used for shoe heels, painting rollers, mallet heads, oil seals, diaphragms, anti-vibration mountings, gears and pump impellers.

Rubber hydrochloride
This material is better known as 'Pliofilm' and is used to form a transparent film for the vacuum packaging of foods and DIY hardware. It is easily identified by its unusually high tensile strength and tear resistance.

Silicone rubbers
Although silicone rubber has a relatively low tensile strength, it has an exceptionally wide working temperature range of −80 to +235°C. Thus it often outperforms other rubbers which are superior at room temperature, but which cannot exist at such temperature extremes. It can be used for mould linings and high-temperature seals. It is also used in space vehicles and artificial satellites.

Styrene–butadiene rubber (SBR)
A general-purpose synthetic rubber used for tyres, belts, floor tiles and latex paints. It is superior to natural rubber in respect of skid resistance, solvent resistance and weathering.

Table 2.3 Thermosetting plastics

Type	Applications
Phenolic resins and powders	The original 'Bakelite' type of plastic materials, hard, strong and rigid. Moulded easily and heat 'cured' in the mould. Unfortunately, they darken during processing and are only available in the darker and stronger colours. Phenolic resins are used as the 'adhesive' in making plywoods and laminated plastic materials (Tufnol)
Amino (containing nitrogen) resins and powders	The basic resin is colourless and can be coloured as required. Can be strengthened by paper-pulp fillers and are suitable for thin sections. Used widely in domestic electrical switchgear
Polyester resins	Polyester chains can be cross-linked by adding a monomer such as styrene, when the polyester ceases to behave as a thermoplastic and becomes a thermoset. Curing takes place by internal heating due to chemical reaction and not by heating the mould. Used largely as the bond in the production of glass-fibre mouldings
Epoxy resins	The strongest of the plastic materials used widely as adhesives, can be 'cold cast' to form electrical insulators and used also for potting and encapsulating electrical components

Plastics

Plastic is the popular name for a group of materials scientifically known as *polymers*. They may be soft and flexible or hard and brittle or anything in between, however they all become *plastic during the moulding process*. They may be classified as *thermosetting* plastics or *thermoplastics*.

- *Thermosetting plastics* are also known thermosets. These materials are available in powder or granular form and consist of a synthetic resin mixed with a filler material. The filler reduces the cost and modifies the properties of the material. A colouring agent and a lubricant is also added. The lubricant helps the plasticized moulding material to flow into the fine detail of the mould. The moulding material is subjected to heat and pressure in the moulds during the moulding process. The heat from the moulds changes the plastic material chemically. This chemical change is called *polymerization* or, more simply, *curing*. Once cured the moulding is hard and rigid, and cannot be softened by heating. It cannot be recycled. If made hot enough it will just burn. Some typical thermosets and their typical applications are summarized in Table 2.3.
- *Thermoplastics*, unlike thermosets, soften every time they are heated. In fact, waste material trimmed from the mouldings can be ground up and recycled. They tend to be less rigid and more 'rubbery' than the thermosetting plastic materials. Some typical thermoplastics and their applications are summarized in Table 2.4. Thermoplastics can be recycled.

The materials described so far are available to manufacturing companies in a variety of forms as shown in Fig. 2.10. They may be supplied as castings,

Table 2.4 Thermoplastic materials

Type	Material	Characteristics
Acrylics	Polymethyl-methacrylate	Materials of the 'Perspex' or 'Plexiglass' types. Excellent light transmission and optical properties, tough, non-splintering and can be easily heat-bent and shaped. Excellent high-frequency electrical insulators.
Cellulose plastics	Nitro-cellulose	Materials of the 'celluloid' type. Tough, waterproof, and available as performed sections, sheets and films. Difficult to mould because of their high flammability. In powder form nitro-cellulose is explosive.
	Cellulose acetate	Far less flammable than nitro-cellulose and the basis of photographic 'safety' film. Frequently used for moulded handles for tools and electrical insulators.
Fluorine plastics (Teflon)	Polytetrafluoro-ethylene (PTFE)	A very expensive plastic material, more heat resistant than any other plastic. Also has the lowest coefficient of friction. PTFE is used for heat-resistant and anti-friction coatings. Can be moulded (with difficulty) to produce components with a waxy feel and appearance.
Nylon	Polyamide	Used as a fibre or as a wax-like moulding material. Tough, with a low coefficient of friction. Cheaper than PTFE but loses its strength rapidly when raised above ambient temperature. Absorbs moisture readily, making it dimensionally unstable and a poor electrical insulator.
Polyesters (Terylene)	Polyethylene-teraphthalate	Available as a film or in fibre form. Ropes made from polyesters are light and strong and have more 'give' than nylon ropes. The film makes an excellent electrical insulator.
Vinyl plastics	Polythene	A simple material, relatively weak but easy to mould, and a good electrical insulator. Used also as a waterproof membrane in the building industry.
	Polypropylene	A more complicated material than polythene. Can be moulded easily and is similar to nylon in many respects. Its strength lies between polythene and nylon. Cheaper than nylon and does not absorb water.
	Polystyrene	Cheap and can be easily moulded. Good strength but tends to be rigid and brittle. Good electrical insulation properties but tends to craze and yellow with age.
	Polyvinylchloride (PVC)	Tough, rubbery, practically non-flammable, cheap and easily manipulated. Good electrical properties and used widely as an insulator for flexible and semi-flexible cables.

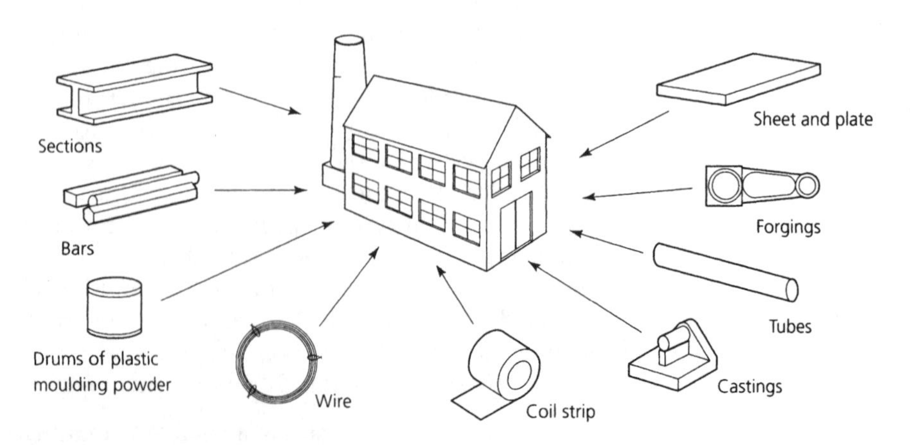

Figure 2.10 Forms of supply.

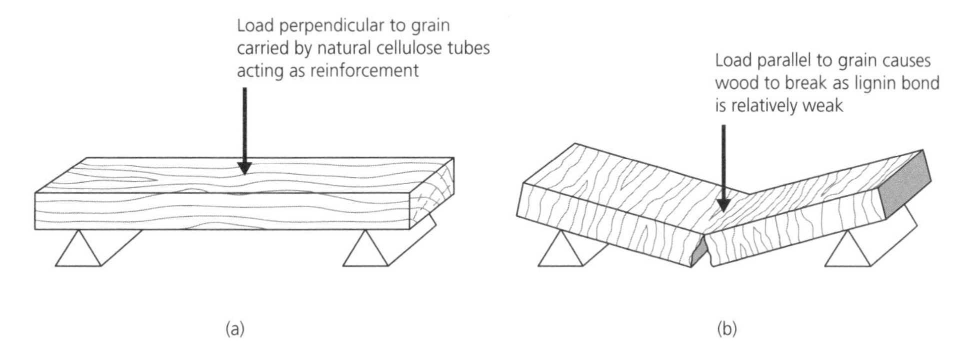

Figure 2.11 Effect of the direction of grain on the strength of wood.

forgings, rolled and extruded sections, bars, coiled strip, wire, tubes, sheets and plates. Plastic materials are often supplied in paper sacks and in drums. All these materials are readily available in large quantities and at relatively low cost.

Composites

Composites are materials in which a relatively weak matrix is used to bond together high strength reinforcing fibres or particles.

Timber (wood) products

Wood is a natural composite. It consists of hollow cellulose fibres through which the sap flows, bonded together by a matrix of lignin. The cellulose fibres are strong but the lignin bond is relatively weak. Therefore it is the cellulose fibres in wood that gives it its characteristic grain. Wood in the form of beams and planks is sawn from the trunks of trees. Most woods used in the construction industry are softwoods. These come from evergreen trees (conifers). Hardwoods come from deciduous trees (oak, elm, mahogany, teak, etc.) which lose their leaves in the winter. Deciduous trees are very slow growing and many are found in the tropical rain forests. For environmental reasons the use of hardwoods should be avoided wherever possible.

Solid wood cut from the trunk of a tree has the grain running in one direction only. This results in its strength being highly directional. Figure 2.11(a) shows a plank being bent in a plane *perpendicular* to the lay of the grain. Loading the plank in this way exploits its greatest strength characteristics. Figure 2.11(b) shows what would happen if a piece of wood is loaded so that bending occurs *parallel* to the lay of the grain. The wood breaks easily since the lignin bond is relatively weak compared with the natural cellulose reinforcement fibres.

Plywood

Plywood overcomes this problem and is a synthetic composite that exploits the directivity of natural wood. Figure 2.12 shows how plywood is built up from veneers (thin sheets of wood) bonded together by a high-strength and water-resistant synthetic resin adhesive. The veneers are stacked so that the direction of the grain of each successive layer is at right angles to the preceding one. When the correct number of veneers have been laid up, the adhesive component of the composite is *cured* under pressure in a hydraulic press.

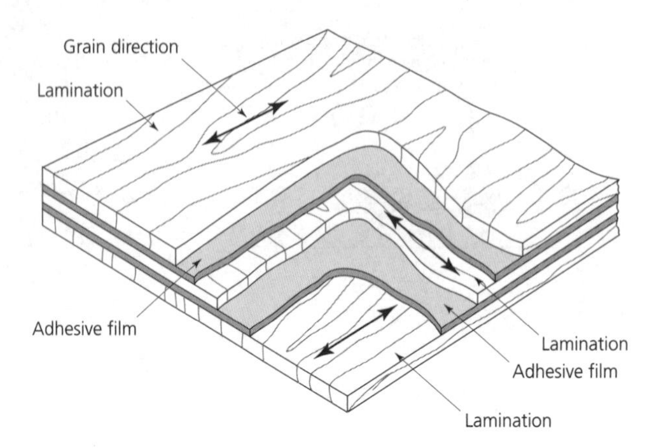

Figure 2.12 Structure of plywood.

Originally, when timber was sawn from the trunks of trees much of the available wood was wasted since only the *heartwood* was used. Nowadays timber technology has advanced so that most of the available wood is utilized in one way or another. The bark and brushwood is ground up for horticultural use. The branches and *sapwood* is ground up and mixed with synthetic resin adhesive to make *chipboard*.

Laminated plastic (Tufnol)

Tufnol is made from fibrous materials such as paper, woven cotton cloth, woven woollen cloth, woven glass fibre cloth, etc., impregnated with a thermosetting resin. The sheets of impregnated material are laid up in powerful hydraulic presses and they are heated and squeezed until they become solid and rigid sheets, rods, tubes, etc. This material has high strength and good electrical properties. It can be machined with ordinary metal working tools and machines. Tufnol is used for making insulators, gears and bearing bushes.

Glass fibre reinforced plastics (GRP)

Woven glass and chopped strand mat can be bonded together by polyester or by epoxy resins to form mouldings. These may range from simple objects such as crash helmets to complex hulls for ocean-going racing yachts. The thermosetting plastics used to bond the fibre reinforcement together are cured by chemical action at room temperature and a press is not required. The glass fibre is laid up over plaster or wooden moulds and coated with resin that is well worked into the reinforcing material. Several layers or *plies* may be built up according to the strength required. When cured the moulding is removed from the mould. The mould can be used over again as many times as is required. *Note*: To aid removal of the complete and cured moulding, the mould is coated with a release agent before moulding commences.

Cermets

These are materials that combine the ductility and toughnes of a metal matrix with the hardness and compressive strength of ceramic particles to provide particle reinforcement. Certain metal oxides and carbides can be bonded together and 'sintered' into a metal powder matrix to form important composite materials called *cermets*. This is short for *ceramic* reinforced *metals*.

Sintering is a process where the compacted powder in the shape of the finished product is heated to a high temperature in a furnace that has a reducing atmosphere to remove any oxygen present. Cermets are used in the production of very hard and abrasion resistant cutting tools that can operate continuously at much higher temperatures than conventional high-speed steel (HSS) tools. Modern CNC machine tools invariably use cermet tipped tooling, i.e. tungsten carbide tipped tools and mixed carbide tipped tools and boron nitride tipped tools.

Test your knowledge 2.3

1. State typical applications for the following materials giving the reason for your choice.

   (a) Low carbon steel
   (b) High carbon steel
   (c) Cast iron (grey)
   (d) Ductile brass
   (e) Phosphor bronze
   (f) Gunmetal bronze
   (g) Duralumin
   (h) A thermosetting plastic
   (i) A thermoplastic
   (j) An elastomer
   (k) A ceramic material
   (l) A cermet.

2. Explain what is meant by a 'composite material' and give an example and an application of

   (a) A laminated composite material
   (b) A fibre-reinforced composite material
   (c) A particle-reinforced material.

2.2.4 Fibres and yarns

Natural fibres

Natural fibres come from plants and animals. *Cotton fibres* come from the seed heads of the cotton plant. *Linen fibres* come from the stem of the flax plant. These are both cellulosic fibres. *Wool fibres* are animal hairs usually from sheep, but also from goats, camels, rabbits and llamas. *Silk fibres* are made by a caterpillar known as a silk worm. The silk worm uses the fibres to make a cocoon in which to change from a caterpillar into a moth. Wool and silk fibres are protein fibres.

Man-made (synthetic) fibres

Man-made fibres do not occur naturally. One group of man-made fibres is known as *cellulosic fibres*. A natural starting point such as wood pulp is treated to extract the cellulose. This is then treated with chemicals to form a thick sticky liquid like treacle. It is then forced through a spinnaret to form long strands that are solidified using a chemical bath, warm air, or cold air. Fibres made in this way are *viscose, modal, acetate* and *triacetate*. Fibres that are entirely made from chemicals are known as synthetic fibres. The chemicals used are generally extracted from coal or oil. Fibres that belong to this group include *polyamide* (*nylon*), *polyester*, *acrylic* and *elastane* (stretch fabrics).

Table 2.5 Performance characteristics of textile fibres

Performance characteristics	Acetate	Acrylic	Cotton	Linen	Polyamide (nylon)	Polyester	Silk	Triacetate	Viscose	Wool
Abrasion resistance	*	***	**	***	****	****	**	*	*	**
Absorbency	**	*	***	***	*	*	****	**	****	****
Crease resistance (stretch)	**	***	*	*	****	****	**	***	**	****
Flame resistance	*	*	*	*	**	**	****	*	*	****
Insulation	**	***	*	*	***	***	***	**	**	****
Moth resistance	****	****	****	****	****	****	**	****	****	*
Mildew resistance	****	****	*	*	****	****	*	****	*	*
Resistance to acids	***	****	**	**	*	****	*	**	**	***
Resistance to alkalis	*	***	****	***	***	****	*	***	***	**
Resistance to bleach	*	***	***	***	***	****	*	*	*	*
Resistance to build up of static electricity	*	*	****	****	*	*	****	*	****	****
Tensile strength	*	**	***	***	****	****	****	*	**	*
Thermal conductivity	*	*	***	****	*	*	*	*	**	*
Thermoplasticity	**	**	n/a	n/a	***	****	n/a	***	n/a	n/a

Microfibres are fibres that are less than one denier thick – this is very fine. When woven into cloths they are lightweight yet strong, crease resistant, yet drape beautifully. Polyester and nylon fibres are particularly suitable for this treatment. Table 2.5 summarizes the performance characteristics of the textile fibres discussed above.

Smart fibres represent cutting-edge technology and respond in various ways to changes in the environment and will be considered in Chapter 4.

Yarns

Yarns are made by a process known as *spinning*. The fibres are arranged to lie roughly parallel to each other and then they are twisted (spun) together. The twisting (spinning) brings the fibres into close contact with each other and the friction between the fibres holds them together. The exact process used to spin the fibres will change the appearance and the performance characteristics of the yarn produced. The spun yarn can then be woven or knitted to make a fabric.

Cotton, linen and wool fibres are quite short and are known as *staple fibres*. A yarn made from this type of fibre has to be tightly twisted to hold the fibres together. This type of yarn has a slightly 'hairy' surface that does not reflect light well and has a matt appearance. It is known as staple yarn.

Longer fibres are known as *continuous filament fibres*. They require less twisting to form the yarn. The surface of the yarn is smoother and reflects light, giving a more lustrous appearance. Yarns made from continuous filament fibres are usually less bulky. This type of yarn is known as *filament yarn*.

Woollen and worsted spinning systems

The woollen system of spinning can be used to make yarn from almost any fibre, not just wool. In this system the fibres can be quite short and not highly combed during the spinning process. This produces a yarn that is coarse and hairy. In the worsted system, longer fibres are used and they are repeatedly combed so that the fibres are parallel. This gives a smooth, regular yarn that is hard wearing.

Complex and plied yarns

A yarn made by twisting fibres together is called a *single yarn*. This can be twisted together with other single yarns to form a *multiple-ply yarn*. Two single yarns twisted together make a *two-ply yarn* and three single yarns twisted together makes a *three-ply yarn* and so on. As many as twelve yarns can be plied together. Plying yarns produces a stronger, more even yarn which is more balanced if a 'Z' twist (clockwise twist) is plied with an 'S' twist (anticlockwise twist). Multiple-ply yarns can be twisted together to make a *corded yarn*. Corded yarn is very strong and has a wide range of uses from decorative embroidery threads to yachting ropes.

Interesting effects can be created by twisting different colours and thicknesses of yarn together. Yarns made by plying together different types of fibre (for example natural and synthetic fibres) combine the different performance characteristics of the various types of fibre. All these yarns are *plain yarns*. That is, they are smooth and regular along their length. Fancy yarns are uneven and vary along their length. A selection is shown in Fig. 2.13.

2.2.5 Woven fabrics

With the exception of sewing, embroidery and knitting, yarns are woven into cloth before being made into garments. Weaving is the most common method of producing fabric from yarn. Two sets of yarn are interlaced at right angles on a machine called a loom as shown in Fig. 2.14

One set of parallel yarns is put into the loom. These are called the *warp* yarns. They have to be strong in order to withstand the strain of the weaving process. They run the length of the fabric and are referred to as the *straight grain*. The *weft* yarns are woven in and out of the warp yarns across the fabric. The weft yarns are carried across the fabric by a shuttle. The *heddle* lifts alternate warp yarns to create a space called a *shed* through which the shuttle carrying the weft yarns can pass. The first heddle is then lowered and the second heddle is raised to provide a new shed for the return of the shuttle. The weft yarns are pushed into place by the *reed*.

When the shuttle reaches the edge of the fabric it passes around the last warp yarn and travels back across the fabric. This provides an edge called a *selvedge*. A selvedge prevents the fabric from fraying. The fabric width is the distance between the selvedge on one side and the selvedge on the opposite side of the fabric. Unfortunately woven fabrics fray at cut edges where no selvedge exists. Woven fabrics are quite firm and do not stretch much. Any stretch that does occur is diagonally across the fabric. This is known as *stretching on the bias*. A woven fabric is strongest along the grain of the fabric, following the line of the warp yarns.

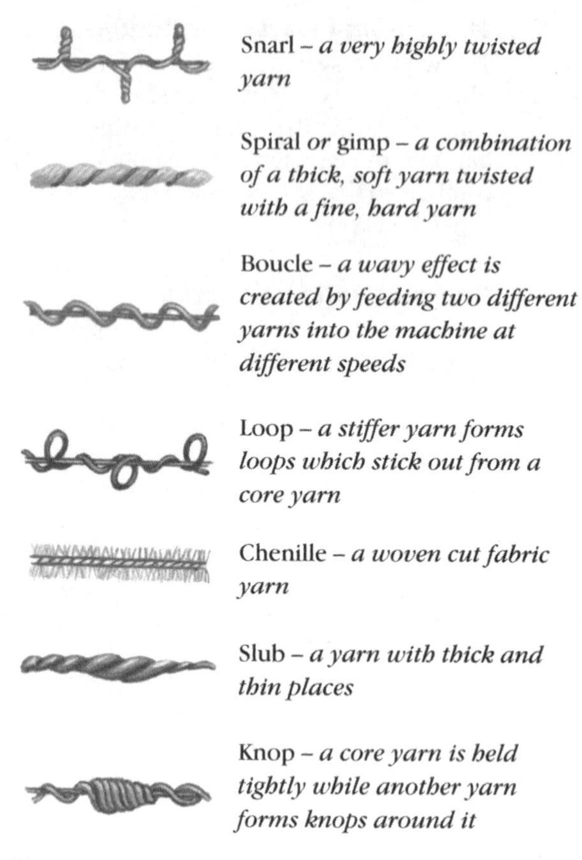

Figure 2.13 Fancy yarns.

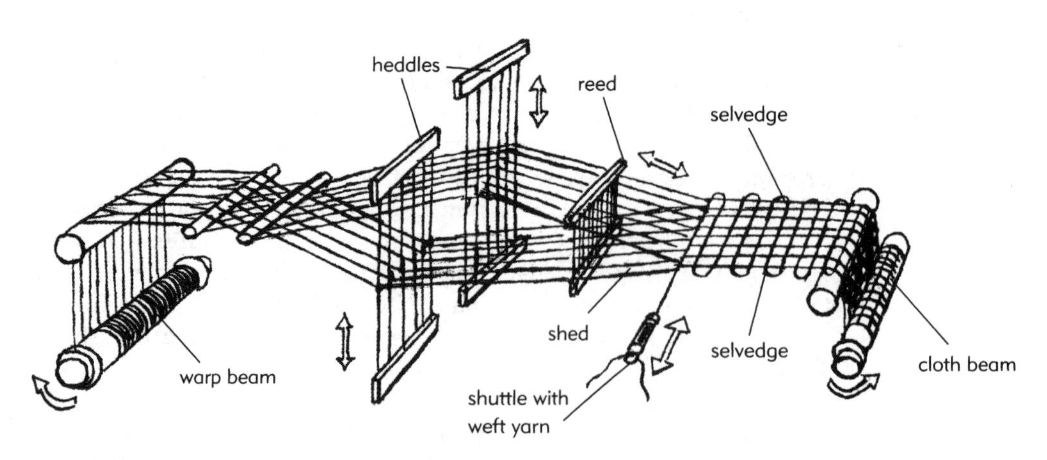

Figure 2.14 A basic weaving loom.

Plain weave

An example of plain weave is shown in Fig. 2.15(a). This is the simplest weave. The weft yarn passes over one warp yarn and under the next. On the return, it passes over the yarns it passed under on the previous row. Plain weave looks

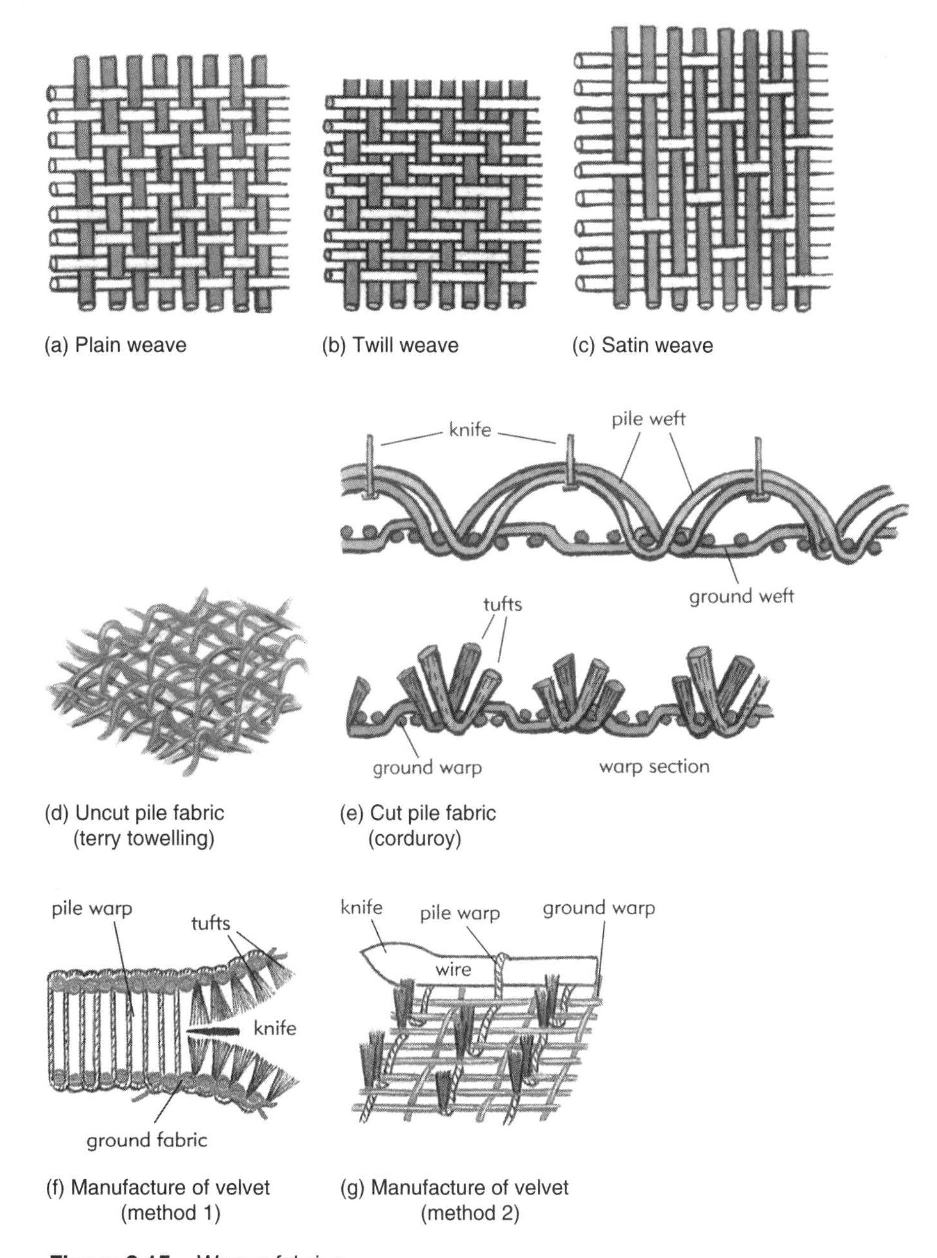

(a) Plain weave (b) Twill weave (c) Satin weave

(d) Uncut pile fabric (terry towelling)

(e) Cut pile fabric (corduroy)

(f) Manufacture of velvet (method 1)

(g) Manufacture of velvet (method 2)

Figure 2.15 Woven fabrics.

the same both sides and the even surface is good to print on to. It tends to crease more readily than some other weaves, but is hard wearing.

Twill weave

This weave makes a pattern of diagonal lines known as *wales*. The weft yarn passes over and under more than one warp yarn at a time as shown in Fig. 2.15(b). In the following row the pattern moves along one thread to create the diagonal characteristic of twill weaves. Twill is the hardest wearing fabric. Unlike plain weave the front and back of the fabric look different.

Satin weave

In this weave the weft yarns pass under between four and eight warp yarns before going over one as shown in Fig. 2.15(c). The right (best) side of the fabric is smooth and shiny since the warp yarns are set close together and almost completely cove the weft yarns. Unfortunately satin weave fabric is not hard wearing since it frays easily as the yarns do not interlace very often. Also the warp yarns lie on the surface of the fabric and can snag easily. Again this prevents the fabric from being hard wearing.

Jacquard weaves

These weaves are very complicated. The jacquard loom can be set to different sequences to produce intricate patterns within the fabric since each warp thread can be lifted individually. Originally controlled by a sequence of punched cards, modern jacquard looms are computer (CNC) controlled.

Pile fabrics

A pile fabric has threads or loops on the surface of the fabric in addition to the warp and weft yarns of the main or *ground fabric*. An extra set of yarn is used to create the loops or threads on the surface. The loops can be left uncut as shown in Fig. 2.15(d) as in the case of *terry towelling* or they can be cut as shown in Fig. 2.15(e) as in the case of *corduroy*. Velvet can be made either by weaving two fabrics face to face with a set of yarns passing between them and then cut to separate them as shown in Fig. 2.15(f). Alternatively, yarns can be left on the surface of the fabric during weaving and then cut as shown in Fig. 2.15(g). Pile fabrics are thicker and harder wearing than other fabrics and the surface is more ornamental. In the case of terry towelling, the increase in surface area improves its ability to absorb moisture. The quality of pile fabrics is determined by the density ratio of the ground weave and pile.

2.2.6 Non-woven fabrics

There are three types of non-woven fabrics in general use, namely *knitted fabrics*, *felts* and *bonded webs*.

Weft knitted fabrics

Hand-knitted fabrics are *weft* knitted. They are made from one long length of yarn. Loops are formed in the yarn which are interlocked with the rows above and below to form the fabric as shown in Fig. 2.16(a). Weft-knitted

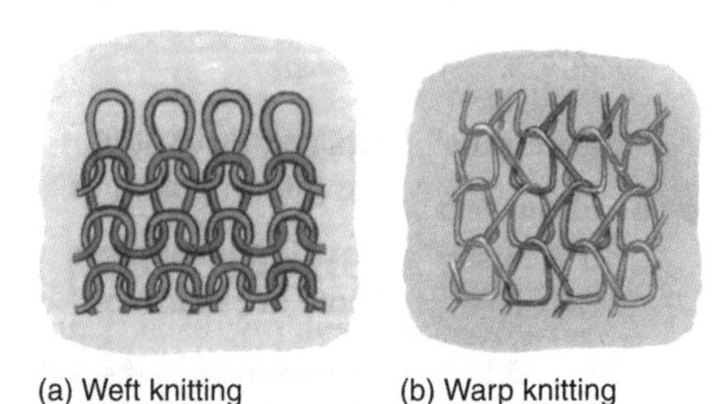

Figure 2.16 Knitted fabrics.

fabrics can also be made by machine. The loop structure of knitted fabrics traps the air, making them good thermal insulators. Plain weft-knitted fabrics, known as *single jersey*, tend to curl at the edges. This makes them difficult to work with. *Double jersey* is made using two sets of yarn and two sets of needles, making a thicker, more stable fabric that lies flat but has less stretch. A wide range of textures and properties can be created in weft knitting, as hundreds of different stitches can be used. Weft-knitting can create straight lengths of fabric or shaped fabrics ready to be sewn up.

Warp knitting

In warp-knitted fabric, loops are linked together from side to side as shown in Fig. 2.16(b). Warp-knitted fabrics can only be made by machine. Warp-knitted fabrics can be firm, like a woven fabric, or stretchy. However, they are not as stretchy as single jersey fabric and keep their shape well.

Wool felts

The first stage in making any non-woven fabric is to create a web of fibres. Natural fibres can either be air blown so that they fall in a random arrangement to produce *random laid* fabrics or the fibres can be combed out to produce *parallel laid* fabrics. If two layers of parallel laid fibres are arranged at right angles to each other they are said to be *cross laid*. Synthetic (man-made fibres) can be pumped from the spinneret directly on to a conveyor belt to produce a *spun bonded* fabric.

Wool fibres have scales on their surface that can become tangled together causing the fibres to stick to each other. The web of wool fibres is treated with an alkaline solution and heated. Pressure and mechanical action cause the fibres to become increasingly tangled together forming the fabric. Wool felts are good heat insulators as air is trapped in the web. They can be supplied as cloths or they can be moulded to shape to make hats and can be recycled from waste fibres, although they do not fray they are not as strong as woven fabrics.

Needle felts

Needle felts can be made from almost any fibre. A web is formed and then repeatedly punched with hot barbed needles. Fibres from the surface of the web are dragged through to the back to hold the web together. The web can be increased in strength by bonding its fibres together, see bonded webs. Needle felts are lightweight compared with wool felts and are generally more elastic. Like wool felts, they do not fray.

Bonded webs

The initial web for these fabrics can be made in the same way as wool felts or needle felts. The fibres can then be bonded together in one of the following ways:

- A binder or bonding agent (glue) can be applied to the web.
- A solvent can be used that softens the fibres causing them to stick together where they touch.
- Thermoplastic fibres can be fused together in small areas by the local application of pressure and heat.

- Fibres that melt or dissolve to bond with the other fibres can be included in the web.

Bonded webs do not fray, nor do they stretch or 'give'. They are not as strong as woven or knitted fabrics but have good crease resistance. Bonded webs are used for interlining and interfacing. Since they are relatively cheap to produce they can be used for disposable items, such as cleaning cloths, napkins, underwear and hospital items. Bonded webs can be impregnated with chemicals, such as antiseptics for wound dressings and bandages.

Test your knowledge 2.4

1. Discuss the differences in terms of their advantages and limitations between natural and synthetic fibres used in fabrics.
2. Describe how fibrous materials are converted into yarns ready for weaving.
3. Explain what is meant by the following terms applied to weaving:
   (a) Weft
   (b) Warp
   (c) Selvedge
   (d) Plain weave
   (e) Twill weave
   (f) Satin weave.
4. Describe the three main types of non-woven fabrics and where they would be used.

2.2.7 Miscellaneous materials

- Hides for the manufacture of leather goods and footwear
- Flour from the milled (ground) seeds of cereal crops for the manufacture of bread, biscuits and confectioneries
- Fruit, vegetables and meat from animal carcasses for food processing.

The list is interminable and would include the raw materials for the manufacture of soft drinks (still and fizzy), alcoholic beverages including wines and spirits, paints and varnishes, pharmaceuticals, household washing powders and soaps, solid, liquid and gas fuels, etc.

Mixtures, compounds and blends

A *mixture* is a close combination of different substances in which no chemical reaction takes place, no new substance is created and no heat is taken in or given out. Further, separation of the ingredients can be achieved using only physical processes. For example, if you mix iron filings and sawdust, you can remove the iron filings with a magnet.

On the other hand, if the reactive metal sodium burns in the poisonous gas chlorine, a new substance called sodium chloride is formed. This new substance is the salt used in cooking and as a table condiment. Since a new substance is formed and since heat is given out during the reaction, sodium chloride (table salt) is a *chemical compound*.

Mixtures are also to be found in combinations of materials or ingredients held together by physical rather than chemical means such as mixing flour, salt, water and yeast into dough for bread. However, chemical changes do take place during the baking process.

Blending, on the other hand, consists of mixing different grades of the same substance. For example the tea in your tea bag will have be a blend of the dried leaves of the tea plants from several different plantations from different locations. They will be selected so that the blend of different flavours will give the required overall flavour at a competitive price. Similarly, the coffee you buy will be a blend ground from the roasted beans from a number of different sources. Tobaccos, wines and spirits are also blended in order to improve the flavour or reduce the cost or both.

Test your knowledge 2.5

1. Describe the essential differences between mixtures compounds and blends.
2. Describe the essential difference between an alloy and a pure metal.
3. State the essential differences between cereals, vegetables and fruit.
4. Explain briefly the difference between knitting and weaving.
5. Explain briefly the difference between a warp yarn and a weft yarn when weaving.

2.3 Properties of materials

All materials possess properties or characteristics that make them suitable for use in manufacturing products. These properties or *key characteristics* vary from material to material. Generally, the key characteristics of a material are dependent upon its *composition*. Before a material can be selected for the manufacture of a particular product the following points must be considered:

- The conditions to which the product will be subjected in service
- The process or processes to which the material will be subjected during manufacture
- The cost of the material and its processing
- The service life of the product for which the material is to be used.

2.3.1 Mechanical properties

Strength

In general, we consider *strength* to be the ability of a material to resist an applied force (load) without fracturing (breaking). It is also the ability of a material not to yield. Yielding is when the material 'gives' suddenly under load and changes shape permanently, but does not break. This is what happens when metal is bent or folded to shape. The load or force can be applied in various ways as shown in Fig. 2.17.

It is important to be careful when interpreting the strength data quoted for various materials. A material may appear to be strong when subjected to a static load but will break when subjected to an impact load. Materials also

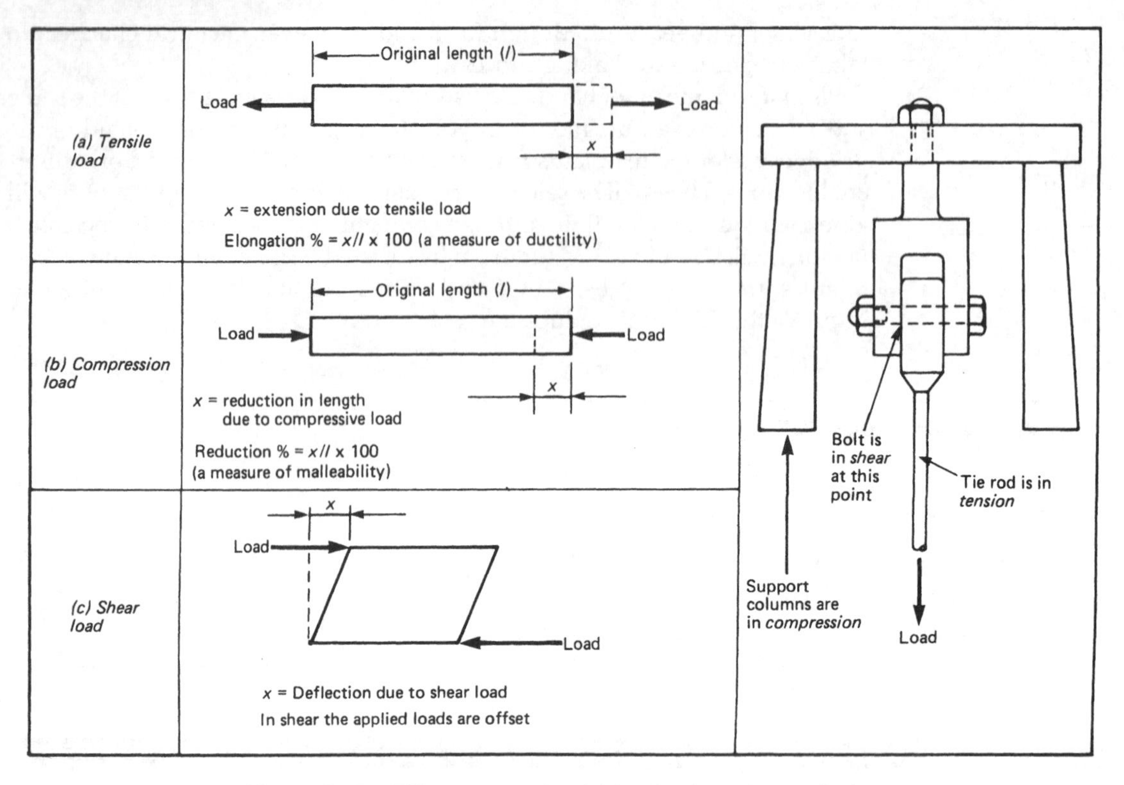

Figure 2.17 Different ways in which a load can be applied.

show different strength characteristics when a load is applied quickly than when a load is applied slowly.

Toughness

Do not confuse strength and toughness. Toughness is the ability to resist impact loads. For example, Fig. 2.18 shows a piece of high carbon steel in the soft (annealed) condition being compared with a piece of the same steel that has been hardened by raising it to red-heat and cooling it quickly (quenching it in cold water). The hardened steel shows a greater strength but it lacks toughness.

Brittleness

Brittleness is the opposite of toughness. Brittle materials lack *impact strength*, that is, they lack *toughness*. Brittleness is often associated with hardness and hard materials. Sintered carbides, quench hardened (untempered) steel, ceramics and glass are examples of brittle materials. Glass breaks easily when hit with a hard object such as a hammer, therefore glass is brittle: it lacks toughness.

Rigidity (stiffness)

Rigidity or stiffness is the ability of a material to resist *distortion*, such as bending and twisting. The frames of machine tools, vehicle chassis and the foundations of buildings require this property. Steel is much stronger than

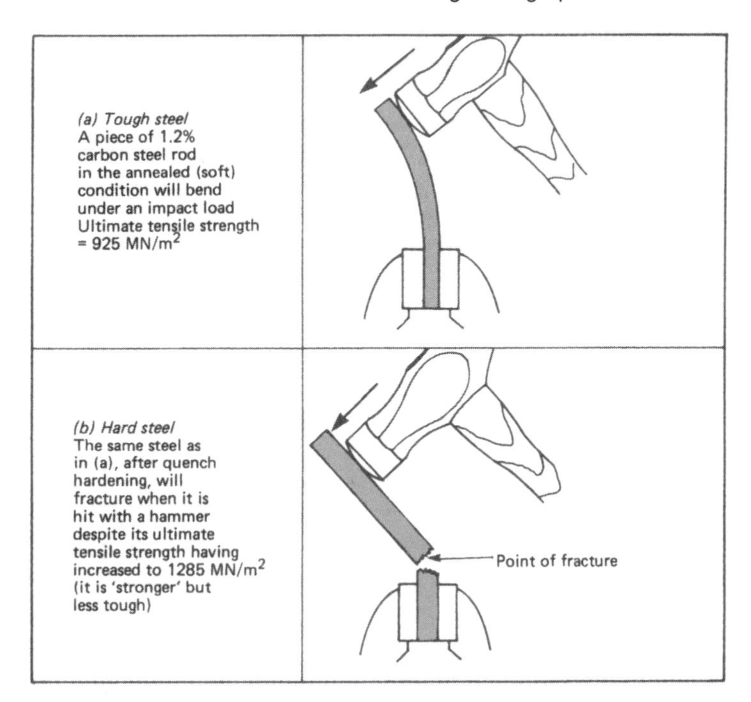

Figure 2.18 Impact loads.

cast iron but it is less rigid. For this reason cast iron is the preferred material for machine tool frames.

Elasticity

Elasticity is the ability of a material to be stretched or deformed and return to its original size and shape. Natural and synthetic rubbers, and metals used for springs and steel rules are examples of materials and products possessing this property. For elastic materials, the elongation (stretch) is proportional to the load provided they are not overloaded. Double the load and you double the stretch. Halve the load and you halve the stretch. Remove the load and the elastic material returns to its original length.

Plasticity

Plasticity is the opposite property to elasticity. It is the ability of a material to be *permanently deformed* or shaped by the application of a force. Most metals have this property if sufficient force is applied. The steel sheets pressed into the shape of car body panels must have the property of plasticity so that they can retain the required shape when taken from the press. If they were elastic, they would spring back into a flat sheet. Plastic materials (polymers) have this property when heated. This enables them to be shaped by a moulding process.

However you have to be careful. Metals tend to show elastic properties up to a limiting load (the *elastic limit*). Beyond this load they either break if they are brittle (e.g. cast iron or quench hardened steel), or they undergo plastic deformation. Thus most metals can show the dual characteristics of elastic or

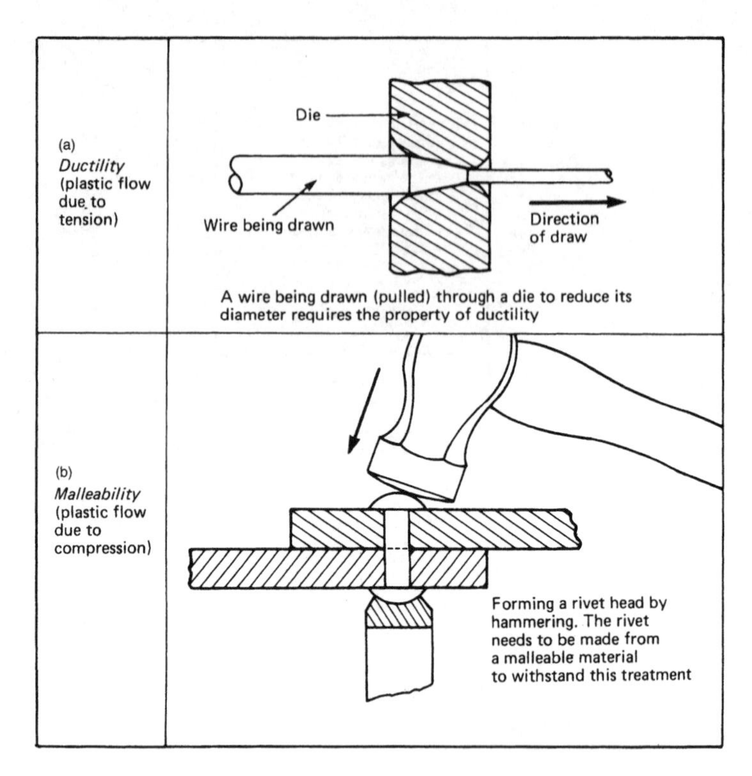

Figure 2.19 Two common engineering processes, drawing and riveting. Drawing exploits ductility whilst riveting exploits malleability.

plastic properties depending upon how severely they are stressed. Metals that can show these dual properties include low-carbon and medium-carbon steel, copper, brass alloys and aluminium and its alloys.

Ductility

Ductility is a special case of plasticity. It is the ability of a material to be *permanently deformed* by stretching. Some metals are very ductile, such as the steel used for the manufacture of car body sections by pressing. The metal copper, as used for wire, has to be ductile so that it can be manufactured by drawing it through dies in order to make it longer and thinner as shown in Fig. 2.19(a).

Malleability

Malleability is also a special case of plasticity. This time it is the ability of a material to be permanently deformed by a compressive force. Products such as forged crankshafts, camshafts, spanners and small valve bodies are produced from malleable materials. The metals used for the production of bars, sheets and sections from metal ingots by hot and cold rolling also need to be malleable. Most metals become more malleable when they are heated. This is why a blacksmith gets a steel bar red-hot before forging it to shape by hammering it on an anvil. Figure 2.19(b) shows a rivet being cold-headed. The rivet has to be made from a malleable material.

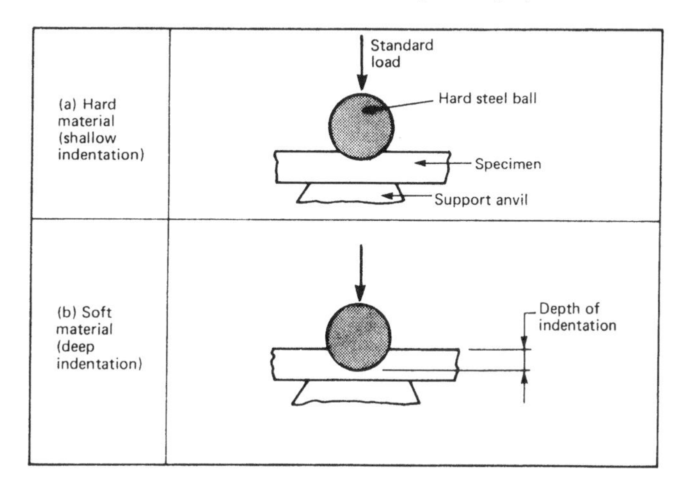

Figure 2.20 The effect of pressing a hard steel ball into two materials with different hardness properties.

Hardness

Hardness is the ability of a material to resist indentation and abrasion by another hard body. The degree of hardness of a material ranges from that of soft polymers to diamonds. A diamond is the hardest known material. Figure 2.20 shows the effect of pressing a hardened steel ball into two pieces of metal with the same force. The ball sinks further into the softer of the two pieces of metal than it does into harder. There are various hardness tests available. The Brinell hardness test uses the principles set out above. A hardened steel ball is pressed into the specimen by a controlled load. The diameter of the indentation is measured using a special microscope. The hardness number is obtained from the measured diameter by means of conversion tables. Obviously, the specimen must be softer than the hardened ball indenter.

For testing hard materials a diamond indenter is used. For example, in the Vickers test, a diamond pyramid is used that leaves a square indentation. The distance across the diagonals is measured and again the hardness number is obtained by the use of conversion tables.

Test your knowledge 2.6

1. Name the main mechanical properties required by the materials used to make the following products:

   (a) A steel rule
   (b) Sewing yarn
   (c) A machine frame
   (d) A squash racket handle
   (e) An axe head.

2. State the properties that you associate with the following materials:

   (a) Glass
   (b) Rubber
   (c) Low-carbon (mild) steel
   (d) Nylon.

2.4 Physical properties

2.4.1 Electrical conductivity

Electrical conductivity is the ability of a material to conduct electricity. Materials that can conduct electricity are called *conductors*. Most electrically conductive materials are metals, the exception being the non-metal carbon. There is also a group of non-metals called semi-conductors used in the manufacture of diodes, transistors and integrated circuits, but these are outside the scope of this book. The most common metal used for electrical conductors is copper. It is a very good conductor of electricity, it is easily drawn out into the finest of wires, and can be easily soldered to make permanent connections. Other highly conductive metals are aluminium, silver, gold and platinum.

Materials that are *very poor conductors* of electricity are called *insulators*. There is no such thing as a perfect insulating material, only materials that are very poor conductors. Typical materials used for insulators are polymers, composites and ceramics. Dry wood is also an insulator. It is possible, however, to make polymers and elastomers conductive for specialized applications by adding a metallic or carbon powder during compounding. Aircraft tyres are made conductive in this manner so that static electrical charges collected during flying can leak safely to earth on landing. This reduces the risk of sparks that could cause a fire and it also reduces the risk of electric shocks for the passengers as they disembark.

2.4.2 Magnetic properties

Only *ferromagnetic* materials show strong magnetic properties. Such materials are iron, cobalt, nickel and gadolinium. Originally iron and steel were the only magnetic materials available. Since the chemical name for iron is the Latin word *ferrum*, such metals were called *ferromagnetic*. The other metals were added later. Other metals and non-metals are added to the ferromagnetic materials to modify their magnetic properties as required.

Soft magnetic materials

Materials such as pure iron and low-carbon steels only become magnetized when placed in a magnetic field. These materials are used for such things as transformer cores and electromagnet cores. They do not retain their magnetism when the energizing magnetic field is removed. They are called 'soft' magnetic materials.

Hard magnetic materials

Materials such as quench hardened high-carbon steels are more difficult to magnetize when placed in a magnetic field. However, once magnetized, they retain their magnetism even when the energizing magnetic field is removed. They are called 'hard' magnetic materials. Hard magnetic materials are used for the permanent magnets found in loudspeakers. Alloy steels have been developed (such as the alloy Columax) that are as much as 30 times more powerful than hard high-carbon steel.

2.4.3 Thermal properties

Thermal conductivity

Thermal conductivity is the ability of a material to conduct heat. Generally the materials most able to conduct heat are the same as those that are most able to conduct electricity. Therefore metals such as copper and aluminium are the most thermally conductive. Such metals are used for the manufacture of cookware, and soldering iron bits.

Thermal insulation

Materials that are poor conductors of heat are called *thermal insulators*. Generally, they are the same materials as electrical insulators. Polymers, composites and ceramics are poor or medium conductors of heat. Air is a good heat insulator if it is trapped and can't circulate. It is the air trapped been two pieces of glass that provides the insulation in double-glazing. Again, it is the air trapped in the fibreglass used for loft insulation that prevents heat loss.

Refractoriness

This is the ability of a material to remain unaffected by high temperatures. That is, it can be raised to high temperatures without burning, softening, or melting. For example, the firebricks that are used to line furnaces are made from refractory materials. Thermal insulators are not necessarily refractory materials. Expanded polystyrene plastic is a very good heat insulator but it melts at quite low temperatures and it will also burn. It is not a refractory material.

Thermal stability

Thermal stability is the ability of a material to retain its shape and properties over a wide or specified temperature range. The metal nickel and its alloys, such as 'Invar', are used in the manufacture of measuring tapes and master clock pendulums. 'Monel' is a nickel alloy that can resist high temperatures without appreciable corrosion or dimensional change (creep) and is used for the manufacture of steam turbine blades and 'K-Monel' alloy is used in the manufacture of pressure gauges.

2.4.4 Density

Density is the mass of material per unit volume of a material. Expressed mathematically:

$$\text{Density (kg m}^{-3}) = (\text{mass}) \div (\text{volume})$$

Manufactured products, such as aircraft, may be made from materials with a low density such as the aluminium alloys. Other products may also be made from low-density materials such as polyethylene. At the other end of the scale, car batteries contain components that are made from lead that has a very high density. Table 2.6 lists the densities of some common materials.

2.4.5 Flow rate

Flow rate is the rate at which a material will flow under pressure and is an important factor in the processing of polymers, health products such as

Table 2.6 Densities of common materials

Material	Density (km m^{-3})
Aluminium	2700
Titanium	4500
Lead	11 300
Tungsten	183 920
Polyethylene	960
Polyamide (nylon 66)	1150
Polyesters	1360
Polyvinylchloride	1400
(PTFE) Polytetrafluoroethylene	2140

toothpaste, confectionery and food products such as dough, sauces, fondants and pastes. The flow rate of a material will depend upon its viscosity. Viscosity is the ease with which a material will flow. For example, an oil with a high viscosity will not flow easily, whilst an oil with a low viscosity will flow easily. Viscosity is affected not only by temperature and pressure, but also by the consistency and composition of a material.

In the polymer processing industry flow rate is of critical importance in extrusion and moulding processes. Flow rate must also be considered in the pharmaceutical and toiletries manufacturing industries. Products, such as toothpaste and ointments, would be difficult to squeeze out of the tube if too thick but, if too thin, it would be difficult to control them. This lack of control would result in waste.

2.4.6 Flavour

Flavour is the property of a food or drink determined by taste. The flavouring is the substance, or substances, that gives these products the unique qualities that attracts consumers to them. Flavours may be natural such as vanilla and fruit juices or essences distilled from natural products. Nowadays, many essences and flavourings are synthesized, artificial products.

2.4.7 Colour

Colour is the visual effect on the eye produced by light of different wavelengths reflected from the surface of the product. The colouration of the surface of the product will determine the wavelength or mixture of wavelengths reflected. The colour of manufactured products is very important to satisfy customer preference. This is particularly the case in the automobile, textile and clothing industries. Colouring matter is often added to processed foods in order to enhance and improve their colour and to give them a more attractive and appetizing appearance for the consumer. Colouring may be natural such as cochineal, a scarlet colour made from dried insects; or synthetic such as tartrazine made from petrochemicals.

2.4.8 Texture

Texture is the property of a product that is responsible for its 'feel'. Feel is very personal and what may be pleasant to one person might be unpleasant to another. This is particularly the case with fabrics and is important in the clothing and upholstery industries.

2.4.9 Resistance to degradation

All materials will degrade over a period of time when exposed to atmospheric conditions and the rate of degradation will vary from material to material. It is virtually impossible to prevent degradation, but it can be retarded by changing the composition of the material (for example, alloying steel with nickel and chromium to make stainless steel) or by protecting any exposed surfaces (for example, painting or electro-plating). In some cases the degree of degradation can be reduced to a level where it is undetectable by the unaided eye.

Corrosion

Corrosion is the process by which a material is progressively damaged by chemical or electrochemical attack. Most metals are less resistant to degradation by corrosion than ceramics and polymers, and the rates of corrosion vary between metals. Iron and steel are the most common ferrous metals and will corrode rapidly (rust) under the influence of oxygen and moisture when left unprotected. Metals such as brass, bronze and stainless steel are alloyed in such a way as to make them very corrosion resistant, but even so in certain environments (e.g. marine environments) brass will corrode more rapidly than bronze and bronze more rapidly than stainless steel.

Weathering

Weathering is a degradation process in which the material is damaged by exposure to atmospheric effects such as sunlight, rain, heat or cold. Sometimes the effect is physical. For example, if water is absorbed into brick or stone, it will expand as it freezes and split the brick or stone into fragments. Wind and rain will also erode (wear away) the surface of softer materials. Sometimes sunlight (ultraviolet rays) will break down the chemical structure of the material. Certain polymers may become brittle, and will craze and crack when exposed to sunlight, heat or immersion in water over a long period of time.

Pigments are added to some polymers to act as stabilizers to absorb the ultraviolet (UV) rays in sunlight. For example, window frames are made from an ultraviolet resistant grade of polyvinyl chloride (UPVC). In the case of polyethylene (polythene) sheet, carbon black is added to retard ultraviolet degradation. In another case, polyethylene detergent bottles are stabilized with polyisobutylene to protect the polymer against the effects of the detergent.

On the other hand, some polymers are deliberately made to degrade. At one time, polymers buried in land-fill sites did not break down. Nowadays many polymers are biodegradable. They will break down as the result of bacterial attack when buried and therefore will not pollute the ground.

Test your knowledge 2.7

1. Before the days of multigrade motor oils, cars were often difficult to start up in very cold weather. Explain the reason for this.

2. Calculate the volume of an aluminium casting if its mass is 54 kg. You will find the density of aluminium in Table 2.5.

3. An electric cable consists of copper wire surrounded by PVC plastic. Suggest why these materials were used in its manufacture.

4. In your own words, explain the difference between materials that are thermal insulators and materials that have refractory properties.

5. If a screwdriver blade is stroked with one pole of a magnet, the screwdriver becomes magnetized, and remains magnetized. State whether the screwdriver blade has 'hard' or 'soft' magnetic properties. Give the reason for your choice.

Key notes 2.2

- Metals containing iron are called ferrous metals (Latin for iron = ferrum).
- Steels contain a small amount of carbon ranging from 0.15 per cent in the case of low-carbon (mild) steels to 1.5 per cent for high-carbon (tool) steels.
- Alloy steels contain other metals such as nickel and chromium to make them tougher, stronger or more corrosion resistant. High-speed steels (HSS) contain cobalt and tungsten so that they can remain hard and sharp at higher temperatures than high carbon steels.
- All the carbon in steels is combined chemically with the iron.
- Cast iron has more than 1.5 per cent carbon present, so that some of the carbon content remains uncombined as flakes of graphite between the crystals of metal.
- Non-ferrous metals do not contain any iron.
- Brass is an alloy of copper and zinc.
- Phosphor bronze is an alloy of copper and tin with a small amount of phosphorus.
- Gun-metal bronze is an alloy of copper and tin with a small amount of zinc present.
- Aluminium is a light but relatively weak metal, however it forms the basis of many much stronger 'light-alloys' used in the aircraft industry.
- Elastomers are natural or synthetic rubbers showing elastic properties.
- Thermosetting plastics change chemically when they are moulded and can never again be softened.
- Thermoplastics can be softened every time they are heated.
- Composite materials are a physical combination of two or more materials that exploit the benefits of both materials, e.g. fibre-glass consists of glass fibres bonded together with a polyester resin to make a strong tough material for small boat hulls.
- Cermets are ceramic powders bonded in a matrix of metal, e.g. tungsten carbide cutting tools.
- Yarns are spun from natural cotton or wool fibres or from synthetic fibres, such as nylon or terylene.
- Fabrics can be woven or knitted from yarns.
- Felts and webs are made from a random lay of cut fibres. In the case of webs the fibres are bonded together by the use of adhesives.
- Mixtures are a close association of material particles that do not normally react with each other and can be physically separated.
- Compounds are chemical combinations of two or more substances to form totally different new materials, e.g. the metal sodium burns in the poisonous gas chlorine to produce sodium chloride (table salt).
- Blending is the mixing of different grades of the same substance to improve the flavour or reduce the cost or both.
- Strength is the ability of a material to resist an applied force (load) without fracturing (breaking). The applied load may be tensile (stretching), compressive (squashing) or shearing (offset).
- Toughness is the ability of a material to resist impact loads.
- Brittleness in a material indicates lack of toughness.
- Rigidity is the ability of a material to resist distortion.
- Elasticity is the ability of a material to be stretched or deformed under load, but to return to its original size and shape when the load is removed.
- Plasticity is the ability of a material to deform under load and to retain its deformed shape when the load is removed.
- Ductility is a special case of plasticity caused by a tensile load.
- Malleability is a special case of plasticity caused by a compressive load.

- Hardness is the ability of a material to resist scratching or indentation by another hard object.
- Conductivity is the property of a material that allows electricity or heat to travel through it.
- Insulators are poor conductors of electricity or heat.
- Soft magnetic materials become magnetized when placed in a magnetic field, but lose their magnetism when that field is removed.
- Hard magnetic materials become magnetized when placed in a magnetic field and keep their magnetism when that field is removed.
- Refractoriness is the ability of a material to be unaffected by high temperatures.
- Density is mass per unit volume.
- Flow rate is the rate at which a material will flow under pressure.
- Flavour is the property of a food or drink determined by taste.
- Colour is the visual effect determined by light of different wavelengths.
- Texture is the 'feel' of a material.
- Corrosion is the degradation of a material by chemical or electro-chemical attack.
- Weathering is the degradation of a material when exposed to sunlight, rain, heat or cold.

2.5 Production details and constraints

Once key design features (Section 2.1.6) and suitable materials have been agreed between client and designer the next area to consider are the *production details and constraints*, that is any factor within the following areas that can affect the ease with which a newly designed product may be profitably manufactured.

2.5.1 Available labour

A newly designed product may require large numbers of production personnel with skills not presently available within the existing workforce. To overcome this problem, manufacturing management may decide to recruit more workers with the necessary skills or retrain (reskill) existing workers. Alternatively, the designer may have to modify the proposed design, so that it can be produced by the existing workforce. If more workers cannot be employed and a design change is not feasible, then the new work may need to be subtracted out. In some instances, where a product is being redesigned to take advantage of new technological developments such as automation it may be necessary to employ a smaller workforce that is multi-skilled so that it can programme and maintain the automated plant but does not have to operate it 'hands-on'. Alternatively, where a process is labour intensive, it may be more economical to transfer production abroad to countries where suitable labour is plentiful at a lower cost than in the UK.

2.5.2 Available materials

The new product should be designed to use readily available materials (standard sizes and specifications) which can be purchased easily and quickly from a material stockholder. Readily available materials should keep material costs to a minimum. Materials that are difficult to obtain are likely to be more expensive and may hold up manufacturing if they are not delivered when required. The designer's choice of materials will influence the manufacturing process that is chosen to manufacture the new product. Modern manufacturing tends to employ a near finished size (NFS) philosophy so that machining processes are kept to a minimum. For example, sintered powder metal compacts may be used in place of forgings. Although basically more expensive, this extra cost is more than offset by the lack of need for expensive machining operations.

2.5.3 Available technology

To maintain their competitive advantage new products should be designed to make as much use as possible of new technology providing it is well proven and appropriate for the product. New technology should not be used just because it happens to be available. It should only be used if it will provide a better quality product and a more economical price. Generally automated plant is more costly than its manually operated counterpart. Therefore the volume of work to be produced on it must be sufficient to keep it fully employed to warrant its purchase, otherwise it may be more economical to subcontract some processes out to specialist manufacturers who have sufficient volume of work to warrant the use of 'cutting edge' technology. Wherever possible standard components should be used as these can be bought in from specialist suppliers and are of guaranteed quality if made to British Standard specifications.

2.5.4 Health and safety requirements

The Health and Safety At Work Act of 1974 places the responsibility for safety equally upon:

- Manufacturers and suppliers of manufactured products and equipment
- Employers
- Employees.

So manufacturers have to ensure that all products are manufactured within relevant Health and Safety laws. Any manufacturing processes that can cause injury to or adversely affect the health of employees or do damage to the environment can give rise to a production constraint in the form of a *Prohibition Order.* Such an order is issued by an inspector of the Health and Safety Executive (HSE) and prohibits the process being used until it can be made safe or replaced by one that is less hazardous. For example:

- Chemical hazards may create burns and toxic fumes that can cause injury
- Electrical hazards can cause burns and shock
- Extreme conditions of heat, cold, humidity and noise cause stress and fatigue
- Equipment giving off radiation can cause tissue damage
- Environmental pollution can be caused by the improper disposal of scrap materials, waste products and excessive noise from manufacturing processes. Exposure to excessive noise can cause damage to employees' hearing.

Safe working practices will be considered in Chapter 3.

Test your knowledge 2.8

1. State what is meant by the term 'production constraint'.
2. Name the persons and organizations who have legal responsibilities under the Health and Safety at Work Act.
3. Find out what constraints are available to an HSE inspector and under what circumstances they are used.
4. Explain briefly under what conditions members of a workforce may require training to re-skill them. Suggest why this may be more satisfactory than replacing them with workers who already have the required skills.
5. Explain under what circumstances manufacturing processes should be subcontracted out.

2.6 Quality standards

Quality can be defined as *fitness for purpose*. The product quality will have been agreed with the customer during the generation of the design brief. The designer will have had to conform to this quality requirement at every stage of the design process. The product itself must also conform to the appropriate British Standards Institute's (BSI) recommendations, where appropriate, for safety in use. It must also conform to the European safety requirements and legislation and carry the appropriate CE markings. Similarly, the production equipment or machines chosen to produce the components to be used in a newly designed product must be capable of consistently meeting the product designer's specified requirements for dimensional accuracy, geometric accuracy and surface finish. In other words the chosen production equipment must have adequate *process capability* consistently to manufacture the required product.

Customers constantly demand increasing quality in the manufactured products that they buy. They want:

- Value for money
- State of the art technology
- Fitness for purpose.

There is no such thing as absolute quality. A customer's ideas of what represents quality will change with time and with competition in choice. A motor cycle that was considered state of the art 50 years ago would not satisfy the requirements of the present-day market except, perhaps, as a collectors' item. Successful companies need to ensure that their products are every bit as good or better than those manufactured by their competitors. Nowadays most companies adopt a strategy of *Total Quality Management* (TQM) where everyone from the boardroom down is responsible for the achievement of quality standards. Quality cannot be 'inspected into' a product, inspection can only prevent faulty products being passed on to the customer. The manufacture of a high level of faulty products is a waste of time and a waste of money. Quality should be 'designed into' the product in the first instance and the production processes used during manufacture should be chosen to ensure that the quality targets are maintained.

2.6.1 Quality assurance

Quality assurance is the result of creating and maintaining a quality management system that ensures all finished products achieve *fitness for purpose* and *conform* to the specifications agreed with the customer.

In the UK, British Standard BS EN ISO 9001/2 lays down the guidelines and rules for a quality management system that is recognized and approved internationally. Since companies that are approved to this quality control standard can only purchase their goods and services from similarly approved companies, it is necessary for most companies to be approved. To obtain the BS kitemark a company must have its quality management system assessed by an independent, external certification organization. If it passes this *audit* and conforms to the requirements of the national standard, its name appears in the Department of Trade and Industry (DTI) *Register of Quality Assessed United Kingdom Companies*. It can then trade with other similarly accredited companies and its customers can be *assured* of the quality of its products.

The benefits of satisfying the requirements of BS EN ISO 9001/2 and being registered with the DTI can be summarized as:

- Greater control over raw materials and bought-in components
- A complete record of production at every stage to assist in product or process improvement
- Cost effectiveness because there is a reduction in scrap and waste and the need for re-working. Products are increasingly *right first time*
- Customer satisfaction because quality has been built in and monitored at every stage of manufacture before delivery. This enhances a company's reputation and leads to repeat orders and increased profitability
- Acceptability of a company's products in global markets.

Let's now see how this can be achieved.

2.6.2 Quality control

Quality control is not only about good design and conformance to quality specifications. Quality should permeate every aspect of a company's business, always focusing upon the needs and care of the customer. This is achieved by setting up a system under the direction of a *quality manager* that embraces all members of a company from the chairman of the board downwards and has the full support and commitment of the board and senior management. This is the concept of total quality management (TQM). Total quality management can be defined as an effective system of coordinating the development of quality, its maintenance and continuous improvement, with the overall objective of the achievement of customer satisfaction, at the most economical cost to the producer and, hence, at the most competitive price. The quality system must provide for:

- *Traceability* so that all products, processes and services needed to fulfil a customer's requirements can be identified in order that they can be traced throughout the company. This is necessary in the case of a dispute with a customer regarding the quality of any particular product supplied or component within that product.

- *Control of design* so that any design meets a customer's requirements through consultation by the design and marketing staff with the customer.
- *Control of 'bought-in' parts* to ensure that they conform to previously agreed specifications. This verification can be achieved by buying only from a BS EN ISO 9001/2 accredited source or by rigorous inspection of the bought-in parts.
- *Control of manufacture* through clear work instructions and documentation together with effective process control and inspection procedures. Accreditation of the company's measuring and testing equipment should be by an external body such as NAMAS (National Measurement and Accreditation Service).

2.6.3 The effect of level of output on quality control

Let's see how assessing the required standards of production are affected by the level of production. First we must consider the number of items being manufactured to a given design. One-off structures such as bridges, oil rigs and ships that are built as a single entity and the finished article cannot be sampled and tested to destruction. However, batches of materials and bought-in components, such as nuts and bolts, can be sampled and tested. The design can be proved using models of such structures. Ship designs can be tested using models towed through large water tanks in which scale type waves can be simulated and the stability of the ship can be assessed under varying conditions of weather, currents and loading. Similarly, models of ships, bridges and high-rise buildings can also be tested in wind tunnels to measure the effect of gale-force winds. Models can also be tested in suitable rigs to study the effect of earthquakes.

In the case of quantity production, it would be too expensive to inspect every component at every stage of its manufacture. Therefore when batch, flow and line production is undertaken it is necessary to use *statistical sampling techniques*, to establish *points of inspection*, and to specify *quality indicators*. Figure 2.21 shows a typical production system and the position of the *points of inspection*. These are also called *quality control points.*

2.6.4 Points of inspection

These must be inserted into the manufacturing chain so that costly time is not wasted on processing parts that are already defective. There is no point in gold plating a defective watchcase. Such inspection points should be:

- Prior to a costly operation
- Prior to a component entering a series of operations where it would be difficult to inspect between stages (e.g. within an automated flexible manufacturing cell)
- Prior to a processing station where failure of a defective part could cause an enforced shutdown of the whole station (e.g. automatic bottling plants, where breakage of a bottle could result in machine damage or failure, in addition to the time lost and cost of cleaning the work station)

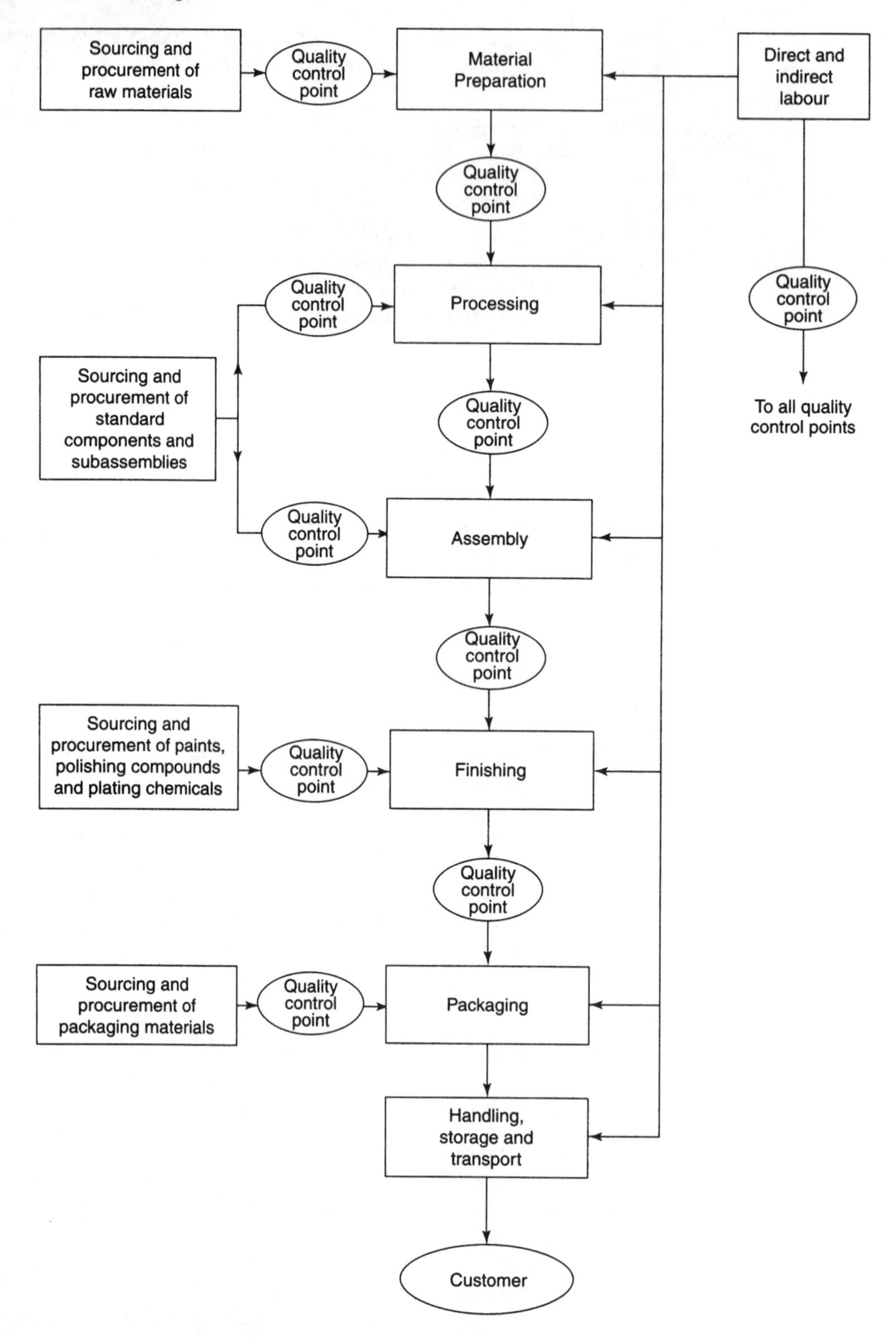

Figure 2.21 Typical production system showing the quality control points.

- Prior to a *'point of no return'* where rectification would be impossible after the operation (e.g. final assembly of a *'sealed for life'* device).

Points of inspection will vary from plant to plant and from product to product, but there are certain points of inspection that have reasonably common ground, so let's now look at them:

- *Raw materials and bought-in parts.* As previously stated these should be bought preferably from a BS accredited source and be manufactured to an established BS/ISO specification. If not, then *acceptance sampling and testing* should be carried out on every batch procured. Some key items and materials, such as those used in vehicle and aircraft control systems, will undergo acceptance sampling irrespective of their source of procurement. Suppliers are usually given a *vendor rating* depending on the percentage of acceptable products received previously and the number of warranty claims made against the supplier. On no account must substandard materials and parts be allowed to enter the production chain as this may lead to a whole batch of products being scrapped after expensive processing has been carried out.
- *Material preparation.* This is the next point of inspection to ensure that the correct quantity of the correct material has been issued for the job. Also that it has been cut to the correct size and any degreasing or other preliminary preparation process has been carried out.
- *Processing.* Modern manufacturing organizations usually aim to produce goods that are *right first time.* You cannot inspect quality into a product. Inspection can only hope to remove defective (faulty or flawed) components from the manufacturing sequence. These defective products then have to be corrected or scrapped. This costs the company money. Therefore it is more economical to ensure than products are made correctly from the start. That is, they are made *right first time.*
- *Assembly.* Even assuming that all the components parts are correctly manufactured, it is still essential that they are correctly assembled. It is important that components have not been omitted or damaged during assembly, or that electric wiring has not been incorrectly connected.
- *Finished product.* The final inspection must include correct appearance (freedom from blemishes), correct functioning of the assembly, total compliance with the customer's specification, total compliance with national and international safety and legal requirements.

The quality control department, under the leadership of the quality manager, and with the full support of the board of directors, needs to establish well-defined procedures at each point of inspection in order to avoid confusion and to ensure that the system works consistently. These procedures should include:

- The method of inspection and the sampling frequency
- The reworking, down-grading or scrapping of components and assemblies
- The recording and analysis of inspection data
- The feedback of information so that faulty manufacturing and assembly processes and techniques can be corrected with the aim to be *right first time.*

2.6.5 Quality indicators

Quality indicators can be categorized as:

Variables. These are characteristics where a specific value can be measured and recorded and which can *vary within prescribed limits*. For example, a particular component dimension is given as 100 ± 0.5 mm. Therefore any component with a measured dimension lying between 100.5 and 99.5 mm inclusive would be acceptable. The British Standards Institute publishes tables giving various qualities of fit for a range of dimension sizes. Care must be taken to ensure that the quality of fit is adequate for the correct function of the part concerned. The specification of unnecessarily accurate components makes them unduly costly. High accuracy is necessary between the pistons and cylinder bores of a car engine, but would be quite inappropriate for the wheel and axle of a garden wheelbarrow. Variables are measurable and can include, for example:

- length, width, height, diameter, position and angles inspected by measurement
- mass
- electrical potential (volts), current (amps) and resistance (ohms)
- fluid pressure
- temperature
- appearance
- taste, sound, smell and touch
- functionality (fitness for purpose).

Attributes. These can only be *acceptable* or *unacceptable* – for example, colour (if you have ordered a blue suit, a brown one will not be acceptable).

2.6.6 Statistical sampling techniques

As stated previously, for 'one-off', prototype and small batch production, every product is inspected. This is 100 per cent sampling and inspection. It is also used for key components where a fault could cause a major environmental disaster or the loss of human life. Under these conditions, the high cost of 100 per cent inspection would be small when compared with the cost of failure in service. For the large volume production of many items, 100 per cent inspection is too time-consuming and expensive if carried out manually. Either fully automated 100 per cent inspection is built into the production line or *statistical sampling* is used.

In statistical sampling, samples are taken from each batch and the quality of the batch is determined by analysis of the sample. If the sample satisfies the criteria of the quality specification, then the whole batch is accepted. If the batch does not satisfy the quality specification the following lines of action may be taken:

- The batch is re-inspected to confirm the original findings;
- The batch is sold at a reduced price to a less demanding market;
- The batch is reworked to make it acceptable;
- The batch is scrapped.

The most difficult decision when using statistical quality control is in selecting the most satisfactory sample size and frequency. If the sample size is too

Table 2.7 Frequency distribution for wooden strips

Length in millimetres	Number of strips
249.3	1
249.4	2
249.5	3
249.6	5
249.7	7
249.8	10
249.9	13
250.0	19
250.1	11
250.2	8
250.3	6
250.4	5
250.5	4
250.6	3
250.7	2
250.8	1
Total	100

large, then the cost of inspection may be too great. If the sample size is too small, then the inspection results may give the wrong impression and faulty goods may slip through the inspection net. The sample may be taken from the whole batch or, for very large batches and continuous production, samples may be taken on an hourly basis.

Let's consider taking 100 wooden strips at random from a batch of 1500 wooden strips all of which are supposed to be 250 ± 0.5 mm long. Since we cannot saw to an exact length strips that vary between 250.5 mm long and 249.5 mm long will be acceptable. We have given the size a *tolerance* of 1 mm. The wooden strips are measured and their actual sizes are recorded and grouped as shown in Table 2.7. This is called a *frequency distribution table* since it tells us how frequently each size occurs. These results can be plotted as a histogram (see Section 2.7.4) as shown in Fig. 2.22. The histogram shows that of the 100 sample strips tested, 91 were within the specified limits and that nine were either oversize or undersize. That is $(9 \div 100) \times 100 = 9\%$ were defective, so it follows that 91 per cent were acceptable. It can be a reasonable assumption that these results will apply to the whole batch so, if these percentages are applied to the whole batch of 1500 strips, then $(91 \div 100) \times 1500 = 1365$ strips should be acceptable and $(9 \div 100) \times 1500 = 155$ strips will be defective (scrap).

We would prefer to have no scrap since scrap represents wasted time, money and materials. We therefore set warning limits inside the action limits. In this example warning limits could be set at ±0.4 mm. Further, instead of waiting until the whole of the batch has been produced, samples could be taken more frequently in order to establish *trends* and *drifts*. If the trend is for the cut sizes to drift towards one of the warning limits then the machine or process has to be adjusted to bring the situation back under control again before any out of tolerance components (scrap) are produced.

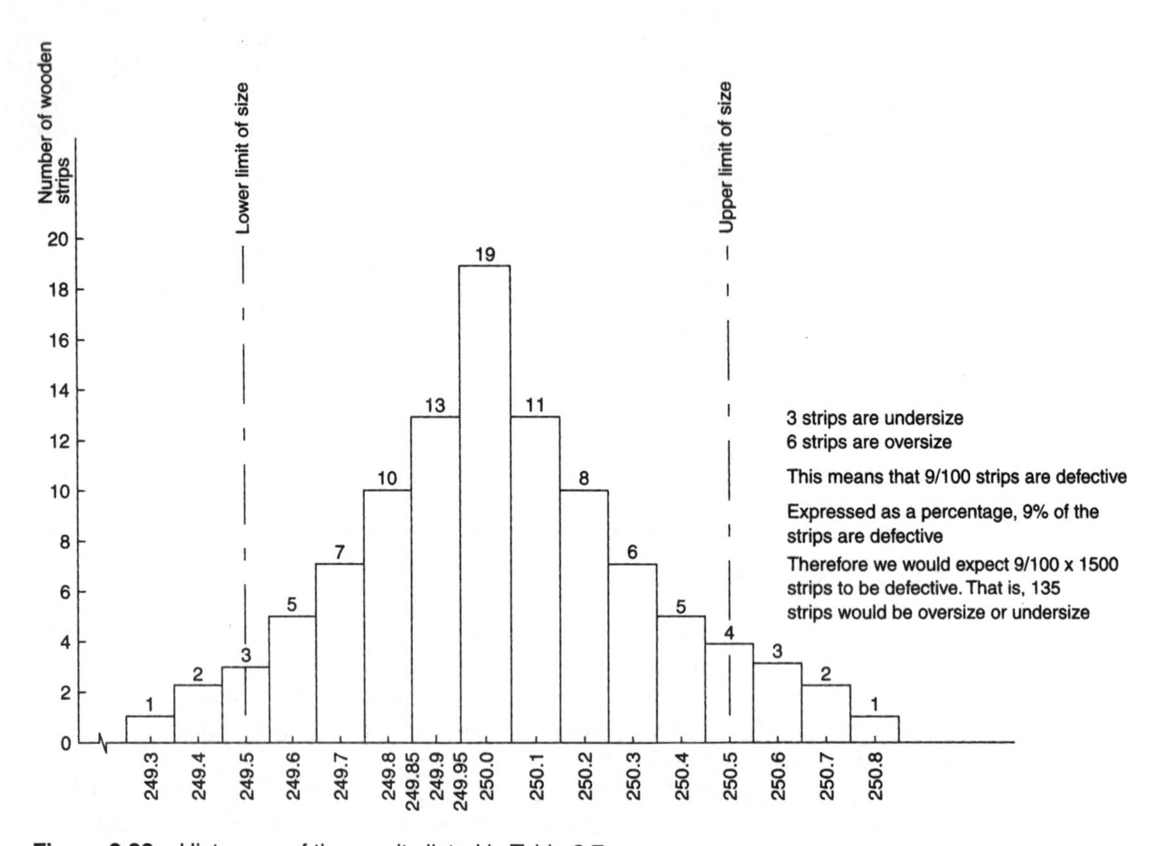

Figure 2.22 Histogram of the results listed in Table 2.7.

Statistical data can be recorded manually or stored on a computer. One advantage of using a computer is that it can be loaded with a software package that will carry out the statistical analysis at the touch of a button and present the outcomes as either a graph, a chart or a spread sheet. A computer can also handle very large numbers very quickly. Computers can also be used for much more comprehensive analysis including such factors as the amount of reworking that has had to be done, the amount of unreclaimable scrap produced, customer complaints, etc., and print out a weekly or monthly report.

2.6.7 Defects

Defects are faults or flaws in products that do not comply with the quality specification.

Critical defects

Critical defects are those likely to result in hazardous or dangerous conditions for anyone using or coming into contact with the product. An example of such a critical defect is a fault in the controls of an aircraft. Not only could this result in the loss of the aircraft, the crew and the passengers, but also the people on the ground at the crash site.

Major defects

Major defects are those likely to result in failure or a reduction in the operating efficiency of a component or product, but not necessarily endangering life. An example of such a defect is the failure of the electric motor in a vacuum cleaner.

Minor defects

Minor defects affect the saleability of a product, but will have only little or no effect on its use. An example of such a defect is a loose button or a stain on a shirt.

Incidental defects

Incidental defects are those which have no effect on the use or function of a product. An example of such a fault is a small air bubble in a glass paperweight.

2.6.8 Inspection and testing methods

Inspection and testing are essential to all quality control systems. However, the inspection and testing techniques adopted will depend upon the product and the quantity being produced. The inspection of a shirt or a pair of trousers will be far less rigorous than the inspection of an aircraft engine. Further, the inspection process used will depend upon the position of the quality control point in the manufacturing system. Let's now consider the *Sector Specific Standards* as they affect the client design brief.

Level of output during production

As previously discussed the techniques, frequency and methods of sampling depends upon the volume of production. It will depend upon whether the process is automated or manual. Automated processes are invariably more consistent and inspection can be in-built.

Level of performance

The level of performance of a motorcycle, car, aeroplane or boat, for example, is essential if it is to conform to the design brief and satisfy the customer. Clothing and fabrics are not judged by power, speed and reliability, but by their ability to hang gracefully and fit properly, to resist deterioration caused by repeated washing, to be shrink resistant and to retain their colours and not fade in sunlight. Food products will be judged by flavour and texture.

Quality of materials

Materials inspection can include:

- Mechanical testing for strength as described previously
- Chemical analysis particularly of foodstuffs and pharmaceuticals to ensure there are no harmful bacterial organisms or toxins present
- Sonic and X-ray testing for cracks and flaws in pressure vessels and high-pressure pipelines that may cause loss of life and/or ecological damage in the event of failure

- Testing for physical properties (electrical, magnetic, and thermal)
- Visual appearance, surface blemishes, poor colour, surface cracks in ceramic products, etc.

Tolerance

It is not possible to make a product to an exact size or weight and, even if it were possible, we would never know since it is impossible to measure exactly. All sizes and weights must have tolerances. We first met tolerances when we were considering statistical sampling techniques in Section 2.6.6. Let's now consider limits and tolerances a little further.

Figure 2.23 shows a metal rod. Its *nominal sizes* are length 100 mm and diameter 20 mm since these are the sizes by which the component is known. These are also sometimes referred to as the *basic sizes* to which the limits of size are applied. The *length* has *bilateral* limits of size in this example since they lie above and below the nominal size. The *diameter* of this example has *unilateral* limits of size since they lie on the same side of the nominal size. Providing all the components lie within these limits of size, the component will function satisfactorily.

The tolerance is the mathematical difference between the upper and lower limits of size. For example, the difference in length between the upper limit of 100.2 mm and the lower limit of 99.8 mm is 100.2 − 99.8 = 0.4 mm. Therefore the tolerance is 0.4 mm. Similarly, for the diameter the tolerance

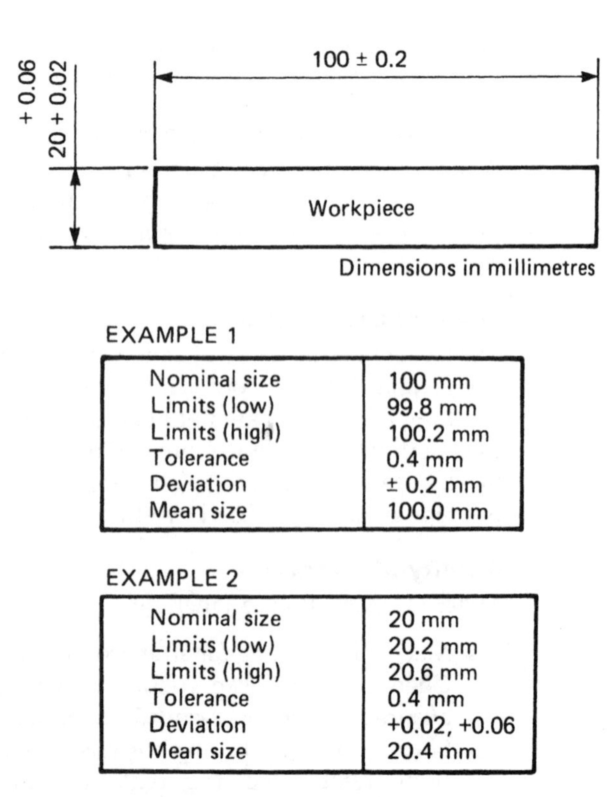

EXAMPLE 1

Nominal size	100 mm
Limits (low)	99.8 mm
Limits (high)	100.2 mm
Tolerance	0.4 mm
Deviation	± 0.2 mm
Mean size	100.0 mm

EXAMPLE 2

Nominal size	20 mm
Limits (low)	20.2 mm
Limits (high)	20.6 mm
Tolerance	0.4 mm
Deviation	+0.02, +0.06
Mean size	20.4 mm

Figure 2.23 Use of tolerances.

is $20.6 - 20.2 = 0.4$ mm. The tolerance is affected by the basic size of the dimension. A tolerance of 0.4 mm would be considered to be very coarse when applied to a diameter of 20 mm, however the same tolerance would be considered unduly fine when applied to a diameter of 20 m. Tables of tolerances and limits of size are published by the British Standards Institute to suit different sizes of component and the quality of fit required between two mating components, such as shafts and bearing bushes. Gauges and measuring instruments are generally chosen so that they are about 10 times more accurate that the dimension they are measuring.

Limits and tolerances also apply to the weight composition of mixtures as in the manufacture of foodstuffs or the volume composition of liquids in the blending of drinks or the dilution of concentrates. Limits and tolerances also apply to the viscosity of liquids, for example motor oils and paints. They also apply to the density of substances, where density is the weight per unit volume.

Product finish, packaging and presentation

The finish and presentation of a product is all important. No matter how well it is made, a product will not attract a buyer unless it is well made and well presented. You would not expect a new car in showroom condition to be presented to potential customers in a dirty condition or with the bodywork scratched or dented.

Similarly the packaging of produce on a supermarket's shelves is important. It must be convenient to handle, attractive in appearance, advertise the maker's brand name clearly and provide the all-important information concerning nutrition, ingredients, cooking information and weight or volume.

Health, safety and hygiene

As stated previously, all manufacturers, wholesalers and retailers have a 'duty of care' to their customers. They are responsible for and must ensure the 'fitness for purpose' for merchandise that they sell. All goods for sale or hire should clearly display the appropriate *British Standard kitemark* and *European Union CE mark*. Goods for sale or hire must in no way contravene the Consumer Protection Act. Foodstuffs must be hygienically packed, stored and presented so that they are uncontaminated at the point of sale and well within a clearly indicated 'sell by' date. Packaging materials should be biodegradable or capable of recycling.

Test your knowledge 2.9

1. Explain briefly what is meant by the terms:
   (a) Quality standards
   (b) Quality assurance
   (c) Quality control.
2. Discuss where the points of inspection should be inserted into the manufacturing chain.
3. Explain the difference between a variable and an attribute when discussing quality indicators.
4. What is statistical quality control and the circumstances under which it should be used.

5. Differentiate between:

   (a) Critical defects
   (b) Major defects
   (c) Minor defects
   (d) Incidental defects.

6. A dimension has a nominal size of 100 mm with an upper limit of size of 100.25 mm and a lower limit of size of 99.75 mm. State:

   (a) The magnitude of the tolerance
   (b) Whether the tolerance is unilateral or bilateral.

Key notes 2.3

- Production constraints refer to the availability of suitable materials, technology, and a suitably skilled labour force.
- Health and safety at work is the responsibility of employers, employees and suppliers and hirers of materials and equipment equally. The Act is enforced by officers of the Health and Safety Executive.
- Quality is fitness for purpose.
- Quality assurance ensures that finished products achieve fitness for purpose and conform to the specifications agreed with the customer.
- Quality control is the responsibility of the quality manager and achieved through total quality management (TQM).
- Points of inspection are strategic points in the manufacturing chain where materials, components and/or assemblies must be inspected so that time and money is not wasted on goods that are already defective, e.g. it is no good gold plating an already defective watch case.
- Variables are characteristics where a specific value can be measured and recorded and which can vary within prescribed limits.
- Attributes are either right or wrong. A car ordered in a red finish is either red (acceptable) or it is some other colour (unacceptable).
- Statistical sampling, a quality control technique used in mass production where a limited number of components from a large batch are inspected and the quality outcome of the whole batch is predicted by statistical techniques. It avoids the cost of inspecting every component in the batch.
- Defects are faults or flaws in a product. They may be *critical* if they lead to hazardous or dangerous conditions, *major* if they are likely to cause failure or reduction of operating efficiency, *minor* if they have little or no effect on the usefulness of a product, *incidental* if they do not affect the function of a product.
- Tolerance is the mathematical difference between the upper and lower limits of a dimension.

2.7 Generating proposals to meet a design brief

Let's now consider how a design brief (Table 2.8) or a design specification can be used to develop initial ideas concerning how the product might be designed and manufactured to satisfy the customer's requirements around the key features, production constraints and quality standards as previously discussed.

For example, a design brief calls for a small hand-held, portable electric fan that can direct a gentle stream of cool air on to the face of the user whilst watching sporting events on hot sunny days. The basic requirements are for the fan to be light in weight, compact, pleasant to hold, reliable, efficient, be powered by a readily available source of energy, and be capable of being manufactured and marketed at a relatively low cost.

Table 2.8 Design brief (key design features)

Product:	Hand-held, portable electric fan
Aesthetic requirements:	• Comfortable to hold by hand for extended periods of time • bright, primary colours for easy recognition • tough, durable and compact • light in weight • quiet running
Contextual requirements:	• to be used out of doors in hot locations • to be used in temperate and tropical countries • equally attractive to men, women and children • safe for children to use • easy to change batteries
Performance:	• to be powered by dry batteries (2 × 1.5 V type AA) or by solar cells • low current consumption for long life • must provide adequate ventilation • reliable, durable, capable of achieving a long service life
Production parameters:	• low cost – under £12.00 • initial batch size 5000 per month • quality standards – minimum battery life of 2 hours' continuous running – moulding finish to be comfortable to hold and free from sharp edges – to be moulded from a tough, impact resistant plastic

Initially there would be a group discussion involving the customer, the technical design team, the stylist and the production engineer in a mind-mapping session involving the whole team. Freehand sketches such as those shown in Fig. 2.24 would be made and discussed. A formal design proposal (Table 2.9) would then be developed for submission to the customer.

Figure 2.24(a) shows one possible design. The fan could be driven by a small direct current (dc) motor powered by two size AA dry cells connected in series to give an output of 3 volts. The fan operates when the device is switched on. Figure 2.24(b) shows an alternative design. The fan is again driven by an electric motor but the energy source is a solar panel built into the case of the fan. No switch is required. To stop the fan motor, the solar panel is folded over so that the active elements are covered. To start the fan the solar panel is simply opened and exposed to light.

Test your knowledge 2.10

1. For a simple product of your own choice (it can be quite simple, such as a confectionery item, a garment, or a simple tool):

   (a) Sketch the product
   (b) Produce a design brief for the product
   (c) Tabulate the key design features
   (d) Tabulate the process constraints for its manufacture.

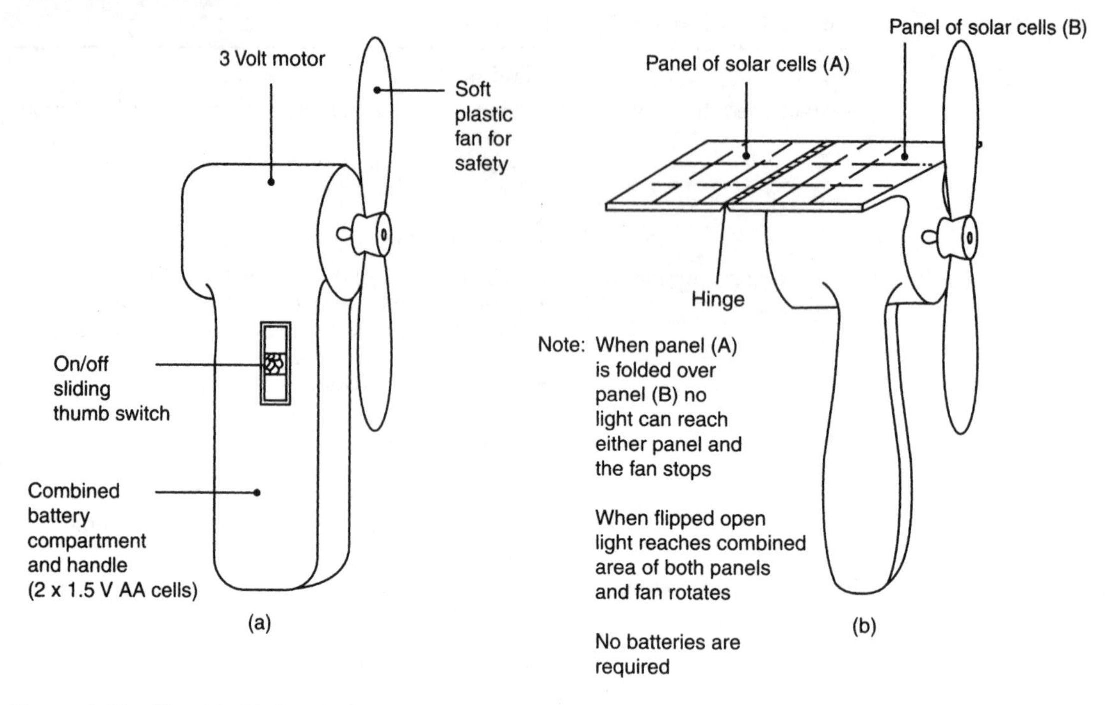

Figure 2.24 Hand-held electric fan.

Table 2.9 Design proposals for a hand-held portable electric fan

Key design features	Battery powered	Solar powered
Size	Comfortable to hold and compact if fan blades are made to fold when not in use	More bulky than when battery powered because of the solar cell panels, particularly when open
Weight	Approximately 150 g with batteries	Approximately 120 g
Power source	2 × 1.5 volt AA size dry batteries	2 solar cell panels
Running time	Minimum of 2 hours specified. Actual time in excess of 2 hours depending on type of battery used: standard/long life/alkaline rechargeable	Continuous and indefinite if light level is sufficient to activate the solar cells
Motor	Permanent magnet 3 V, d.c. low consumption motor	Permanent magnet 3 V, d.c. low consumption motor
Ventilation	Constant output from the fan over the life of the batteries	Varies depending on level of illumination. Output generally lower than for batteries
Ease of operation	• Thumb-operated slide switch on handle • Batteries inserted into handle by removing 'click-on' base	• No switch required. Fan starts immediately solar panel is flipped open. Fan stops when panel is closed • No batteries to change
Reliability	• Indefinite life. Oil impregnated bearings require no maintenance	• Indefinite life. Oil impregnated bearings require no maintenance

(Continued)

Table 2.9 *(Continued)*

Key design features	Battery powered	Solar powered
	• Batteries should be removed when not in use for long periods • Proven technology	• Proven technology • No switch contacts to wear out
Durability	• Moulded from tough, high impact resistant plastic such as acrylonitrile-butadiene-styrene (ABS)	• Moulded from tough, high impact resistant plastic such as acrylonitrile-butadiene-styrene (ABS)
Process constraints	• No reskilling required – plastic mouldings and motor units will be outsourced • ABS to be used as it is impact resistant • Plastic moulding is an available technology • Health and safety. Fan blades are designed from soft plastic and the rotational speed is limited for safety • Safety clothing will be worn by process workers during manufacture	• No reskilling required – plastic mouldings, motor units and solar panels will be outsourced • ABS to be used as it is impact resistant • Plastic moulding and solar panels are now available technologies • As for battery powered fan • Safety clothing will be worn by process workers during manufacture

2.7.1 Assessing the feasibility of design proposals

The first step is to assess the alternative design proposals such as those outlined in the previous section. The factors to be considered will vary from product to product, but the main factors common to most products are:

- Cost
- Marketability
- Materials
- Production requirements
- Quality standards
- Suitability for purpose.

Table 2.10 compares the feasibility assessment for both of our fan designs. Both have their own advantages and limitations and it could be difficult to make a decision.

If the feasibility assessment does not give a clear-cut indication as to which design should be selected, then the next stage is to make prototypes for testing to provide data for a more detailed evaluation. At the same time the stylist can model various shapes using 3D software on a computer and producing examples by rapid prototyping techniques (see Case Study 4.5, p. 245) driven by the computer model. This would give the assessment team an opportunity to evaluate alternative shapes and material textures to establish which would be the most comfortable to hold.

This assessment enables the products to be subjected to a range of tests incorporating the extremes of service conditions the product is liable to meet in use. It is also possible for prototypes to be tested not only by the manufacturer and by the customer originating the design brief, but also by members of the public to assess their reactions to the alternative designs. Sometimes

Table 2.10 Feasibility assessment of the design proposals

Assessment	Battery powered	Solar powered
Production	• The plastic casing can be an injection moulding using available technology • Assembly methods need to be considered at the design stage as assembly is labour intensive • Use 'snap-together' components and adhesives, avoid screwed fastenings wherever possible • Use compression joint connections rather than soldering • It is feasible to produce	• The plastic casing can be an injection moulding using available technology • Assembly methods need to be considered at the design stage as assembly is labour intensive • Use 'snap-together' components and adhesives, avoid screwed fastenings wherever possible • The hinge connecting the solar panels could be a problem • Use compression joint connections rather than soldering • It is feasible to produce, but some further work is needed on the hinge connecting the solar panels
Materials	• The motor and switch are standard components that can be outsourced cheaply from the Far East • ABS high impact plastic will be used for the mouldings • It is feasible to obtain parts and materials	• The motor and solar panels are standard components that can be outsourced from the Far East • ABS high impact plastic will be used for the mouldings • It is feasible to obtain parts and materials
Cost	• It is feasible to manufacture for less than £12	• The feasibility of manufacturing for less than £12 will depend upon being able to source low cost solar panels of suitable quality. Also on simplifying the design of the hinge
Marketability	• There is a market for the battery powered fan in the UK and abroad • Some difficulty in obtaining batteries in Third World countries • Can be used indoors and in poor light – an advantage in the UK • It is feasible to market this produce	• There would be little demand for the solar powered fan in the UK • There could be a demand in sunnier and hotter countries • Would be popular in Third World countries where batteries are scarce and the added bulk of the solar panels would not be a problem • It could be feasible to market abroad if price could be reduced, possibly by licensing manufacture to a Far East low cost country
Quality standards	• The mouldings can be produced to the required finish and toughness • Reliable motor units and switches are available • Battery life exceeds the requirement of the design brief • It is feasible to produce this battery operated fan to the specified quality standards	• The mouldings can be produced to the required finish and toughness • Reliable motor units are available • The solar cells are prone to damage. The design needs to be revised to give them greater protection • Subject to some design modifications it should be possible to meet the specified quality standards
Overall assessment	• The design for the battery operated fan is acceptable • The market is available • This product is approved for prototype production	• It was felt that this was an innovative design that required further development • In view of the size of the potential overseas market further design and marketing studies were commissioned

even the most favoured design has to be modified before it is finally considered fit for production.

2.7.2 Selecting preferred designs

It must be emphasized that it is no good generating an ideal design if it is too expensive for the public to buy. As well as the technical merit of the design, the following information should also be provided:

- **Production** – Ideally, when producing assemblies for test, it is advisable to use the plant and the equipment that will eventually be used in full-scale production. This way, the design team can evaluate the adequacy and technical feasibility of the product prior to full-scale manufacture. It also enables them to carry out a *process capability study* of the equipment to identify if the allocated machinery is capable of holding selective tolerances and providing the required quality of finish. A *process capability study* is a practical assessment carried out under controlled conditions to determine whether the machinery allocated is capable of working to and holding the required tolerances.
- **Materials** – The materials to be used should be fully specified together with the sizes required, properties such as strength and durability, resistance to environmental attack, together with its characteristics and suitability for the chosen manufacturing processes.
- **Cost** – The cost of materials, resources and production processes and an estimated selling price for each item that the customer will pay.
- **Market** – The type and size of the market and the level and cost of pre-production and post-production publicity campaigns required.
- **Quality standards** – The quality of the manufacturing tolerances, materials used, finish, and performance (fitness for purpose) must be specified.

Having assessed the design proposals and, possibly, built completed prototype assemblies in order to carry out feasibility tests, it is necessary to select the preferred design, by carrying out a range of *design reviews.*

- Design reviews should be conducted by a team of specialists not normally associated with the development of the design.
- The review team should include specialists such as field engineers, reliability and quality engineers, method study engineers, materials engineers, and representatives from the procurement, marketing, costing, and packing and shipping departments.
- The review team should work to clearly defined criteria, including customer requirements, industry standards, national and international safety, environmental and hygiene regulations.
- The review team will evaluate areas of reliability, performance, ease of maintenance, interchangeability, installation, safety, appearance, cost, value to the customer and ergonomics. Ergonomics is defined as 'the study of economics in the human environment'. Put more simply, is the product convenient to use? For example, the flight deck of an aircraft or the driving position of a modern car is *ergonomically designed* so that the controls fall readily to hand and are easy to use with the minimum of movement and effort.

Test your knowledge 2.11

1. You are a member of the design review team assessing our two prototype handheld fans. Write a brief report on your findings and make a recommendation for manufacture giving the reasons for choice.

2.7.3 Presentation techniques

At various stages during the design process it is necessary for individual designers to present their ideas to the rest of the team and for the team leader to present their ideas to the customer. For such presentations to be successful they must be carried out in a planned and professional manner.

Your presentation technique must be organized to suit your target audience. A presentation suitable for your fellow professionals using the appropriate technical vocabulary would be largely meaningless if given to members of the general public. Similarly, a presentation given to the engineering design team at your firm would be unsuitable for representatives of the accounts department. They are no less knowledgeable, but they would require very different information. It is also important to mix your delivery techniques to avoid monotony. Let's now consider the techniques available to the presenter.

Verbal

Rehearse your presentation thoroughly so that you do not have to read it from the script, try to use the headings as prompts. The details are only there in case you have a lapse of memory.

- Don't use words you find difficult to pronounce.
- Try to be enthusiastic and let your voice show it.
- Be in command of your subject. The less you have to refer to your notes the more impressed will be your audience.
- Speak up so that everyone can hear, and speak clearly. Don't rush, pause at the end of each sentence for a moment.
- Don't hesitate and don't repeat yourself.
- Don't fidget with your hands and don't pace about. Have all your audio-visual aids organized conveniently so as to keep movement to a minimum.
- Arrange the verbal part of your presentation so that there are 'natural breaks' where you can introduce your audio-visual material without interrupting your flow and spoiling the continuity. Remember that most people's concentration span is quite short.

Overhead projector

To use an overhead projector to the best advantage:

- Make sure to set it up properly before the start of your presentation so that it is the correct distance from the screen. The whole of your transparency should be visible without material near the edges falling off the screen. Mounting your transparencies in cardboard frames helps to avoid this happening since the projected area is a constant size.
- Make sure the projector is focused before you start. A coin placed on the illuminated panel provides a sharp outline for focusing.

- Don't leave the projector on all the time. Only turn it on when you need it. To leave it on distracts the attention of your audience from you, the speaker. Remember it's your presentation. You're the star of the show.
- If there is more than one piece of information on the transparency don't show it all at the same time. This can be distracting and confusing. Cover the transparency with a sheet of thin card and reveal each item of information as and when you require it.
- Remember that if the audience want to reinforce a particular piece of information you can easily return to an earlier transparency. For this reason make sure all the transparencies are numbered sequentially. Make sure numbers relate to the prompts in your notes. Make sure you stack them in the correct order as you remove them from the projector.
- Overhead projectors can be used with other devices apart from transparencies. You can obtain measuring instruments such as voltmeters and ammeters with transparent scales.
- Finally, copy your OHP transparencies as handouts and make these available to your audience at the start. It will be too late after they have struggled to take notes in dimmed light. You do not want them leaving your presentation feeling annoyed at this simple oversight.

Slides

Despite the advances in computer technology and digital cameras, 35 mm colour transparencies are still useful for big screen presentations. Make sure that they are strictly relevant and not just 'pie-filling'. Also make sure they are loaded into the magazine of the projector in the correct sequence and the correct way up!

Flip chart

A flip chart is a large pad of white paper supported on an easel. Felt pens of various colours are used to draw or write on the paper. This should not form part of your main presentation but is useful for quick diagrams and notes during a question and answer session following the main presentation. When a sheet is full, you merely turn the paper over and carry on to the next sheet. You can, therefore, turn back to an earlier sheet whenever the need arises.

Video recording

Video recording has largely taken over from ciné-film, it is an excellent method of presenting material particularly if sound is to be incorporated. Unfortunately, although it is easy to make a video recording with modern lightweight camcorders, it is very difficult to make good quality and properly edited recordings. Unfortunately, our efforts tend to be compared with the highly professional material we watch on television.

Computerized presentation

With the availability of LCD projectors that can be driven from a laptop computer, large screen presentation can be achieved using CAD generated images, digital camera images, and material scanned in from photographic and published sources, even material down-loaded from web sites (check on copyright first). All this material can be saved to disc and edited before presentation. Your notes can be held on a second laptop computer acting as a 'portable teleprompt'.

Technical vocabulary

The technical vocabulary you use should be appropriate for your audience and for the product or design you are promoting. Your audience will often contain persons of:

- Different technical and professional backgrounds
- Different social and educational backgrounds
- Different ethnic groups for whom English may not be their first language.

If you are presenting a new product incorporating new technology, the jargon terms associated with it may not yet have filtered through into general usage even amongst professional persons. Therefore it is often useful to issue a *glossary of the terms* used in your presentation. This particularly applies to *acronyms*. Acronyms are words made up from the initial letters of other words, for example NATO stands for the North Atlantic Treaty Organization.

2.7.4 Feedback

In order to finalize the design, it is necessary to obtain customer feedback at the end of the presentation. Two levels of assessment (judgement) have to be made.

Quantitative judgements

These relate to measurable features of the design such as size, weight, performance, strength and cost. In the case of foodstuffs judgements may have to be made relating to such dietary matters as fat, salt (sodium) and sugar content to comply with health requirements.

Qualitative judgements

These relate to matters such as colour, finish and fitness for purpose. That is, the quality of the finished product, whether it represents value for money for the customer, and whether it complies with national and international safety and hygiene standards.

Once the client has carried out their analysis of the proposed design, user panels are frequently set up across the country to assess customer reaction. User panels consist of small groups of people of selective age and sex groups. They are given the opportunity to try out prototype samples of the design and comment upon it. They may also be invited to comment upon the proposed packaging and advertising literature. This information provides feedback to the manufacturer, the marketing personnel and the design team. Final amendments are then made to the design, packaging and advertising literature and the design is put into production. Figure 2.25 shows a simple feedback report form. Feedback forms should be well constructed with unambiguous questions and be easy to complete by means of boxes that can be ticked or by the use of one-word answers.

Test your knowledge 2.12

1. Explain the importance of feedback as part of a presentation.
2. Explain the essential differences between quantitative judgements and qualitative judgements.

Product	Design Team				
Functions	Performance				
	U	M	A	G	O
Operating instructions Ease of operation Packaging Advertising Material					
U = Unacceptable M = Marginal A = Acceptable G = Good O = Outstanding					
General comments					

Figure 2.25 Feedback form.

2.7.5 Support material for presentations

Nowadays, people are often called upon to give *presentations* to senior management, customers, financial sponsors, the media, and other interested parties. The more professional the presentation, the more influence it will have and the more successful it will be. Having discussed the various techniques available to you when making a presentation, let's now consider the material you may have to present.

Reports

Reports are essential forms of communication in business today:

- *Individual reports* are used for internal routine communication of day to day issues and are written in the *first person* (I would be grateful if you will attend etc. ...).
- *Formal reports* are used to communicate with such persons and organizations as your superiors and customers. They are written in the *third person* (The sales figures for this month indicate that ... it is unfortunate little improvement has been shown etc. ...).
- *Legal reports*, such as accident reports, are often completed on preprinted forms which only requires the respondent to fill in the sections indicated in response to a series of formal prompts. The style of writing (first person or third person) will depend upon and reflect the nature of the prompts and questions.
- *General reports* are usually written in the third person and are commissioned by an official body or person. Examples are shareholders reports from the chief executive of the company, design reports to the technical director or a customer, student reports to his tutor.

Whichever type of report is required, it should be written to a basic set of rules to ensure that it is logically constructed in the accepted format. For a lengthy report, it should be divided into numbered sections that should be prefaced by a *list of contents*. The items in the list of contents should be numbered to agree with the section numbers. An *index* of key points should be included at the end of such a report. The basic format and order of presentation for a report is as follows:

- The title which must convey the contents of the report (e.g. design for an electrically propelled saloon motor car).
- The person(s) or body commissioning the report and to whom it should be addressed. In reports for public distribution the address is normally omitted.
- If the report refers to a meeting (e.g. minutes of a meeting), a list of the persons attending together with their titles should be included. A list of persons apologizing for non-attendance should also be included.
- A brief reference to the terms of reference under which the report is written. For short internal reports to a single specific person this can often be omitted.
- A summary of what the report contains and its conclusions should be inserted near the front of the report as an overview of the entire report.
- The main body of the report detailing the investigations carried out, the facts discovered, assumptions, arguments and opinions arising from the investigations.
- The conclusions of the writer set out in a clear and logical manner. These should be kept as brief as possible.
- The recommendations of the writer that need to be carried out in order to meet the needs of the report.
- An appendix to the report which contains charts, statistics, graphs and any other information that would be a problem if included in the main body of the report. Such material should be numbered so that it can be referred to in the text of the report.
- For a technical report, a glossary of the terms used may be necessary if the intended readership is broad and contains persons with a non-technical background.
- Finally the names and signatures of the authors of the report and the date of its preparation should be added.
- Write out your presentation in full – not only as a starting point to make sure you have covered all the important information in the correct order but as a useful prop if your nerves fail you and your mind goes blank!
- Prepare 'prompt cards' from your detailed presentation. Only one topic area on each card with the heading in bold type.
- The prompt cards should also remind you which visual aid you intend to use and the point where you intend to use it. It is useful for these to be colour coded.
- The prompt cards should also carry brief and clear notes summarizing your full write-up with the points highlighted.

Visual aids

At one time or another we have all suffered the 'lecture' where some erudite visiting speaker stands up and drones on and on reading from his or her notes in a monotonous voice, whilst the chair we are sitting on seems to get harder and harder.

We should always try to liven up our presentation with various visual aids. This not only relieves the monotony by introducing a new interest from time to time, but also because 'one picture is worth a thousand words'. A picture of an elephant is much more informative than a written description. There are various types of visual aid available to speakers and first we will consider the overhead projector (OHP) transparency.

Overhead projector transparencies

For a quick, one-off presentation for your colleagues OHP transparencies can be produced on acetate sheets using special transparent colour pens. However, for more formal presentations something better is required. 'Letraset' dry transfer lettering looks a lot more professional, but is only available in black. However a transparency can be 'livened up' by using coloured acetates for the background. For the most professional presentation computer graphics can be pressed into service using a colour printer. The possibilities are endless using readily available modern software and colour printers.

Visual aids (35 mm transparencies)

Colour transparencies made on 35 mm camera film are a quick and simple way of showing pictorial material, such as machines, cars, furniture, clothes, etc. These can be copied from previously published sources (with the permission of the copyright holder) or from the actual objects shown. Remember to check that they are correctly inserted in the projector magazine prior to the presentation. Nothing looks more slipshod than slides that are upside down, reversed left to right or in the wrong order. With the advent of digital cameras the complete presentation can be recorded on disc, so that the whole presentation is under the direct control of the speaker and the projector can be linked to his or her computer.

Visual aids (video recordings)

With the advent of high quality portable camcorders, video recordings have virtually superseded cine-film where moving pictures are required, particularly where sound is required as well. Professionally produced, they are widely used for advertising and for staff training. Compared with ciné-film, video recordings are:

- Quick and relatively cheap to produce
- Easy and cheap to copy
- Easy to re-dub the sound so as to produce various foreign language versions to suit particular overseas markets
- Before the advent of video recordings, a company designing and manufacturing large plant and heavy machines had to bring its overseas customers to this country to see their products or transport expensive and heavy working models to the customer. Nowadays a sales representative can carry a video recording in his or her briefcase recorded in the language of the customer
- Nowadays video recordings can compete with ciné-film using projection equipment linked to a computer or DVD player for large scale presentations. For example, the launch of a new car where cinema-quality large screen viewing is required.

Having reviewed the systems available for illustrating presentations, let's now look at the sort of material for which the various systems can be used. We will start by considering what we can show on OHP transparencies. These can be either hand-drawn or computer generated.

Test your knowledge 2.13

1. Compare and contrast the different types of reports that you may be called upon to write in a manufacturing situation.
2. Compare and contrast the following visual aid techniques and state where each could be used to advantage:
   (a) Overhead projector
   (b) 35 mm transparencies
   (c) Video recordings
   (d) Computer.

Key notes 2.4

- Visual aids are techniques for illustrating a lecture or verbal presentation.
- Overhead projector (OHP): a device that enables the speaker to project hand drawn or computer generated transparencies on to a screen whilst still facing his/her audience.
- Slides are 35 mm photographic transparencies that can be projected on to a screen.
- Flip chart is a large pad of white paper supported on an easel and which can be written or drawn on with multicolour felt pens. It has largely superseded the blackboard and chalk.
- Video recording is a sound and vision recording on tape that can be played through a television set or an LCD projector and has largely replaced ciné-film for illustrating lectures and presentations.
- Quantitative judgements relate to the assessment of measurable quantities.
- Qualitative judgements relate matters such as colour, feel, taste, finish and are largely subjective.
- Reports are a written or spoken (verbal) method of making a presentation.
- Reports need to be drawn up in a logical manner so that they are easily understood.
- The language used should be suitable for the target audience.
- Notes are personal to the speaker and should only be used as an 'aide memoire'. Avoid reading from your notes. They are only there to prompt you if your memory momentarily fails.
- Feedback should be by well-constructed forms issued at the end of a presentation. They are essential for design assessment.

2.8 Graphical communication

2.8.1 Graphs

Graphs are used to represent numerical data in pictorial form. The data are represented by a line (curve) varying in position between two axes (plural of axis). The horizontal axis is usually called the X-axis or the *independent* variable axis. The vertical axis is usually called the Y-axis or the *dependen*t variable axis. This is because the value of Y is related to and *depends* in some way on the value of X. Therefore only two values can be represented on a simple graph. The scale of the axes can influence the appearance of the graph. They can be arranged either to exaggerate the changes being depicted or they can

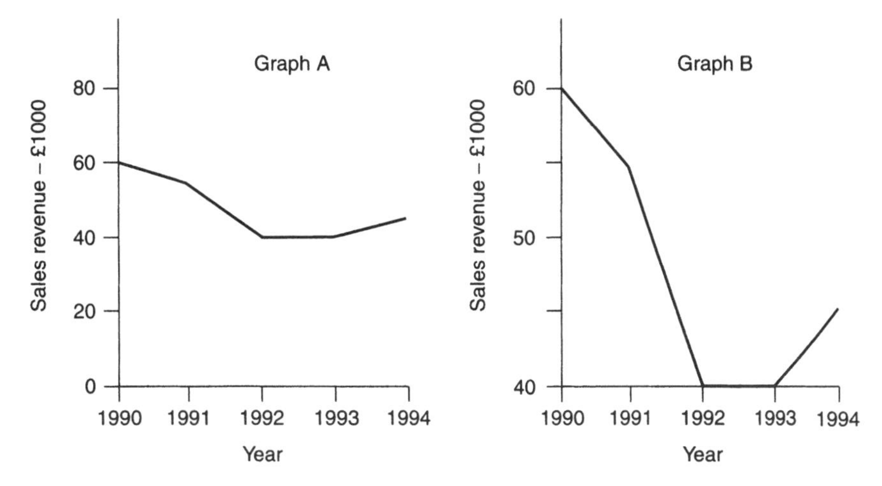

Figure 2.26 Graphs.

be arranged to minimize the apparent changes. It all depends upon how we wish to influence our audience. Let's now consider the graphs shown in Fig. 2.26.

It is the first impression that people tend to remember. So by changing the scales we can influence that first impression. Mathematically, both Fig. 2.26(A) and (B) are identical. However, at first sight, the sales figures shown in graph (A) appear to remain reasonably constant with a slight downward trend, but rising again after 1993. On the other hand graph (B) shows what appears to be a catastrophic fall in sales. However, remember that both graphs are numerically the same. This distortion of the scales is widely used by the media to emphasize a particular point in order to influence public opinion. The use of colour can also affect interpretation. Bright colours such as red against a paler pastel coloured background can be used to focus the viewer's eye on a point of emphasis.

2.8.2 Pie charts

Pie charts are one of the simplest ways of representing numerical data, as they are easy to draw and understand. The numerical value is represented by the angle of the segment. For example, if the total sales for a range of goods is £1000 the whole circle (360° represent £1000). If sales of garment 'A' amounted to £200 over the same period of time then its segment would have an angle of $360° \times (£200 \div £1000) = 72°$. This is shown in Fig. 2.27(a). Figure 2.27(b) shows how the sales revenue from Fig. 2.26 can be shown as a pie chart.

2.8.3 Bar charts

Bar charts show numerical data by changing the heights of the bars when they are drawn vertically (histograms) or the lengths of the bars when drawn

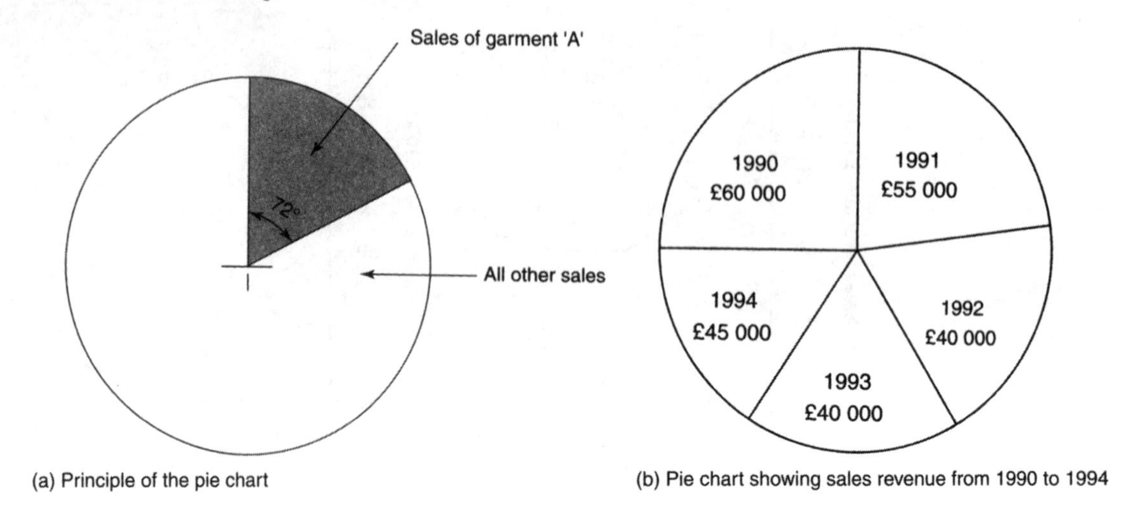

(a) Principle of the pie chart

(b) Pie chart showing sales revenue from 1990 to 1994

Figure 2.27 Pie charts.

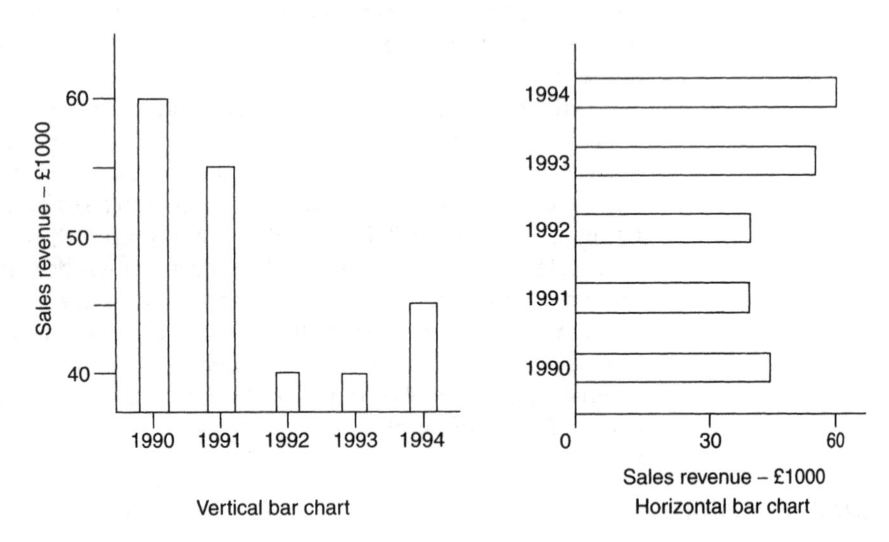

Figure 2.28 Bar charts.

horizontally (Gantt charts). A typical histogram is shown in Fig. 2.28. This shows the same numerical data as Fig. 2.26.

2.8.4 Gantt chart

The *Gantt chart* is widely used in business to compare expected performance against actual performance. It is a horizontal bar chart and an example is shown in Fig. 2.29. It is possible to compare the results of three sales persons against each other, as well as against the company sales target. Gantt charts are also used for comparing actual production against target times and delivery dates.

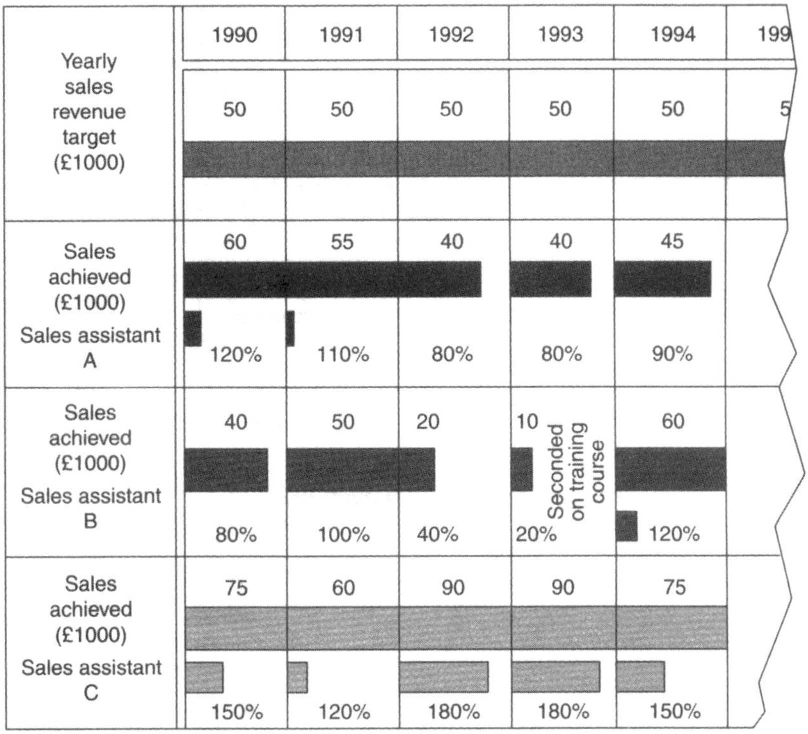

	1990	1991	1992	1993	1994	199
Yearly sales revenue target (£1000)	50	50	50	50	50	5
Sales achieved (£1000) Sales assistant A	60 120%	55 110%	40 80%	40 80%	45 90%	
Sales achieved (£1000) Sales assistant B	40 80%	50 100%	20 40%	10 20% Seconded on training course	60 120%	
Sales achieved (£1000) Sales assistant C	75 150%	60 120%	90 180%	90 180%	75 150%	

Figure 2.29 The Gantt chart.

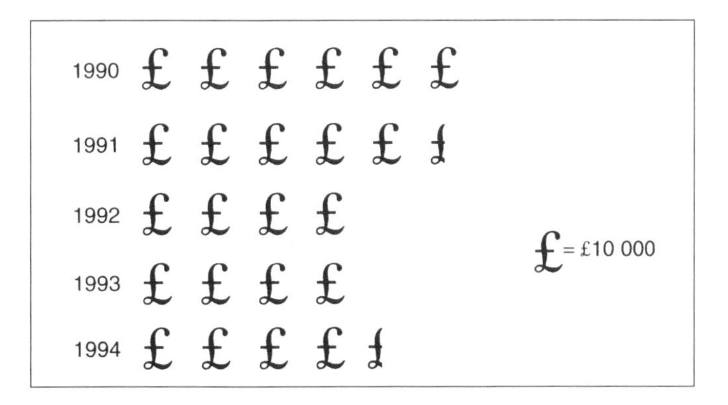

Figure 2.30 Pictogram showing yearly sales revenue.

2.8.5 Pictograms (ideographs)

The pictogram is another method of representing numerical data and is often used when communicating with the general public and other non-technical groups. An *icon* is used to resemble the dependent variable. For example, Fig. 2.30 shows our original sales figures in the form of a pictogram. In this example the icon is the £ symbol and represents £10 000. Smaller amounts can be shown by using only part of the icon. Pictograms are intended for giving an overall impression rather than for detailed analysis.

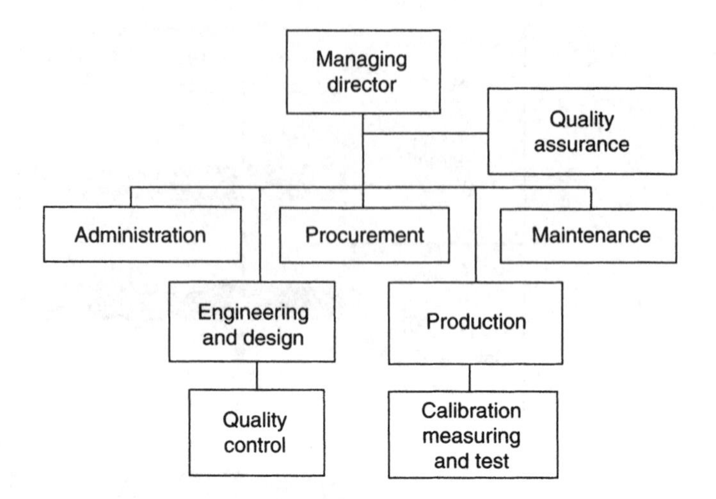

Figure 2.31 Organization chart.

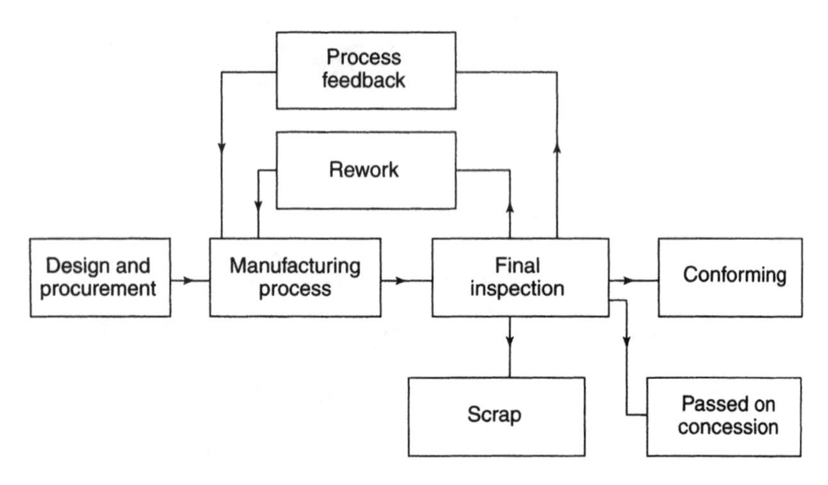

Figure 2.32 Flow chart.

2.8.6 Charts

These are used to show the relationship between non-numerical data such as the organization chart shown in Fig. 2.31. This shows the relationship between the various departments in a company.

2.8.7 Flow charts

These are logical diagrams showing the direction of flow required in order to determine a sequence of information. If you want to make an appointment with your doctor, you ring up your doctor. The request *flows* from you to your doctor. The request for an appointment does not start by flowing from the doctor to you. A typical example of a flow chart or diagram for a quality control system is shown in Fig. 2.32.

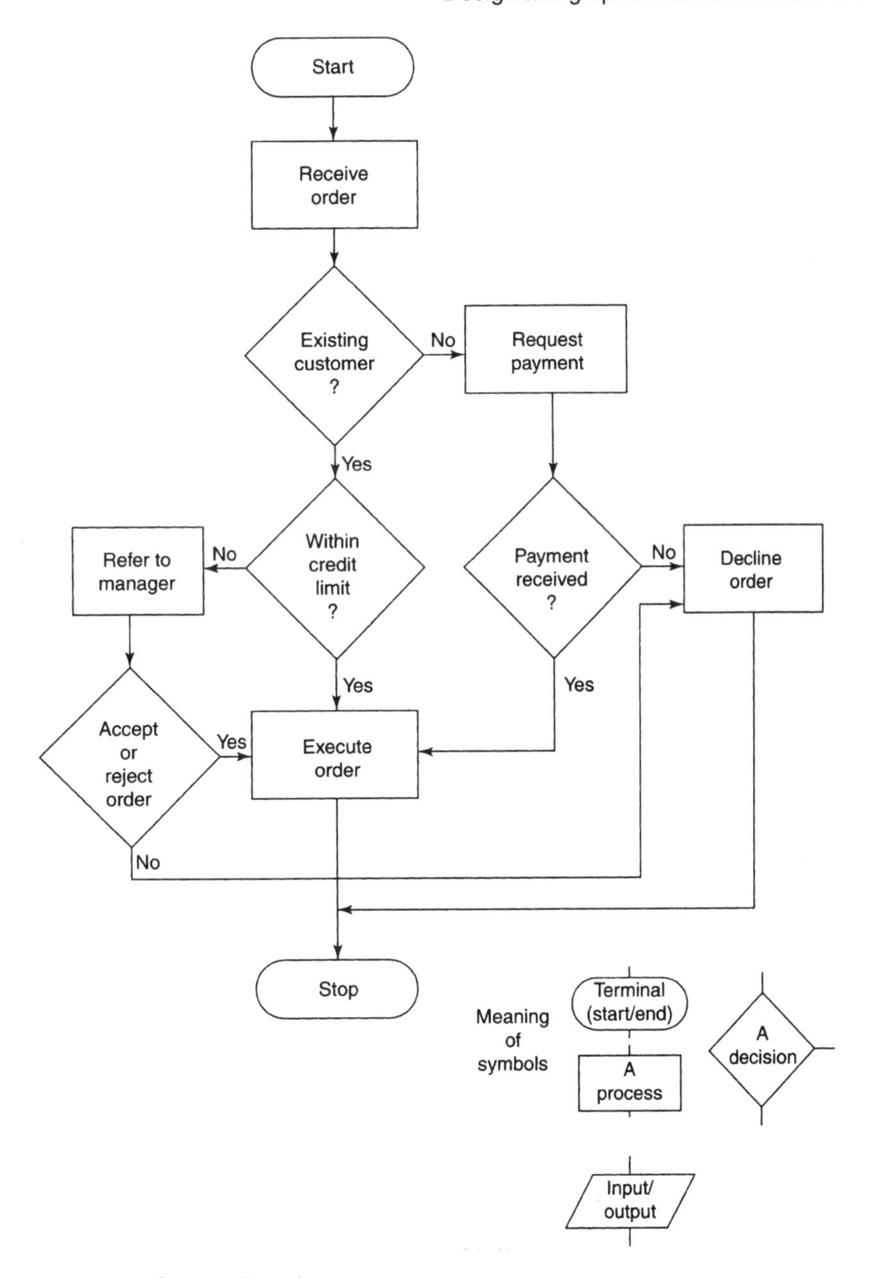

Figure 2.33 System flow chart.

2.8.8 System flow charts

System flow charts are used where procedures can be broken down into logical steps with a simple yes/no decision. Such charts are very popular with computer systems and systems analysts since the computer works in the same logical manner. System flow charts are also known as *algorithms.* Four basic symbols are used to build up diagrams and the use of three of them is shown in Fig. 2.33.

Pipe sizes for the range of MQ boilers				
Boiler type	Pipe diameter in millimetres for no. of radiators			
	1–3	4–8	9–12	13–20
M-001	4.0	5.0	6.0	8.0
M-002	5.0	6.0	8.0	10.0
M-003	6.0	8.0	12.0	14.0
M-004	10.0	15.0	20.0	25.0
M-005	15.0	20.0	30.0	40.0

Data sheet no. MQ./ 123
Issue no.3
Date November

Figure 2.34 Data sheet.

2.8.9 Data sheets

These are used to tabulate data for easy reference. The information contained in such sheets may be distributed as single sheets, or several sheets may be bound together to make a booklet, or the data may be printed out as a wall chart. A typical example is shown in Fig. 2.34. For a given type of boiler and the number of radiators it is to feed, you can immediately see the manufacturer's recommended pipe size for coupling to the boiler. If you wish to feed seven radiators from a type M-004 boiler, you look down the 4–8 radiator column and, opposite the type M-004 boiler you can see that you require 15 mm diameter pipework.

When using data sheets it is important to establish that they are up to date. On receipt of a new issue, all previous data sheets should be destroyed or be returned to the point of issue. Data kept in a computer system should be on a read only memory (ROM) status disk so that unauthorized alterations cannot be made.

Test your knowledge 2.14

1. Explain briefly what steps you would take to make a verbal presentation as professional as possible.
2. Explain how you could illustrate your presentation with easily produced visual aids.
3. Explain what steps you would take to use an OHP effectively.
4. Discuss the use of flip charts in a presentation.

2.9 Technical drawings

Having agreed the design with the customer it is necessary to produce working drawings of the product and the tooling that is to be used in its manufacture. Drawings may also be needed:

- To accompany sales literature.
- For installation and commissioning.

- Maintenance.
- Repairs.

2.9.1 Orthographic drawings

The type of drawing used will depend upon which of the above applications it is intended for. Technical drawings for manufacture should be produced in accordance with the appropriate national standard. For example, mechanical engineering drawings in the UK are produced in accordance with British Standard conventions. Important standards used in mechanical and electronic/electrical engineering are:

- BS 308 Engineering Drawing Practice.
- BS 4500 ISO Limits and Fits (these are used by mechanical and manufacturing engineers).
- BS3939 Graphical symbols for electrical power, telecommunications and electronics diagrams.
- BS 2197 Specifications for graphical symbols used in diagrams for fluid power systems and components.

In addition, abridged and simplified versions of two of these standards are available at a substantially reduced price for students in schools and colleges.

- PP8888 Engineering Drawing Practice for Schools and Colleges
- PP7307 Graphical Symbols for use in Schools and Colleges.

Examples of the drawings used by engineers for showing three-dimensional solids on a two-dimensional sheet of paper or on a computer screen are shown in Fig. 2.35. They are referred to as *orthographic* drawings and can be presented in:

- First Angle or English projection (Fig. 2.35(a))
- Third Angle or American projection (Fig. 2.35(b)).

2.9.2 Pictorial drawings

In addition, pictorial views are often used, particularly for DIY applications where the user may not be familiar with the interpretation of orthographic drawings shown in Fig. 2.36. The most commonly used pictorial drawing techniques used are:

- *Isometric* drawing where all the lines are drawn true length and the *receding lines* (the lines appearing to be 'going into' the page) are drawn at 60° to the horizontal as shown in Fig. 2.36(a).
- *Cabinet oblique* drawing where only the vertical lines are true length and the receding lines are half the true length. The receding lines are drawn at 45° to the horizontal as shown in Fig. 2.36(b).

2.9.3 General arrangement drawing (GA)

Figure 2.37 shows a typical general arrangement (GA) drawing. You can see that this drawing:

- Shows all the components in their assembled positions
- Lists and names all the components and lists the reference numbers of the detail drawings needed to manufacture them

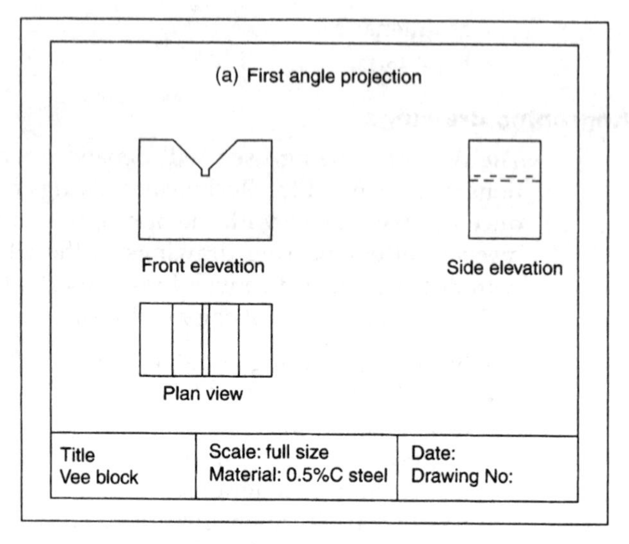

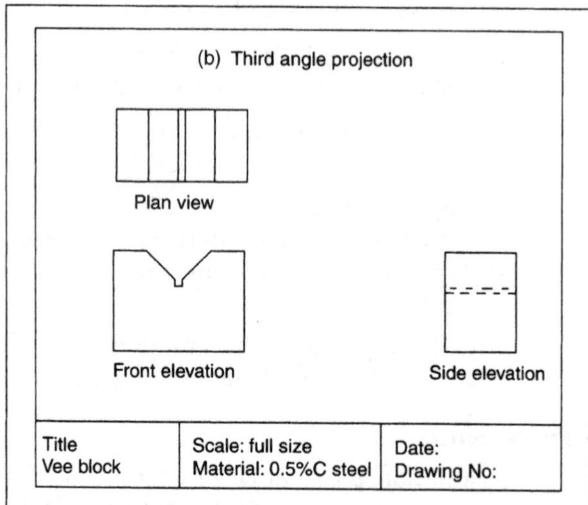

Figure 2.35 Illustration of first angle and third angle orthogonal drawings.

- States the quantity needed of each component
- Lists the 'bought-in' components and indicates their source
- Lists any modifications and corrections.

2.9.4 Detail drawings

As an example, Fig. 2.38 shows one of the detail drawings listed in Fig. 2.37. A detail drawing gets its name from the fact that it provides all the *details* and dimensions necessary to make the component shown.

2.9.5 Exploded drawing

Exploded (assembly) drawings are widely used in maintenance manuals. They show the parts correctly positioned relative to each other, but spaced

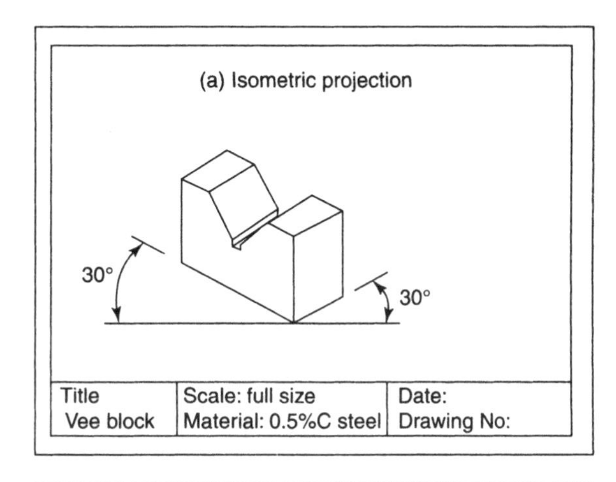

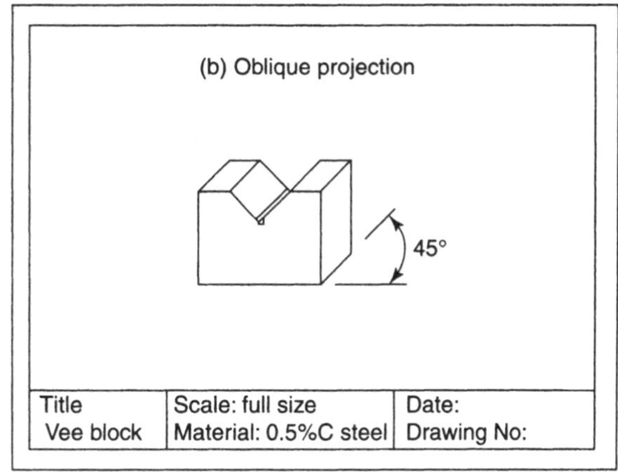

Figure 2.36 Pictorial drawing.

apart so that the parts are easily recognizable. Generally a table of parts (parts list) is included giving the necessary information for ordering spares. An example is shown in Fig. 2.39. Similar drawings are used for the assembly of 'flat-pack' items of domestic equipment and furniture purchased from DIY stores. It makes identification of the parts easy and shows the order in which they should be assembled.

2.9.6 Circuit diagrams

These are drawn using standard symbols for the various components. A typical example is shown in Fig. 2.40(a). This is an amplifier circuit. Nowadays most circuits are built up on a *printed circuit board* (pcb) and, as well as the circuit, the designer also has to provide a suitable layout for printing the circuit board. Figure 2.40(b) shows a suitable layout for the amplifier circuit.

2.9.7 Using computer-aided design (CAD)

Computer-aided design (CAD) has now largely replaced manual methods for producing formal engineering drawings. CAD software is used in conjunction with a computer and the drawing produced on the computer screen is saved

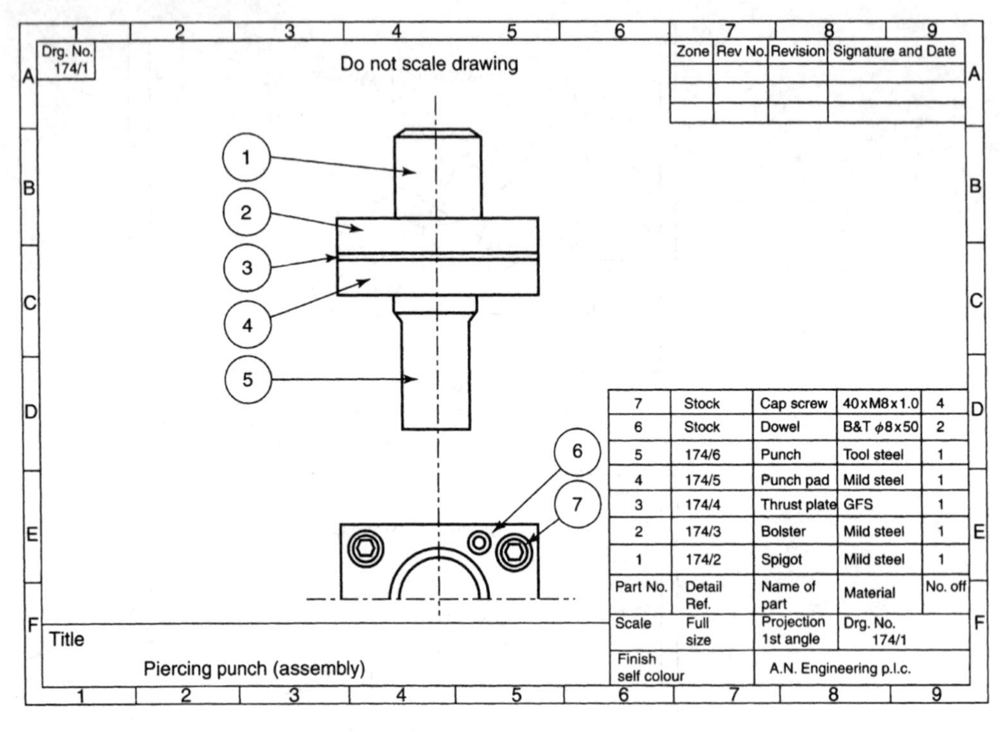

Figure 2.37 General arrangement (GA) drawing.

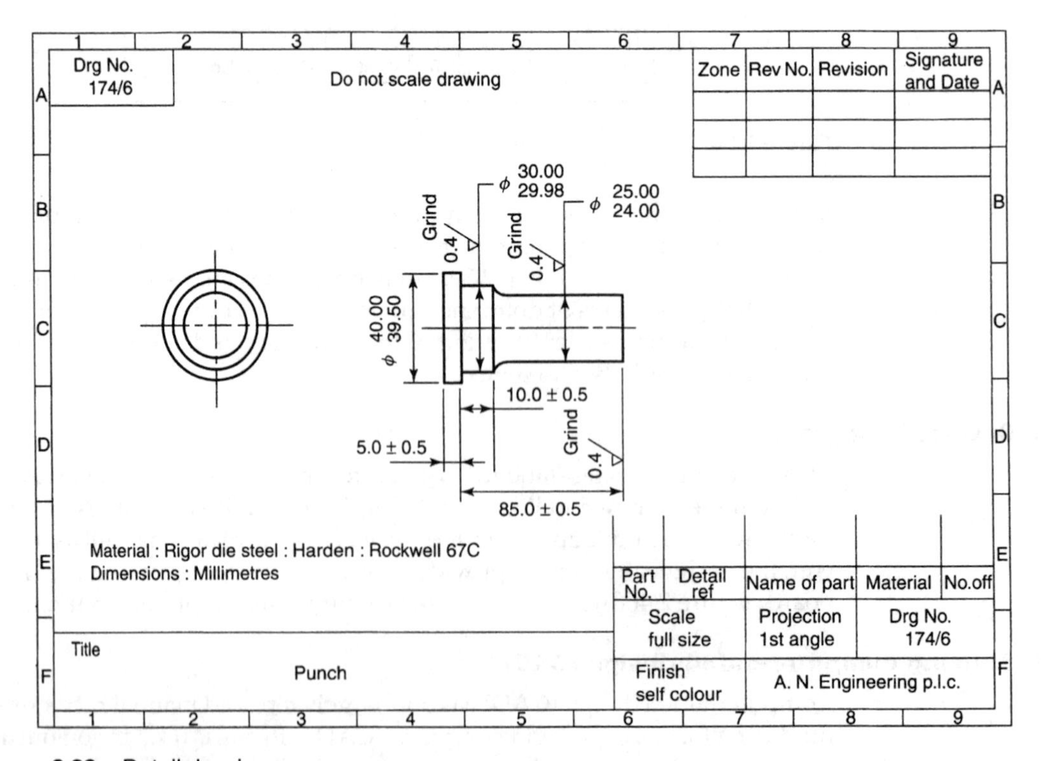

Figure 2.38 Detail drawing.

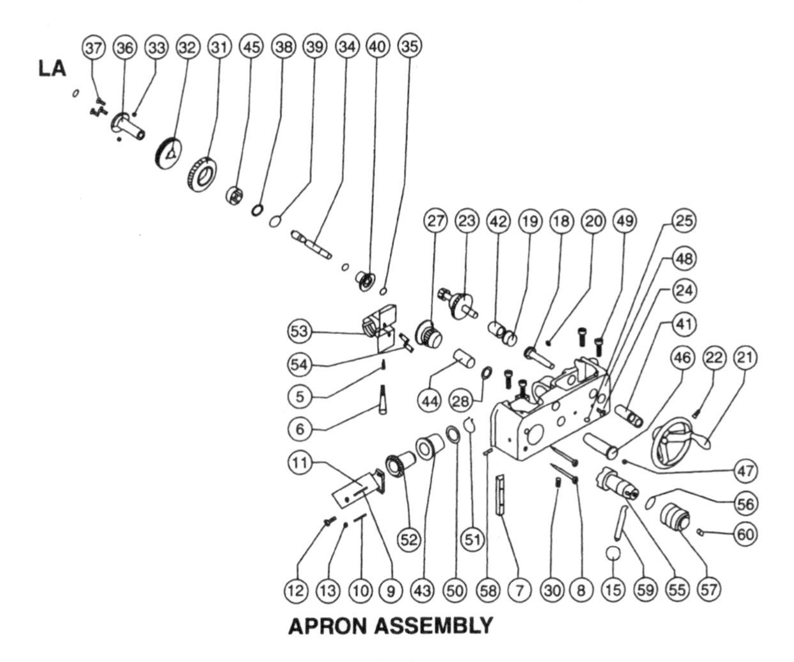

SECTION LA
APRON ASSEMBLY

Drg. Ref.	Part No.	Description	No. Off/Mc.
LA5	A4729	Spring – Leadscrew Nut	1
LA6		Cap Hd. Screw – Leadscrew Nut (2 B.A. $\times 1\frac{3}{4}$")	1
LA7	A2082	Glb Strip – Leadscrew Nut	1
LA8	A9193	Ch. Hd. Screw – Strip Securing	2
LA9	A9194	Adjusting Screw – Glb Strip	1
LA10	A9195	Adjusting Screw – Glb Strip	1
LA11	A9196	Leadscrew Guard	1
LA12		Hex. Hd. Set Screw (2 B.A. $\times \frac{1}{2}$")	1
LA13		Hex. Locknut (2 B.A.)	2
LA15	80002	Ball Knob (KB5/100)	1
LA18	A9198	Hand Traverse Pinion	1
LA19	65004	Sealing Plug – Apron (AQ330/15)	1
LA20	70002	Woodruff Key (No. 404)	1
LA21	A2087	Handwheel Assembly	1
LA22		Socket Set Screw ($\frac{1}{4}$" B.S.F. $\times \frac{1}{4}$") (Knurled Cup Point)	1
LA23	A9199	Rack Pinion Assembly	1
LA24	A2531	Oil Level Plug	1
LA25	65000	Oil Nipple (Tecalemit NC6055)	1
LA27	A9201	Bevel Gear Cluster Assembly (Includes LA44)	1
LA28	A9202	Thrust Washer	1
LA30		Socket Set Screw ($\frac{1}{4}$" B.S.F. $\times \frac{1}{2}$") (Knurled Cup Point)	1
LA31	A9204	Clutch Gear Assembly (Includes LA45)	1
LA32	A9205	Drive Gear	1
LA33	73010	Ball – Clutch (S mm ϕ)	2
LA34	A9206	Operating Spindle	1
LA35		Circlip (Anderton 1400 – $\frac{3}{8}$")	3
LA36	A9207	Drive Shaft	1
LA37		C's'k Hd. Socket Screw (2 B.A. $\times \frac{1}{2}$")	3
LA38	A9782	Washer – Drive Shaft	1
LA39		Circlip – Drive Shaft (Anderton 1400 – $\frac{5}{8}$")	1
LA40	A9280	Knob Operating Spindle	1
LA41	A9210	'Ollite' Bush	2
LA42	A9211	'Ollite' Bush	1
LA43	A9212/1	'Ollite' Bush – Flanged	1
LA44	A7595	'Ollite' Bush	1
LA45	A9220	Clutch Insert	1
LA46	A9203/1	Stud – Gear Cluster	1
LA47	65001	Oil Nipple (Tecalemit NC6057)	1
LA48	10025/1	Apron Assembly (Includes LA41, LA42, LA43)	1
LA49		Cap Screw (M6 ×1× 25 mm)	4
LA50	10217	Thrust Washer	1
LA51	10431	Circlip	1
LA52	A9200/1	Bevel Pinion	1
LA53	A1975/3	Leadscrew Nut – set	1
LA54	10508	Cam Peg	2
LA55	10528	Cam	1
LA56	65007	'O' Ring (BS/USA115)	1
LA57	10529	Eccentric Sleeve	1
LA58		Socket Set Screw ($\frac{3}{14}$" B.S.F. $\times \frac{3}{8}$" Half Dog Point)	1
LA59	10530	Lever	1
LA60		Socket Set Screw (2 B.A. $\frac{1}{4}$", Cup Point)	1
LA61	10424	Guard Plate (not illustrated)	1

Figure 2.39 Exploded view and parts list. Courtesy of Myford Ltd.

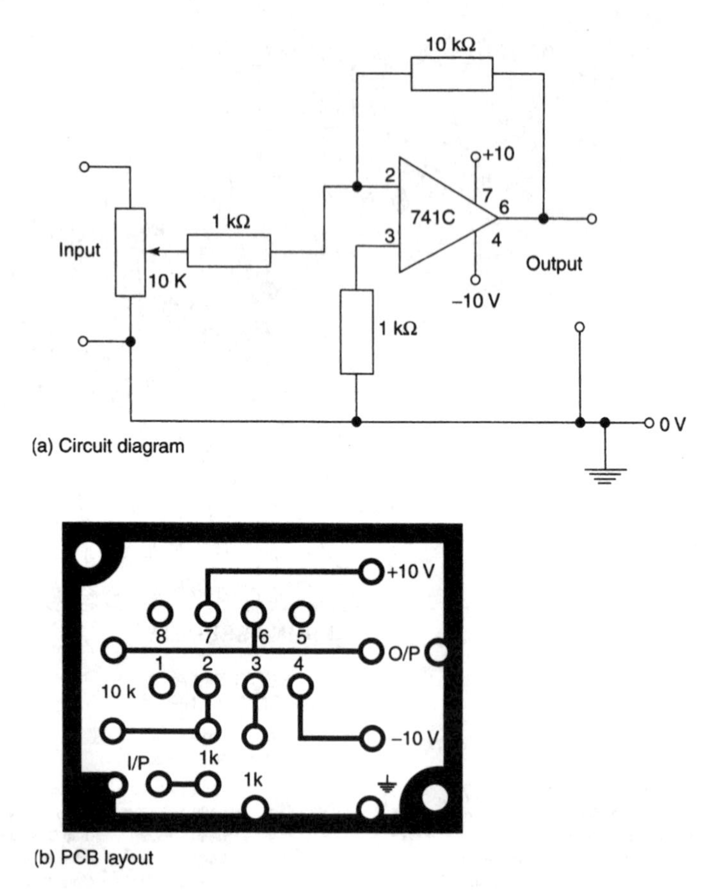

Figure 2.40 Amplifier circuit.

in a computer file on disk. Networked computer-aided design and computer-aided manufacture (CAD/CAM) and computer-aided engineering systems (CAE) have made it possible to share data and drawings over a network and also made them available to computer numerically controlled (CNC) machine tools that carry out automated manufacturing operations. Your school or college will be able to provide you with access to CAD equipment as part of your GCSE course. Figures 2.41–2.43 inclusive show screen shots of three popular screen packages.

Figure 2.41 shows a simple 2D drawing of a gasket using DesignCAD. The drawing is made to scale and the dimensions are still to be added. Figure 2.42 shows a more complicated general arrangement (GA) drawing produced using AutoSketch. Some dimensions have been included. Figure 2.43 shows a 3D assembly drawing produced using AutoCAD.

2.10 Patterns, samples and swatches

Whilst technical drawings are the usual means of transmitting design information when discussing engineering products, joinery products and furniture

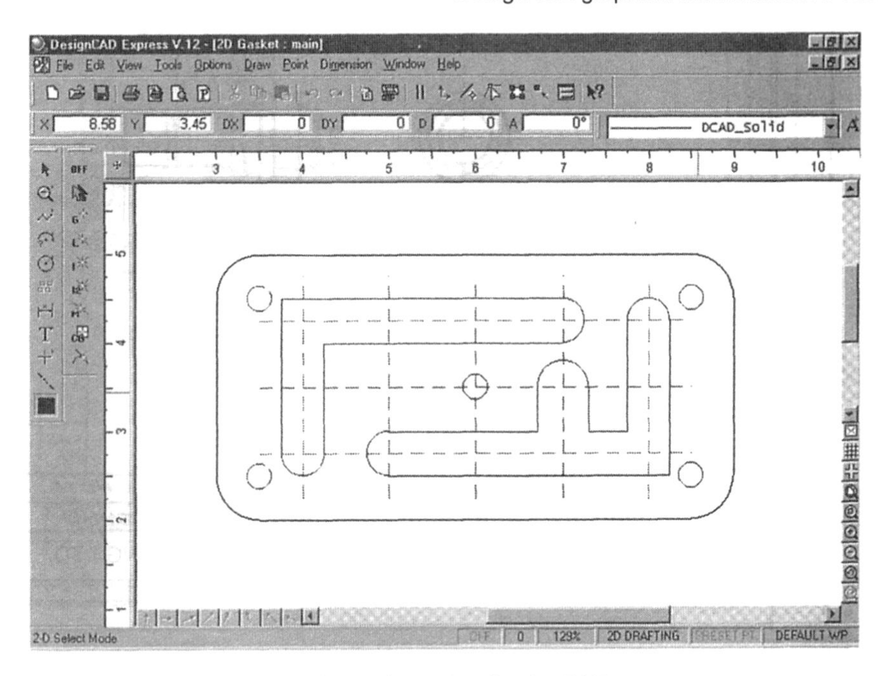

Figure 2.41 A 2D drawing of a gasket using DesignCAD.

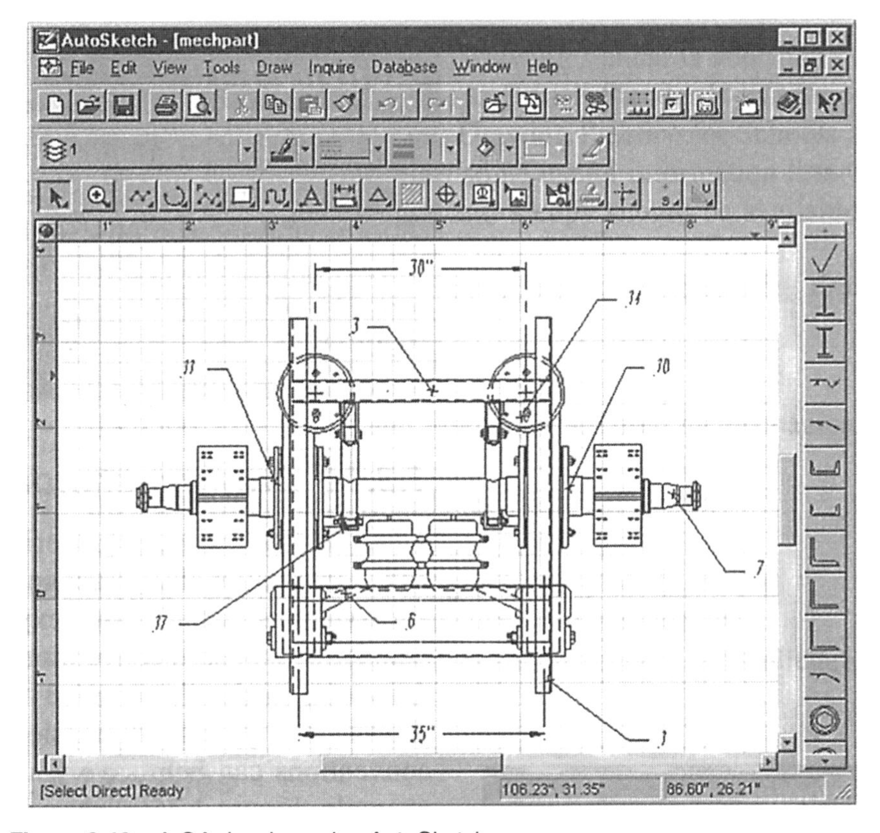

Figure 2.42 A GA drawing using AutoSketch.

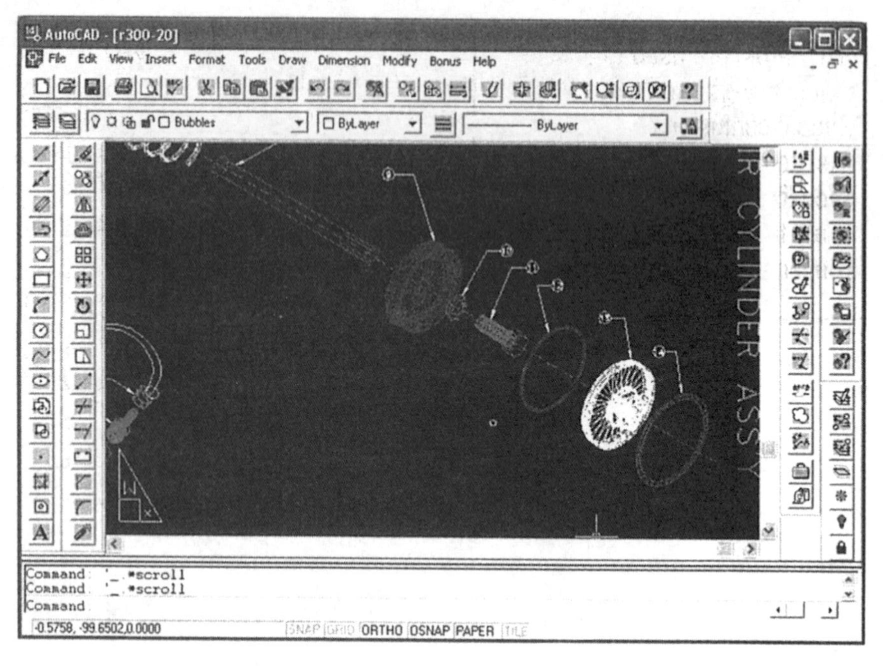

Figure 2.43 A 3D assembly diagram using AutoCAD.

products, they would not be appropriate for, say, convenience foods. For pre-packed convenience foods for supermarkets the product design would be in the form of a *recipe*. This would provide a list of ingredients with the quantities involved together with the processing and packaging information. The packaging would also provide a list of contents, nutritional information and recommended cooking method. Similarly paints, pharmaceutical products, household cleansing products and similar manufactured goods would also rely on a description of the ingredients together with the quantities and methods of mixing and packaging.

Fabrics are usually presented in pattern books in which small samples of the actual cloths of the same type, but varying in design and colour, are bound together. These are called *swatches* and are a convenient method of comparing alternative fabrics.

Clothing can be either 'bespoke' or 'off the peg'. Bespoke garments are individually hand crafted by highly skilled tailors to the customer's individual measurements. As such, bespoke clothing is very expensive. Clothes sold 'off the peg' are mass-produced. Traditionally, the material was cut out using patterns (templates) as shown in Fig. 2.44(a). The layout of the patterns on a piece of woven cloth is shown in Fig. 2.44(b). Note how the pattern lay minimizes waste material (off-cuts) and how they are arranged so that the straight grain line arrows lie parallel to the warp yarns of the fabric. Nowadays mass produced clothing is more likely to be computer designed using a CAD system and downloading the data direct to a CNC cutting machine.

The design is held on a computer file and can be called up every time a repeat order is received. Designer clothing is also produced in bulk, but to the individual design of an internationally famous fashion house. Usually

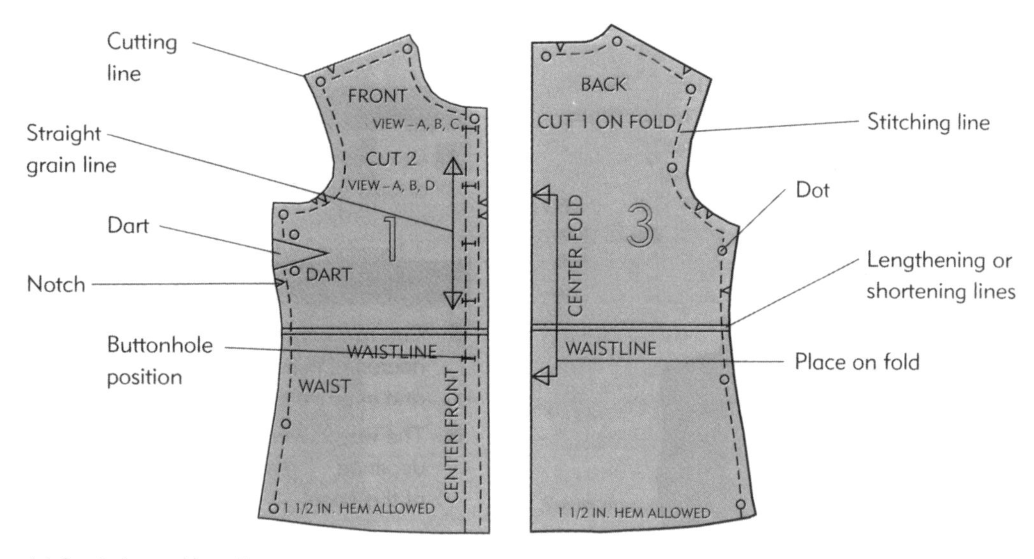

(a) Symbols used in patterns

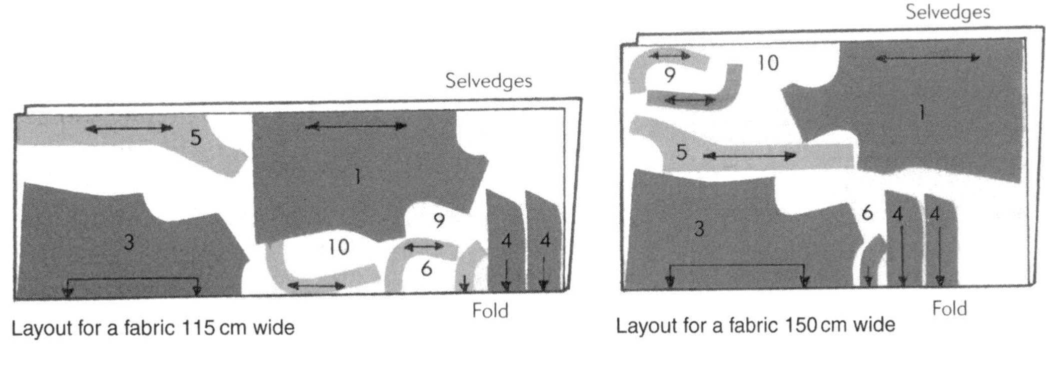

(b) Pattern lays

Figure 2.44 Garment patterns.

the finish is better and the garments carrying the logo of the fashion house will be much more expensive. Knitwear is also computer designed and the data are downloaded on to a CNC knitting machine. Several machines can be controlled from the same computer but the garments are made individually. One advantage of knitwear is that there is no waste. All the yarn is made up into the garment and there are no off-cuts as when cutting out a suit from a length of cloth.

Test your knowledge 2.9

1. Figure 2.45 shows a simple object in first angle projection. Redraw it in:
   (a) Third angle projection
   (b) Isometric projection
   (c) Cabinet oblique projection.

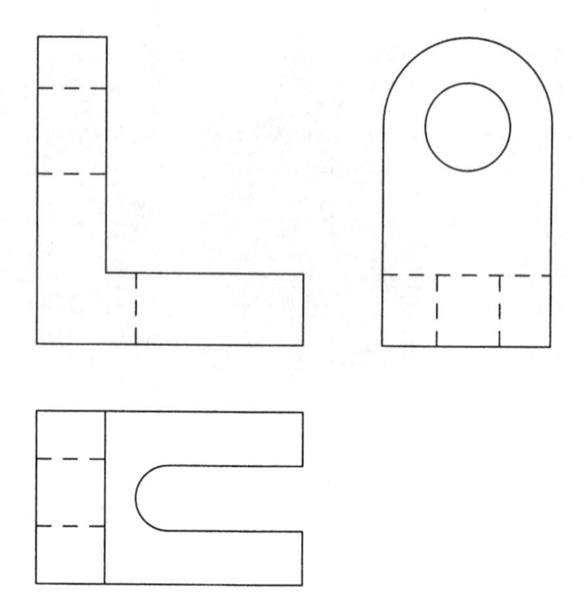

Figure 2.45

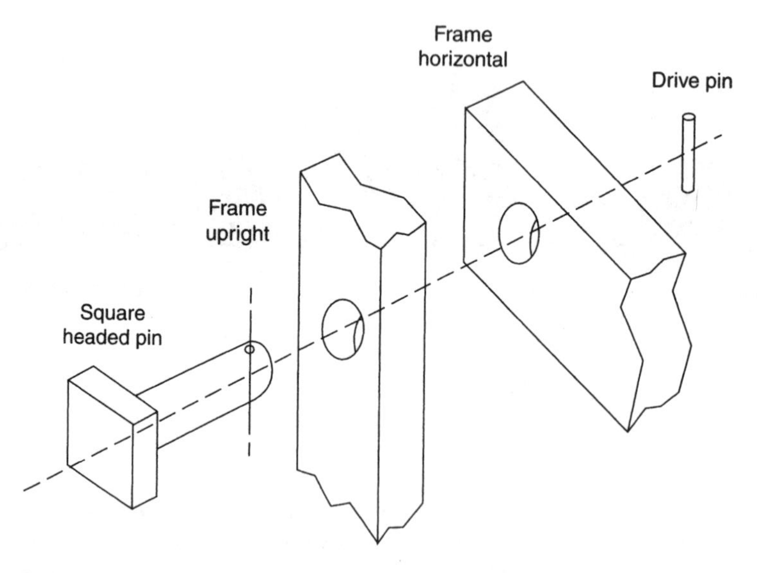

Figure 2.46

2. Figure 2.46 shows an exploded view of a simple assembly. Redraw it as assembled ready for use.
3. Discuss the use of samples, swatches and patterns in garment manufacture.

2.11 From design brief to finished product

Finally, let's now consider a simple product from the initial request to the design team to the finished product.

2.11.1 Design brief

The client is a furniture manufacturer producing popular low cost furniture items for the mass market. Market research has shown that there is a demand for a simple, low-cost stool to seat one person.

- It is to be made from wood
- It is to have a conventional polished finish
- It is to be light in weight, but stable
- Any cushion to be supplied as a separate item from a specialist supplier
- It must be strong enough for a heavy man to stand on whilst changing a light bulb or reaching down articles from a high shelf
- It must be simple to manufacture and competitively priced.

2.11.2 Design proposal

Two designs were prepared. One was for a traditional bench type stool as shown in Fig. 2.47(a) and one for a more modern concept in laminated wood as shown in Fig. 2.47(b). It was felt that the second design might be acceptable in view of the fact that the manufacturer had the technical resources and

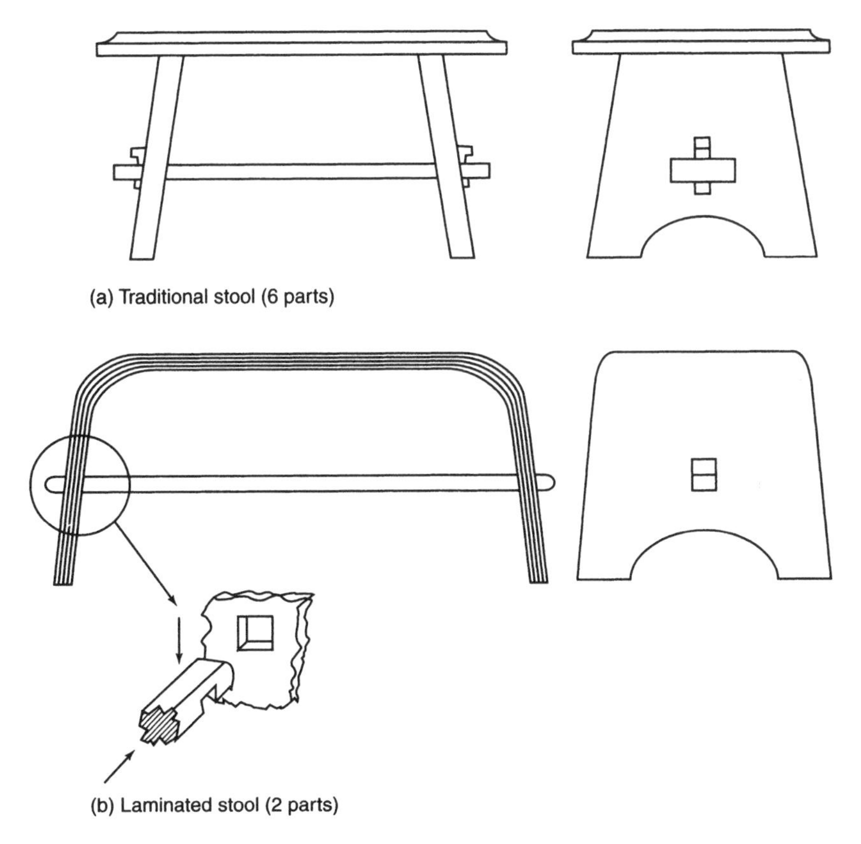

Figure 2.47 Design proposals for a stool.

experience to work in this mode. Comparative, specimen cost analyses were produced for both designs. The second design saved on assembly costs but required higher tooling costs to set it up. In view of the fact that large batch production would be employed if the market research was correct, there should be no difficulty in recovering the set-up and development costs.

2.11.3 Presentation

This was kept quite simple as only a small audience would be present consisting of representatives of the company commissioning the design proposal. These included the managing director, the chief accountant, the marketing manager and the production manager.

A verbal presentation was given by the leader of the design team. Prototypes of the stools were available as visual aids for examination and a cost analysis was presented with the aid of an OHP. The presentation was supported by a printed handout of all the presentation material used together with copies of the manufacturing drawings.

2.11.4 Feedback

The client company congratulated the design team on their work. After discussion it was decided to adopt the second design (Fig. 2.47(b)) for the following reasons.

- The design was in the modern idiom and unlike anything being offered by the competition.
- The design was easy to produce with few parts and little assembly.
- The rounded form and smooth lines would make it easy to maintain and keep clean.
- It satisfied all the key features of the design brief.

Subsequently, however, the client company pointed out that they had a *'legal duty of care'* to their customers and having carried out a *risk assessment* it was considered that the rounded ends could cause the user to slip and fall. The risk would be greater if the user was standing on the stool. The fact that the stool was not intended to be stood upon would not exempt the manufacturer from being sued for damages in the event of an accident. Also it was found that the shape made it difficult to keep a cushion in place.

2.11.5 Finished product

The design team accepted the comments as fed back to them and revised the design as shown in Fig. 2.48. By adding a top 'tray' to the design, the risk of slipping off the rounded ends was removed. The 'tray' also helped to keep the cushion in place. Cut-outs were incorporated in the ends of the 'tray' as shown and these made useful carrying handles. As with the prototype design, modern high strength adhesives were used throughout and there were no metal fastenings required. It was considered that the extra cost involved in the extra work required for the final design was worthwhile in view of the fact that it resulted in a safer and better product.

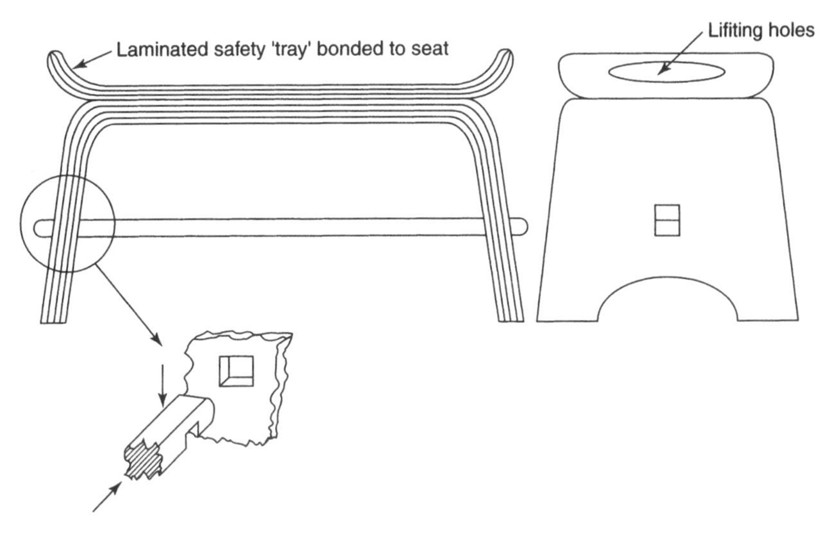

Figure 2.48 Finalized design.

2.12 Calculating the cost of a product

The cost of a product is made up of three main elements. These are made up as shown in Fig. 2.49 and consist of:

- *Direct costs*, sometimes called prime costs, are such items as wages and materials that can be assigned directly to a particular product.
- *Indirect costs,* sometimes called *overhead costs* or *on-costs*, are those items that cannot be directly assigned to a particular product, but have to be shared between all the different products manufactured by the company.
- *Profit.* This is superimposed on all the other costs and provides for such items as dividends, investment, reserve funds and taxation.

Therefore:

Total cost (selling price) = direct costs + indirect costs + profit.

2.12.1 Direct costs (materials)

Material resource costs include raw material costs such as:

- Sheet metal, bars, forgings and castings in the engineering industry
- Rolls of cloth in the clothing industry
- Sawn timber, sheets of plywood and chipboard in the woodworking industries.

Material resource costs also include component costs and the subassembly costs that are used for and can be assigned to a particular product, for example:

- Individual components such as nuts and bolts in the engineering industry, together with subassemblies such as water pumps and gear boxes in the automobile industry

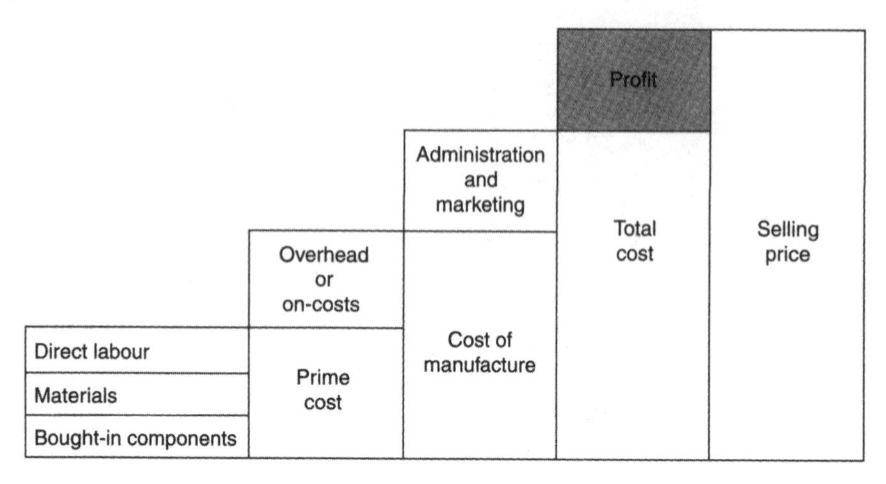

Figure 2.49 Cost structure.

- Buttons and zip-fasteners in the clothing industry
- Nails, screws, adhesives and fittings, such as hinges in the woodworking industries.

2.12.2 Human resource costs

These costs consist of the wages being paid to production workers for the manufacture of particular components, that is:

$$\text{Human resource cost} = \text{operation time} \times \text{hourly rate.}$$

If piecework is involved then the worker will be paid a set rate per item made based on the time taken, that is:

$$\text{Human resource cost} = \text{number of components manufactured} \times \text{rate per item.}$$

2.12.3 Service resources cost

The cost of the services, such as gas, water and electricity, assigned directly to the manufacture of a particular product must be charged directly to that product. Services directly related to production must not be confused with the gas, water and electricity used to light and heat the factory and the water used for personal hygiene, such as washing.

2.12.4 Indirect costs

These are the costs that are incurred by a firm through its very existence. All the direct costs considered earlier were *variable costs*. They varied in proportion to the number of products made. For example, twice as many wooden carrying boxes would have required twice as much wood and double the wages bill. However, indirect costs may be divided into *fixed costs* and *variable costs*:

- *Fixed costs* such as rent, business rates, insurance, bank loan interest, maintenance costs, management salaries, etc., will have to be paid even when the firm is closed down for its annual holiday and no manufacturing is taking place.

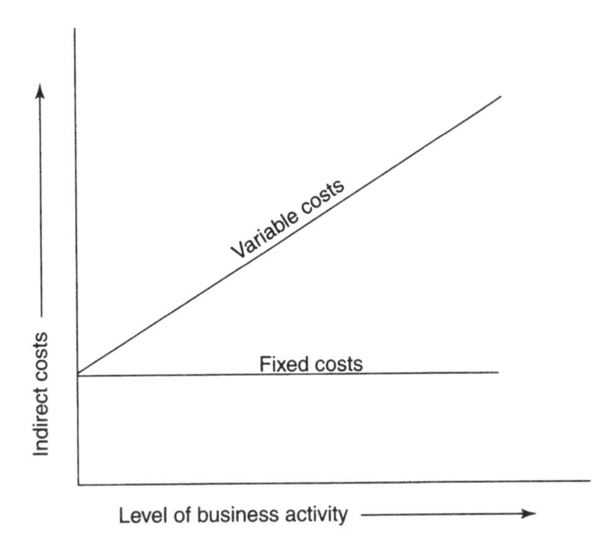

Figure 2.50 Fixed and variable costs.

- *Variable costs* will change with the level of business activity – for example, telephone and fax charges, stationery, postal charges, marketing, etc. The more active the company becomes, the greater will be these charges. The relationship between fixed and variable costs are shown in Fig. 2.50.

$$\text{Total indirect costs} = \text{Fixed indirect costs} + \text{variable indirect cost}$$

where the variable indirect cost is dependent upon the level of business activity.

Indirect costs vary widely from industry to industry and from firm to firm. For example, research and development costs in the pharmaceutical industry represents a major component in the selling prices of its products. In industries manufacturing relatively 'low-tech' products such as drain pipes the research and development charges are very much less. Let's now look at some representative indirect costs and see how they can be allocated to manufactured products.

Research and development costs

These can cover two aspects of manufacturing activity:

- The product being developed
- The method of manufacture.

All manufacturers must constantly examine the market into which they are selling in order to find out the needs of their customers and how their competitors are responding to these needs in an attempt to win business.

Product research and development must be a continuing process so that new and improved designs are always available to satisfy changes in the demands of the market. In order to keep ahead of its competitors, a successful firm tries to lead the market rather than follow it.

In addition, manufacturers must be constantly aware of changes in production technology and materials that can be adopted to make a better product more quickly and cheaply. The Research and Development Department must be constantly assessing new developments in production technology and materials in order to advise the management when changes need to be made to keep the company at the forefront of technology. It is then up to the senior management to assess whether the benefits from the improved technology warrant the expenditure involved.

Management costs

These involve management salaries and their secretarial support salaries. It also involves the cost of the so-called 'perks', such as a company car and health insurance. Many managers have to travel on a global basis in the conduct of the company's business and their travel and subsistence expenses must also be taken into account. Many managers also have to attend conferences and symposiums on new techniques and new national and international legislation and regulations. Conference fees are increasingly expensive.

Administration costs

These costs include the wages and salaries of clerks, typists and general office staff, together with telephone and postal charges, fax and photocopying charges, and stationery. They also include the cost of maintenance personnel, caretakers and security staff, and cleaners. Nowadays, these latter costs tend to be 'outsourced' (contracted out) to specialist firms.

Marketing costs

These costs include advertising and product promotion, sales and marketing staff salaries and commissions. Again, market research, advertising and product promotion is increasingly 'outsourced' to specialist firms.

General expenses

These include rent, business rates, bank charges and interest, heating, lighting, depreciation, insurance, training, and carriage.

Allocation of indirect costs

As already stated, indirect costs are those costs that cannot be associated directly with a particular product. Therefore it is difficult to allocate such costs to the selling price of the product. Various methods of allocating these costs are used. Some are relatively simple and some are extremely sophisticated. We are only concerned with two of the more simple possibilities.

Method 1

The direct human resource costs for a particular company come to a total of £100 000 per year. The indirect costs come to a total of £75 000 per year. Therefore the ratio of the indirect costs to the direct human resource costs is £75 000 ÷ £100 000 = 0.75. This ratio is called the *scaling factor*. We have already seen that:

$$\text{Total cost (selling price)} = \text{direct costs} + \text{indirect costs} + \text{profit}$$

The direct costs (*prime costs*) are the sum of the direct human resource costs plus the direct material costs plus the service costs. The indirect costs are the direct human resource costs $\times$ 0.75. For example, if the human resource costs for a particular product is £10 000, the material costs are £450, and the service resource costs are £50, then we can build up the selling price as follows:

$$\begin{aligned}\text{Total cost} &= £10\,000 + £450 + £50 + (£10\,000 \times 0.75) + \text{profit}\\ &= £10\,500 + £7500 + \text{profit}\\ &= \mathbf{£18\,000 + profit}\end{aligned}$$

The profit margin will depend upon a number of factors but must enable the company to put aside sufficient funds to cover taxation, plant replacement, dividends, and to build up a reserve fund against unforeseen emergencies.

Method 2

Where the material costs are relatively small compared with the direct human resource costs, many firms apply their scaling factor to the total prime cost. They argue that all materials have to be handled in and out of the stores, and that the rent, rates, heating and lighting that are required in servicing the stores area, as well as the wages for the stores personnel, all have to be paid for. The prime cost also includes the service resource costs but, as this is relatively small, it will be left in the prime cost in this example.

Further, the working capital (money) tied up in material stocks is losing the interest that it could earn if invested. Worse, the money to pay for the materials in the stores may have to be borrowed from the bank and loan charges will have to be paid to the bank.

Let's now look at the previous example again. The direct human resource costs for a particular company come to a total of £100 000 per year. The indirect costs come to a total of £75 000 per year. Therefore ratio of the indirect costs to the direct human resource costs (scaling factor) is again £75 000 $\div$ £100 000 $=$ 0.75. We have already seen that:

$$\text{Total cost (selling price)} = \text{direct costs} + \text{indirect costs} + \text{profit}$$

The direct cost is the sum of the direct human resource costs plus the direct material costs plus the service resource costs. As previously, the human resource cost for a particular product is £10 000 and the material cost is £450, and the service cost is £50. However, this time we will apply our scaling factor to the *total prime costs* instead of just to the direct labour costs.

We can now build up the selling price as follows:

$$\begin{aligned}\text{Total cost} &= £10\,000 + £450 + £50 + (£10\,500 \times 0.75) + \text{profit}\\ &= £10\,500 + £7875 + \text{profit}\\ &= \mathbf{£18\,375 + profit}\end{aligned}$$

Therefore we have gained £375 towards the cost of storing and handling the raw materials.

The profit margin will depend upon a number of factors. However, as stated previously, it must enable the company to put aside sufficient funds to cover taxation, plant replacement, dividends, and to build up a reserve against unforeseen emergencies.

In some instances, method 2 would inflate the price of the product unduly and make it uncompetitive. For example, to apply method 2 to the manufacture of jewellery using precious metal and precious stones would substantially increase the selling price compared with method 1. For many applications much more sophisticated methods for apportioning the indirect costs have to be used.

Test your knowledge 2.10

1. Explain the difference between direct and indirect costs.
2. Calculate the total cost of manufacturing 750 products of your own choice, applying the indirect costs by method 2.

Key notes 2.5

- Graphs are a means of showing numerical relationships and data in pictorial form.
- Pie charts are also a simple type of graph much used in presenting numerical information to non-technical audiences. They are frequently seen in the newspapers.
- Bar charts (histograms) use vertical bars or columns to show numerical data.
- Gantt charts are also a form of bar chart, but the bars are arranged horizontally. They are often used to compare actual performance with expected performance.
- Pictograms (ideographs) are a pictorial method of presenting numerical data for non-technical persons. Icons are used to resemble the dependent variable.
- Charts are used to show non-numerical data, such as company organization and relationships.
- Flow charts are used to show the direction of flow of sequenced information.
- System flow charts are used where procedures can be broken down into logical steps with a simple yes/no decision. They are widely used by computer systems analysts. System flow charts are also called *algorithms*.
- Data sheets are use for presenting numerical data for easy reference. For example, inch to millimetre conversions.
- Handouts are a printed record of the presentation for the audience to take away with them. They save having to take notes under dimmed light conditions.
- Technical drawings are used to communicate with the manufacturer of the product at the production management and shop floor level.
- Technical drawings produced in the UK are drawn to BS 308 conventions.
- Drawings may be orthographic and in first angle or third angle projection, or they may be pictorial and in isometric or oblique projection. Exploded views are used for DIY assembly drawings and machine maintenance drawing.
- General arrangement (GA) drawings show all the parts of an assembly in their correct positions and lists the detail drawings needed to manufacture the components used in the assembly together with any 'bought-in' parts required.
- Detail drawings show only one component on each drawing together with all the information needed to manufacture that component.
- CAD is computer-aided drawing and design and has largely superseded manual draughtsmanship in professional design offices.
- Patterns are templates used in garment manufacturing when cutting out material.
- Swatches are small samples of cloth bound together for comparison and choice.
- Direct (prime) costs are items such as wages and materials that can be assigned directly to a product.
- Indirect (overhead) costs are associated with the ownership, maintenance and running of the factory premises, management costs, etc., that cannot be assigned to any particular product, but must be shared amongst all of them.
- Profit is the final cost item superimposed on all other costs and provides for such items as dividends, investment, reserve funds and taxation.

Assessment activities

1. With the aid of a critical friend acting as your customer, prepare a design portfolio for a manufactured product. Your portfolio should include documentary evidence of all the steps involved from the initial design brief to the finalized design.
2. Prepare a presentation including notes, visual aids, handouts and feedback forms for presenting your product proposal to your customer.
3. Prepare representative costs for your product proposal in 2 above using method 1.

Chapter 3 Manufacturing products

Summary

When you have read this chapter you should be able to:

- Produce a production plan and schedule for the manufacture of a product
- Plan the preparation of materials and components
- Plan the critical production and control points. Understand the need to modify the production plan and specification as and when the manufacturing circumstances change
- Appreciate the need for effective teamwork
- Appreciate how to create an effective working environment
- Select suitable methods for processing materials and components
- Select suitable methods for combining, assembling and finishing materials
- Apply the principles of quality and production control techniques during manufacture
- Understand the need for health, safety and hygiene issues when manufacturing products
- Apply the principles of health, safety and hygiene to manufacturing processes.

3.1 Production planning and the schedule for manufacture

3.1.1 Introduction

Production planning is an essential organizational exercise that is carried out before commencing manufacture. We do this regularly as part of the organization of our lives. You may decide that you want a cooked breakfast in the morning. It is no good waiting until the morning comes and find that you do not have suitable ingredients or equipment. So, you think ahead.

- Transfer bacon and sausages out of the freezer and into the fridge so that they can thaw out for the morning.
- Check that the frying pan, spatula and plate are clean and ready for use.
- Check that you have eggs, tomatoes, mushrooms, suitable cooking oil and bread.
- Is the tea, milk, sugar, teapot, and cup and saucer ready? OK, you can now go to bed in the sure knowledge that, barring a power cut, you are all set for your cooked breakfast.
- Estimate the cooking times and set your alarm-clock so that you have time to cook and eat the meal before setting out for work. That is, *schedule* your start to the day.

You have planned the production of your breakfast. You have thought ahead. This is production planning on a small scale but, in principle, it is exactly what has to happen in industry no matter whether you intend to manufacture a nut and bolt or a motor car.

Let's now see what is involved in the production of a typical industrial *production plan*. We can assume that the product specification has been agreed with the customer and that the component and tooling drawings have been prepared as discussed in Chapter 2. Production plans are used to provide all the essential information about the type and quality of the product to be manufactured, such as a single unit, batch or volume production. The production plan will consist of a number of elements. These are listed below.

- Material sourcing and procurement so that suitable material will be available in sufficient quantity for production to commence and continue without interruption. Use standard materials and 'bought-in' components wherever possible. They will be manufactured to BSI specifications and the quality will be guaranteed
- The product specification
- Analysis of the manufacturing process so that the individual tasks can be listed
- Assign the appropriate resources to each task
- Estimate the time that each task will take
- Arrange the tasks in a logical sequence so that the work will flow smoothly through the factory
- Finally, all the above elements have to be brought together into a *production plan* which then has to be communicated to all the appropriate personnel. The production plan must be drawn up in the conventional format for the manufacturing sector concerned.

Having produced a production plan, the next step is to assign starting and finishing dates and times to the individual tasks. That is, *schedule* the production. The production schedule for manufacture will list the following information:

- All preparation and assembly stages
- The sequence and timing of production stages to ensure that the right quantity of components become available at the right time to avoid delays in the next stage of manufacture. At the same time space and money must not be used up by manufacturing too far in advance. Modern production scheduling is based on a *just-in-time* (JIT) philosophy
- Critical production and quality control points must be assigned to strategic points in the schedule
- Suitable and realistic production and quality control procedures must be developed and applied at the critical control points
- Flexibility must be built into the production plan and the production schedule so that they can be easily modified as circumstances change.

Let's consider an actual example.

3.1.2 Product specification

A product specification can vary according to the information needed. For example, a sales person would be concerned with details of size, colour and performance. A production planner is more concerned with those features that are going to influence the materials used and the manufacturing processes required. A product specification normally contains the sort of information shown in Fig. 3.1. Figure 3.2 shows a wooden carrying box and Fig. 3.3 shows a product specification for the box.

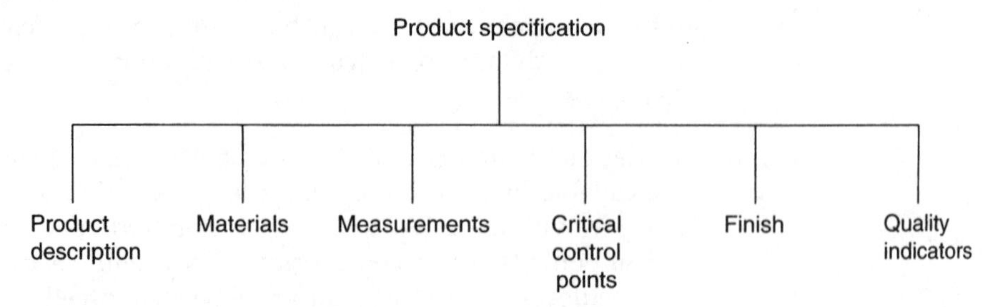

Figure 3.1 Product specification.

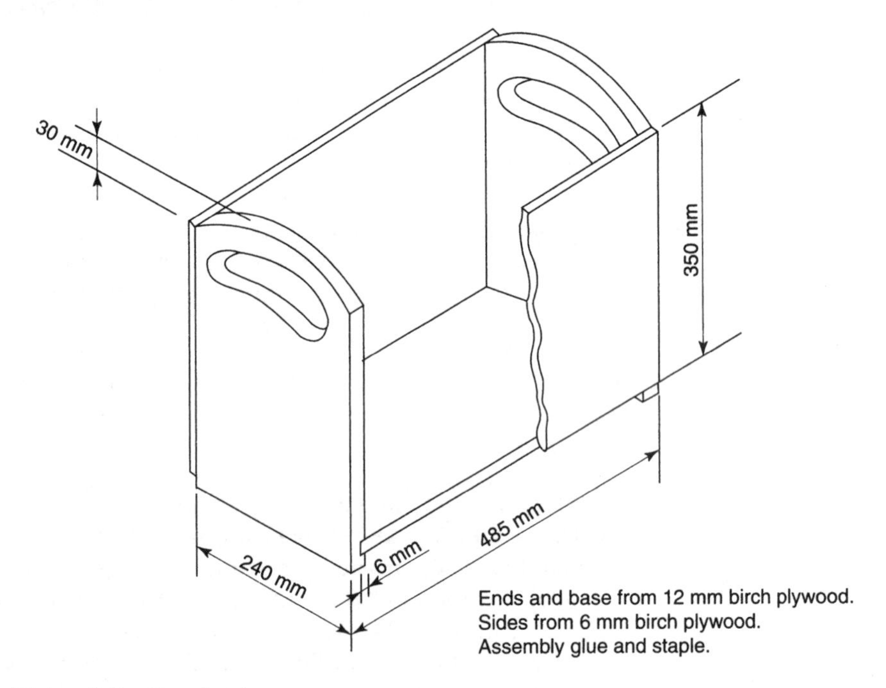

Figure 3.2 Carrying box.

Critical control points

These are the points when, during the course of manufacture of a product, the various components and subassemblies must be checked to make sure that they are correct to specification. This is to ensure that any defective items are rejected before they receive expensive processing or become built into the final assembly. The critical control points for our box are as follows:

- The materials must be checked for type, size and quality before they are issued to the workshop
- The adhesive must be checked to ensure that it is the correct type for the process involved and will produce a sound joint. Sample joints should be tested for strength
- The wood product components must be checked for quantity, size and shape after cutting
- After gluing, power stapling and assembly, the boxes must be inspected to ensure sound joints and correct assembly

Product specification	Job number: 0078/98
Product description	Carrying box for batteries
Materials	Ends and base from 12 mm birch plywood Sides from 6 mm birch plywood Adhesive: Stickall Lacquer: coverite in red, yellow, green, blue, gloss
Measurements	All dimensions as drawing ±1 mm
Critical control points	1. Check materials 2. Check sizes of blanks after cutting 3. Check sizes after assembly 4. Check finish after painting
Finish	Gloss lacquer in bright colours. Spray or brush on depending on quantities
Quality indicators	Materials: thickness and appearance Dimensions: measured for correct size Finish: appearance

Figure 3.3 Product specification for carrying box.

- After finishing (sanding smooth and coating with a spray lacquer), the boxes must be finally inspected for damage or defects before they are finally packed in such a manner as to protect them from damage during storage or transit.

Quality indicators

These have been introduced previously in Section 1.2.5. Quality indicators can be *variables* or *attributes*. These can be applied to the critical control points for our carrying box as shown in Table 3.1.

3.1.3 Key production stages

The key stages of production that are common to all manufactured goods were introduced in Section 1.11. As a reminder, these are summarized in Table 3.2.

Let's now expand this table to include the key stages of production for our box. These are shown as a flow chart in Fig. 3.4. Remember there is no single correct way to make these boxes. The processes chosen and their sequence will depend upon the quantity being manufactured and the manufacturing facilities available. The processes will also reflect the personal preferences of the production manager. Figure 3.4 only shows one typical sequence as an example.

Test your knowledge 3.1

1. Explain why 'production planning' is necessary in any manufacturing industry.
2. For an article of your choice, draw up a typical 'product specification' for:
   (a) A production planning engineer;
   (b) A sales manager.
3. For an article of your choice produce a flow chart showing the key production stages.

Table 3.1 Quality indicators

Critical control point	Quality indicators	
	Variables	Attributes
Materials	Check plywood sheets for correct thickness	Check plywood sheets for: (i) Correct type of wood (birch) (ii) Freedom from surface blemishes
Cut blanks	Check for correct dimensions as stated on the drawing ±1.0 mm	–
Assembly	Check assembled boxes for correct overall dimensions ±1.0 mm	Check assembled boxes for: (i) Correct gluing and tacking (ii) Squareness of corners (iii) Edges and joints sanded smooth
Finish	–	Check finished boxes for: (i) Freedom from blemishes (ii) Correct paint colour (iii) Correct logo/decal if required

Table 3.2 Key stages of production

- Material preparation
- Processing
- Assembly
- Finishing
- Packaging

3.1.4 Resource requirements

We have already looked at many of the resources used in manufacturing in Chapter 1. First, however, let's remind ourselves of the main resource requirements and then see how some of them can be applied to the manufacture of our box.

Capital resources

These are the resources purchased by the company for the manufacture of the company's products. They represent a major part of the company's assets. Ignoring the premises (buildings) in which the company is housed, typical capital resources are:

- Process plant
- Machinery
- Manufacturing equipment.

None of these items are *consumable*. For example, a lathe is an item of capital plant. It is a *capital resource*. The cutting tools used in it are *consumable*. They wear out and have to be resharpened or replaced frequently in the course of manufacturing the company's products. Similarly a sewing machine is also a *capital resource*. The needles wear out or break and have to be replaced from time to time. They are *consumable*.

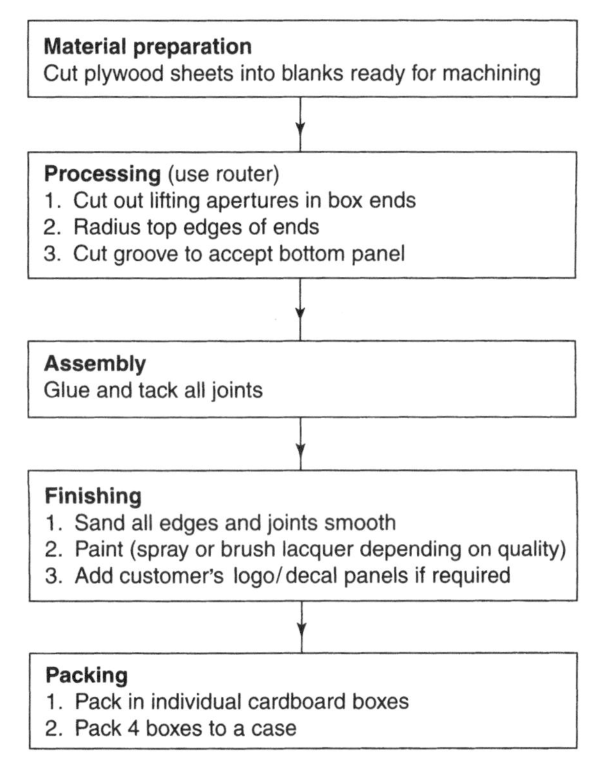

Figure 3.4 Key stages of production for wooden carrying box.

Even capital resources wear out eventually or become obsolete, but this takes place over a very long period of time compared with consumable items. This is referred to as depreciation and money must be put aside on a regular basis from the profits of a company for the eventual replacement of its capital resources (plant).

Tooling resources

These can be consumable items of a general nature applicable to a variety of jobs. Examples of such items are:

- Cutting tools (drills, lathe tools, milling cutters, router blades, etc.)
- Sewing machine needles
- Spot-welding electrodes
- Sanding belts and discs
- Polishing mops.

Tooling resources can also be *job specific* such as:

- Casting patterns
- Plastic moulds
- Templates
- Machining and welding jigs and fixtures.

Although frequently very expensive, these are used only for one specific product. They are charged directly to that product, and are discarded as valueless when that product is no longer made.

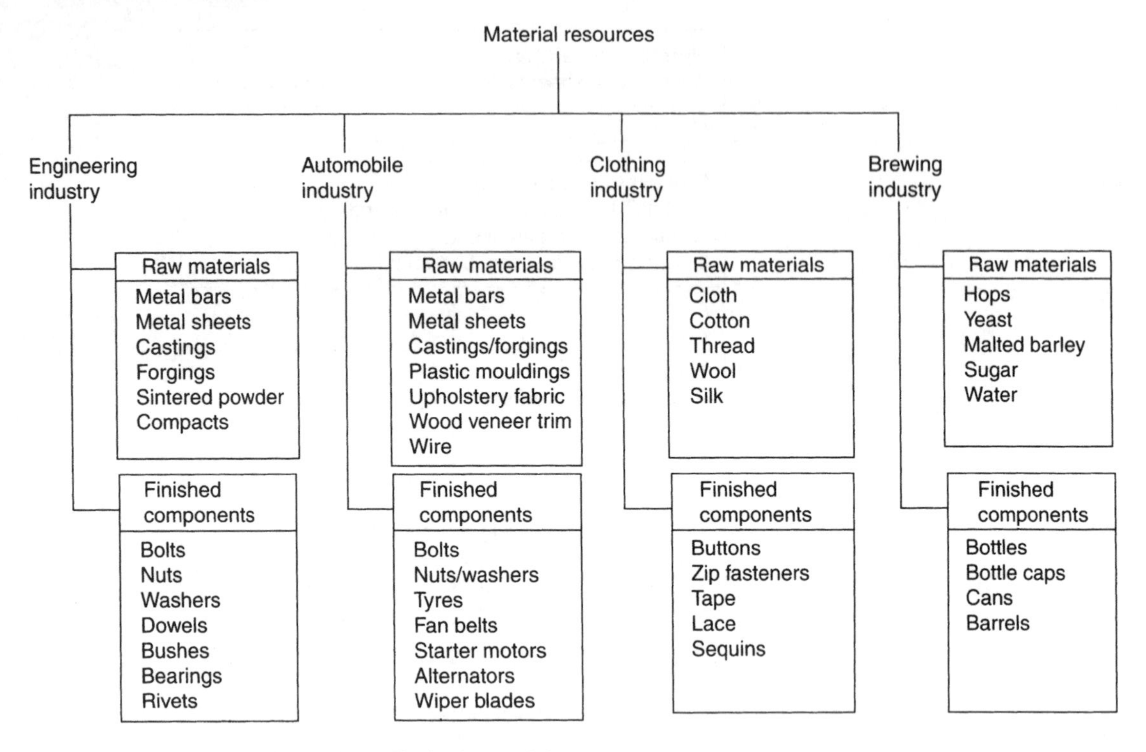

Figure 3.5 Typical material resources.

Material resources

These are the materials required for the manufacture of various products and will vary from one industry to another. Material resources can be broken down into:

- Raw materials such as metal in bar form, forgings or castings in the engineering industry, or rolls of cloth for the clothing industry
- Standardized, finished components such as nuts and bolts in the engineering industry or buttons and zip fasteners in the clothing industry
- Subassemblies such as electric motors, pumps, and valves for the engineering industry.

Some examples of the material resources required for manufacturing are shown in Fig. 3.5.

Human resources

Human resources refer to all the people working in the various manufacturing industries.

Direct labour

This refers to persons employed directly in the manufacture of a product and whose wages can be charged directly to that product.

Indirect labour

This refers to persons involved in the management of the company, in marketing and sales and in various clerical activities essential to the running of the company. Their wages cannot usually be charged to a specific product, but have to be shared amongst all the products manufactured by the company.

Outsourced human resources

This refers to persons servicing a company but not directly employed by it. Increasingly companies are *outsourcing* the fringe services they require on a franchise basis, rather than use employees on their own payroll. For example, the use of outside caterers to run the works' canteen services. This enables costs to be more closely controlled.

Further, with the communication systems now available and the power of personal computers, it is possible for more and more persons to work from home on a freelance basis. For example, book illustrators and copy editors often work from home for the publishing industry. Technical support and maintenance is also an area that is nowadays franchised out to factory-trained individuals who are self-employed.

Service resources

These are the service resources used in manufacturing. In this context they do not include telephones, drinking water, sewerage, electricity and gas for general lighting and space heating, etc. Examples of service resources as applied to manufacturing are:

- Fuel gas for heating process plant
- Process water
- Process drainage and the safe storage and disposal of hazardous liquids
- Electricity for driving process plant
- Compressed air.

Typical resource requirements for our wooden carrying box are summarized in Table 3.3.

3.1.5 Processing times

From the key stages of production we next move on to the resources we would require to make our carrying box. However, before we can start considering the cost of manufacture, we need to work out the time taken to carry out each production stage.

Production stage time = set-up time + (operation time per item × number of items)

Using this formula we will now work out some examples. Note that we must always work in common units. Usually these are either minutes or seconds.

Example 3.1

A joiner takes 2 minutes to set up a circular saw and 15 seconds to cut each piece of wood. Calculate (a) the time to cut 10 pieces of wood, (b) the time to cut 500 pieces of wood.

(a)

Production stage time = set-up time + (operation time per item × number of items)

Table 3.3 Resource requirements: wooden carrying box

Production stage	Operation	Type of resource	Resource required
Material preparation	Cutting plywood sheet to size	• Material • Capital • Human • Service	• Sheets of birch plywood • Circular saw • Operator • Electricity
Processing	Forming box ends	• Capital • Tooling • Human • Service	• Router machine • Templates • Operator • Electricity
Assembly	Glueing and tacking joints	• Material • Capital • Capital • Human • Service	• Staples and glue • Powered staple gun • Hot glue applicator • Operator • Electricity
Finishing	Sanding and painting	• Material • Capital • Capital • Capital • Capital • Human • Human • Service	• Lacquer • Orbital sander • Air compressor • Spray paint gun • Spray booth • Operator for sanding • Painter • Electricity
Packaging	Packing the finished boxes	• Material • Material • Human	• Cardboard boxes • Self-adhesive tape • Packer

$= 2 \text{ minutes} \times 60 + (15 \text{ seconds} \times 10 \text{ items})$

$= 120 \text{ seconds} + 150 \text{ seconds}$

$=$ **270 seconds** or

$=$ **4.5 minutes**

We could have worked in minutes, in which case the example is as follows.

Production stage time $=$ set up time $+$ (operation time per item $\times$ number of items)

$= 2 \text{ minutes} + \{(15 \div 60 \text{ minutes}) \times 10 \text{ items}\}$

$= 2 \text{ minutes} + 2.5 \text{ minutes}$

$=$ **4.5 minutes**

The answer is the same in both cases as, of course, it should be.

Now let's work out the time for the larger quantity.

(b)

This time we will again work in minutes.

Production stage time $=$ set-up time $+$ (operation time per item $\times$ number of items)

$= 2 \text{ minutes} + \{(15 \div 60 \text{ minutes}) \times 500 \text{ items}\}$

$= 2 \text{ minutes} + 125 \text{ minutes}$

= **127 minutes** or

= **2 hours 7 minutes**

Similarly, if this was part of the mass production of window frames then possibly 5000 identical pieces of wood might be required. In this case the total time taken would not be ten times greater because the setting time is the same no matter how many pieces of wood are cut. This is the time saving of large batch production over jobbing production. Try working it out for yourself, you should arrive at a time of 20 hours 52 minutes.

This might be too long a time and would slow down the overall production of window frames. If two circular saws and two operators are available then *concurrent* production can take place. That is, two saws cutting identical pieces of wood at the same time. This time we have an additional formula.

Number of items per operator = total number of items required ÷ number of operators

So, if we cut 5000 pieces of wood using two operators then:

Number of items per operator = 5000 ÷ 2 = **2500**

Reverting to our original formula the time taken would be:

Production stage time = set-up time + (operation time per item × number of items)

$= 2 \text{ minutes} + \{(15 \div 60 \text{ minutes}) \times 2500 \text{ items}\}$

$= 2 \text{ minutes} + 625 \text{ minutes}$

= **627 minutes** or

= **10 hours 27 minutes**

You might have thought that having two operators would have halved the overall time but you have to allow for setting up the extra machine, hence the extra 2 minutes. Table 3.4 shows how we can build up the time for manufacturing 500 of our carrying boxes.

Test your knowledge 3.2

1. Analyse the resources for your school or college under the headings considered in Section 3.1.4.
2. A machinist takes 10 minutes to set up a milling machine and then takes 5 minutes to machine each component. Calculate the time taken to produce:

   (a) 50 components, (b) 500 components, (c) 1000 components if two machines and two machinists are used in this particular instance.

3.1.6 Production schedules

Think of all the parts that go to make up a car. They all have to arrive at the assembly line in the correct quantities at the correct time. Think of the chaos

Table 3.4 Time to manufacture: 500 wooden carrying boxes

Production stage	Machine set-up time (min) (A)	Number of operators (B)	Number of parts (C)	Time per operation (min) (D)	Total operation time (C) × (D)	Production stage time (A) + (C) × (D)	Aggregate time (min)
Cutting base blanks	2.0	1	500	0.25	500 × 0.25 = 125	2.0 + 125 = 127	127
Cutting end blanks	2.0	1	1000	0.25	1000 × 0.25 = 250	2.0 + 250 = 252	127 + 252 = 379
Cutting side blanks	2.0	1	1000	0.20	1000 × 0.20 = 200	2.0 + 200 = 202	379 + 202 = 581
Radiusing ends	5.0	1	1000	0.15	1000 × 0.15 = 150	5.0 + 150 = 155	581 + 155 = 736
Cutting hand grips	5.0	1	1000	0.40	1000 × 0.40 = 400	5.0 + 400 = 405	736 + 405 = 1141
Cutting base slots	5.0	1	1000	0.20	1000 × 0.20 = 200	5.0 + 200 = 205	1141 + 205 = 1346
Glueing and tacking	–	2	500 sets	3.00	500 × 3.0 = 1500	1500	1346 + 1500 = 2846
Sanding	–	2	500 assemblies	2.00	500 × 2.0 = 1000	1000	2846 + 1000 = 3846
Painting	–	2	500 assemblies	3.00	500 × 3.0 = 1500	1500	3846 + 1500 = 5346
Packing	–	2	500 assemblies	–	–	250	5346 + 250 = 5596
					Total production time = 5596 min = 93 hrs 16 min		

if only body shells arrived but no engines, or if only three wheels arrived for each car. Consider the clothing trade. A factory making shirts must have regular supplies of material and buttons. If the wrong quantities arrived at the wrong time the flow of production would break down resulting in huge stocks of unfinished goods that could not be sold. The result would be dissatisfied customers and no *cash flow* to pay the wages and suppliers' accounts. Production management is concerned with preventing these sorts of situations arising. Let's see how this is done.

The production has to be organized so that the sequence of operations enables the work to flow through the factory in the most efficient and cost effective way possible. To do this a *production schedule* has to be produced. This is often in the form of a Gantt chart. These charts were introduced in Section 2.8.4. Let's now draw up a Gantt chart scheduling the production of our carrying boxes. In order to draw up the Gantt chart we have to use the operation times that we have just calculated in Table 3.4. In drawing up our Gantt Chart we will make the following assumptions:

- The factory operates a five day week (Monday to Friday inclusive)
- The daily hours are 08.00 to 12.30 hrs and 13.30 to 17.00 hrs = 8 hours per day
- Work movement (transit) times between operations are ignored for simplicity in this example
- Critical control point inspection times are also ignored for simplicity. In any case they should be organized so as not to interrupt production if no faults are found that require correction
- Sanding down and painting operations are relatively slow and can commence before all the boxes have been assembled.

Let's start our chart on Monday morning at 08.00 hrs. The first 2 minutes will be spent in setting up the circular saw to cut the blanks for the bases. Cutting the blanks only takes 125 minutes so it should be complete by 10.07 hrs. The machine can then be reset for the next operation, namely, the blanks for the ends. Assuming no snags and allowing for the 12.30 to 13.30 hrs break, the blanks should all be cut by 15.19 hrs.

Now comes the clever bit. The operations to finish the end panels (rounding the top edge, cutting the lifting holes, and slotting are done on a different machine called a *router*. Providing it is available, these operations can be commenced immediately whilst the sawyer is cutting the side panels. However life is never so convenient so let's assume the router does not become available until 16.00 hrs on Monday. All routing operations will be completed by 11.50 hrs on Wednesday and, in the meantime, the side panels will be cut and waiting.

Assembly is a relatively slow operation so it can start once about half the final routing operations are complete, say 10.30 hrs on the Wednesday. Note that if two persons are carrying out the assembly, the actual time taken is 1500/2 = 750 minutes. So, if assembly commences at 10.30 on Wednesday it will finish at 16.00 hrs on Thursday.

Sanding down the rough edges and joints is quicker than assembly, so a pool of finished work needs to be built up before this operation can commence. It would be safe to start sanding down at 08.00 hrs on Thursday.

Painting is a slower process than sanding so it can start almost straight away, say, 09.00 on Thursday.

Two operators perform the sanding process so the actual time taken will be 1000/2 = 500 minutes. So if sanding starts on Thursday at 08.00 hrs, it will finish at 08.20 on Friday. Similarly painting can start at 09.00 hrs on Thursday and finish at 14.30 hrs on Friday. The boxes would then be left to dry thoroughly over the weekend and packing would commence at 08.00 hrs on the following Monday. This would be completed by 12.10 hrs. Therefore the boxes could be loaded and ready for despatch immediately following the lunch break. Figure 3.6 shows the finished Gantt chart scheduling the manufacture of our carrying boxes.

3.1.7 Producing a production plan

A production plan brings together all the factors discussed so far in the form of a table. Such a plan includes:

- Identification of the product specification details, including a description of the key production stages and critical control points
- Identification of the resources required
- The estimated processing times for each key production stage
- A production schedule in the form of a Gantt chart.

Figure 3.7 shows the layout of a typical form for a production plan. The Gantt chart scheduling the production would then be derived from this form.

Test your knowledge 3.3

1. Draw up a Gantt chart scheduling a typical day in your life from the time you get up until you go to bed.
2. Draw up a production plan and schedule for a simple component of your own choosing.

Key notes 3.1

- Production planning is essential to ensure that the machines and raw materials all become available at the correct time to ensure a product is manufactured to specification and delivered on time.
- A product specification summarizes all the information about a product that needs to be known for its manufacture.
- A production schedule sets out the time taken for each operation and relates them together usually in the form of a Gantt chart so that the progress of manufacture can be seen at a glance.
- Critical control points are those points during the manufacturing process when it is essential to make quality checks before the work is passed forward for further processing or is sold to the customer.
- Quality indicators may be *attributes* that can only be right or wrong, or *variables* that can vary between specified limits.
- Capital resources are such things as buildings, plant and machinery.
- Tooling resources are consumable items such as cutting tools, jigs and fixtures.
- Material resources can be both raw materials from which the product is manufactured as well as items that are bought in finished and ready for incorporation into the product being manufactured, e.g. buttons for a shirt.
- Human resources refer to all the people involved, directly and indirectly, in the manufacture of products.
- Service resources refer to the gas, water, electricity, compressed air and process drainage required solely for manufacturing processes.
- Processing time is the time taken to set up and perform a particular operation on either an individual product or a batch of products.

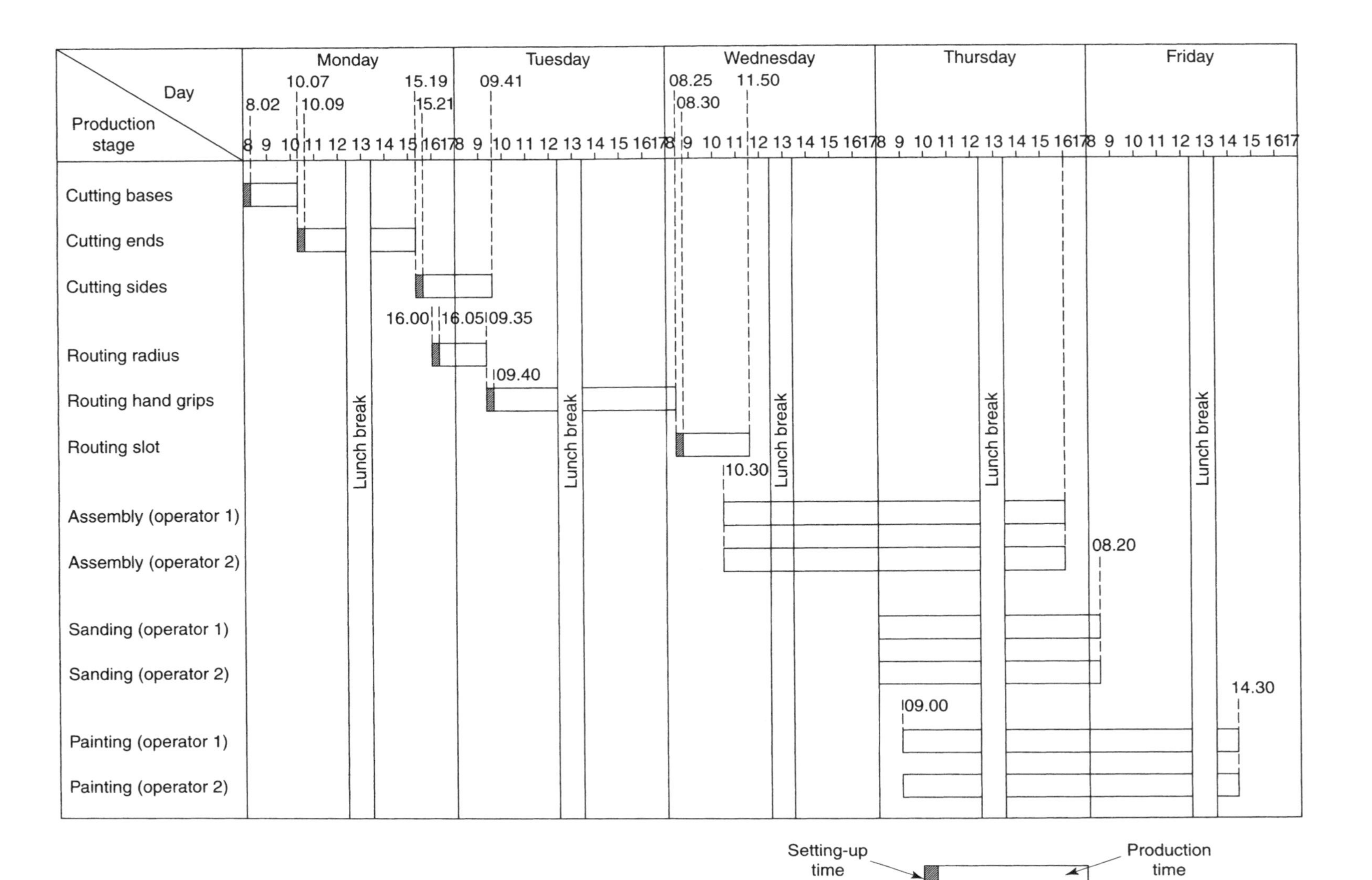

Figure 3.6 Gantt chart for scheduling the manufacture of the wooden carrying boxes.

Customer:		**Product:**		**Design No.**		**Order No.**		**Quantity:**		**Date required:**
Production stage	Capital resource	Human resource	Material resource	Service resource	Tooling resource	Machine set-up time	Number of operations	Operation time per-unit	Total operation time	Production stage time
Notes:							Total production time			

Figure 3.7 Production planning form.

3.2 Teamwork

As soon as the task in hand grows in complexity to be beyond the skills and knowledge of one person, or the amount of work is more than one person can carry out, there is a need for people to work together towards a common goal. This involves *working as a team*.

There are many ways of organizing teams in manufacturing. One method is for each person to specialize in carrying out one particular task as in assembly line working. For example on a car production line, one person fits the wheels to one side of the car and tightens the wheel nuts. Another person performs the same task on the opposite of the car. These two people carry out the same task all day and every day. The advantage of such specialization is expertise at the task. The disadvantages include boredom and lack of personal identification with the whole process and the finished product. Wherever possible repetitive tasks are automated by means of *industrial robots*. Robots do not become tired or bored. The work performed by robots is of consistent quality.

Modern manufacturing favours the use of *manufacturing cells*. Each 'cell' consists of a small group of workers who are responsible for a range of processes. This might be assembling a photocopier from its component parts and subassemblies. Each member of the 'cell' is trained to carry out any of the tasks required in the assembly process. The team is responsible for assembling a predetermined quantity of the required quality in a given time. They can rotate tasks, work hours agreed amongst themselves and organize the work in whatever way they like. They have to produce working photocopiers at a steady and consistent rate, so that they meet their 'quota' every month. The advantages include variety at work, a greater involvement in the product and the process and increased job satisfaction. The disadvantages include the need for the workers to be more skilled and knowledgeable concerning the product they are assembling.

Effective teamwork is only achieved through team building. Team building depends upon:

- Allocating and agreeing roles and responsibilities based upon the strengths and weaknesses of the team members
- Setting and agreeing team targets
- Ensuring good communication between the team members
- Ensuring the team members are motivated
- Creating an appropriate working environment
- Choosing team members who can work together with no clash of personalities.

3.3 The preparation of materials and components

An important stage in manufacturing is the preparation of materials and components prior to processing. Preparatory stages can be difficult to identify when it can be said that all stages of manufacture are steps along the road to the finished product. This is especially so in the food industry where, for example, the grain from a field of wheat is the finished product for the farmer but only the raw material for the miller. The flour from the grain is the finished product for the miller but only the raw material for the baker. Similarly, in engineering, a subassembly may consist of many individual components such as nuts, bolts, washers, screws and springs, etc. All these are the finished products for their individual manufacturers, but only the starting point for the manufacturer of the subassembly. Again, the subassembly is only one part of the complete assembly.

There are many different ways in which materials and components are prepared. However, in the interests of quality control and the avoidance of production delays, there are a number of checks that should be made no matter what materials are involved and no matter what is to be manufactured from them. All materials need to be checked for:

- Type (is it the correct material?)
- Quality (is it to specification?)
- Damage (has it been damaged or blemished prior to or during delivery?)
- Size (is the material the size ordered?)
- Quantity (has the correct quantity been ordered so that production can proceed?)

Once the raw materials, bought-in components, and subassemblies have passed their acceptance checks, their preparation for production can commence. The preparation of materials and components ready for processing can include such activities as those shown in Fig. 3.8. The preparatory processes in this figure are not manufacturing sequences, but merely lists of possible preparatory processes.

Test your knowledge 3.4

1. Draw up flow charts for two examples of material preparation for products of your own choice. The products should be from different sectors of manufacturing industry.

3.4 Machinery, tools and equipment in the working environment

Prior to commencing the manufacturing process required, all equipment, machinery and tools must be correctly and thoroughly prepared. Apart from safety considerations this is very important in the manufacture of quality products. There are two categories of machines *manually operated* and *automated*. First let's consider manually operated machines.

3.4.1 Manual (non-automated) machinery

Machines that are operated by a machinist or an operator are said to be *manually controlled* (non-automated). This method of control depends upon the skill and concentration of the operator and can lead to variations in product quality. The sense of sight of the operator is the major part of his control system and in a company employing *total quality management*, the operator inspects, gauges or measures the components or products produced on his or her machine. Manual machines can be linked to form an integrated manual

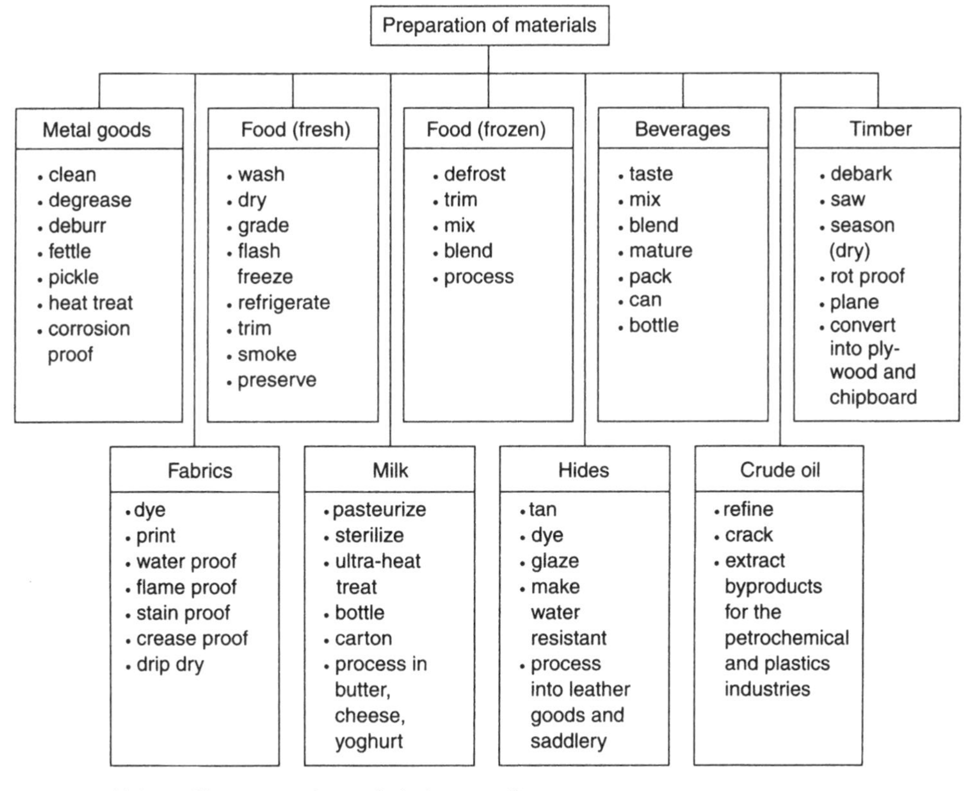

Figure 3.8 Examples of material preparation.

production system. The relationships between the inputs, outcomes and information flow for a manually controlled machine are shown in Fig. 3.9.

3.4.2 Automated machinery

Automated machinery is not new. Automatic screw manufacturing machines controlled by steel cams were also introduced in the nineteenth century. Mechanically controlled automated machines were, and still are, widely used throughout all sectors of manufacturing industry. It is only recently, with the development of powerful microcomputers that the use of computer numerically controlled (CNC), automated machine tools linked with industrial robots has become so widespread. CNC machines are controlled by a dedicated microcomputer. The operator has to program the computer using the data on the component drawing. The machine is then set and adjustments made for any small variations in tool and cutter sizes. Once this has been done the machine will produce identical parts of identical quality. It will not get tired and if it is fitted with automated equipment to load and unload it (industrial robots), it can carry on working in the dark when everyone has gone home. The relationships between the inputs, outcomes and information flow for automated (CNC) machines are shown in Fig. 3.10.

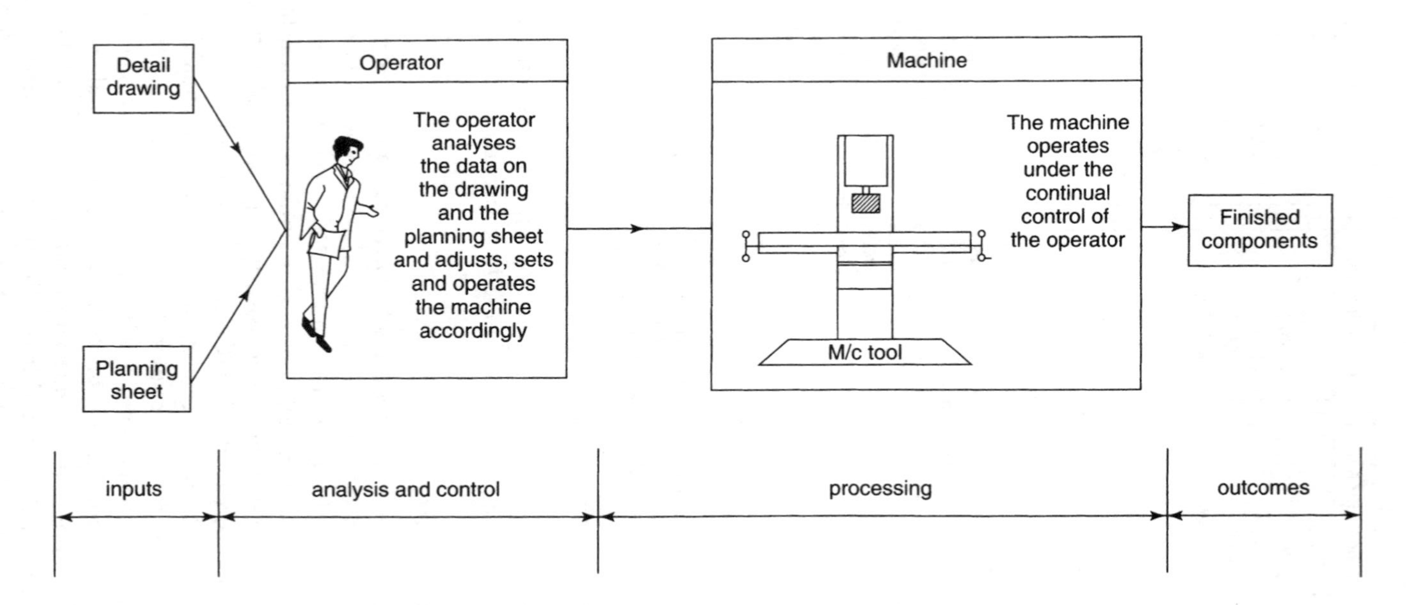

Figure 3.9 Manual (non-automated) manufacture.

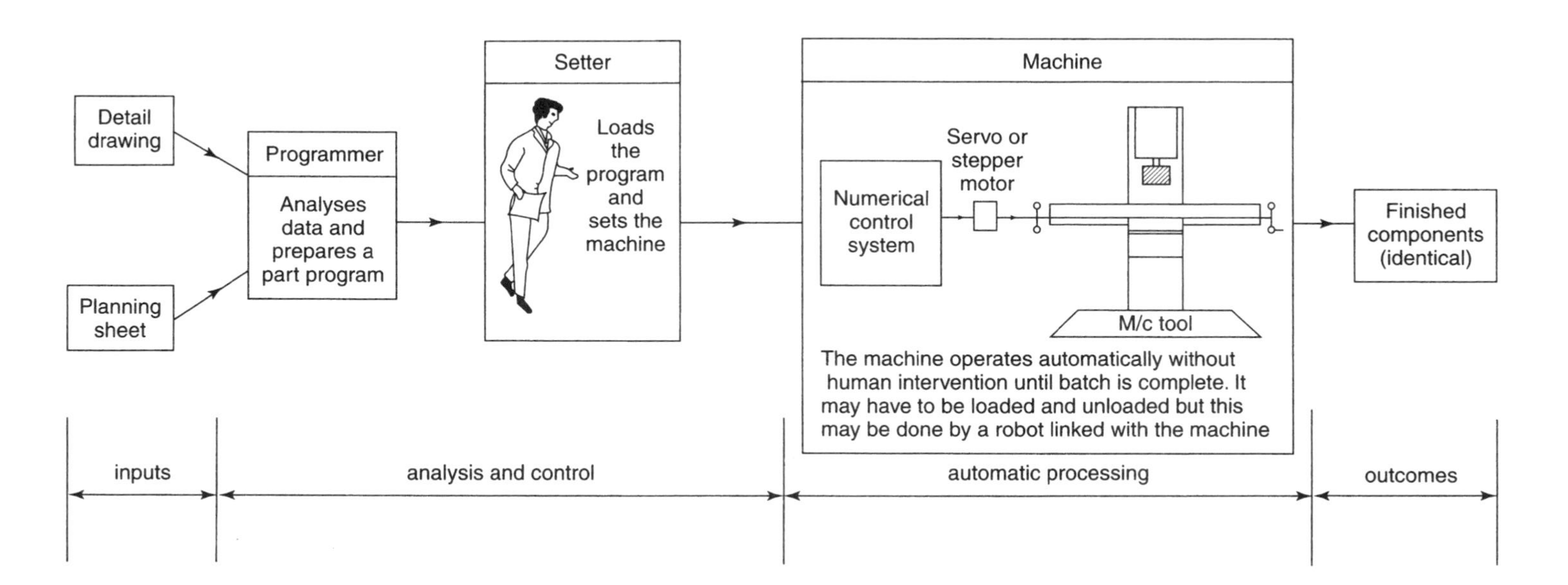

Figure 3.10 Automated manufacture.

With computer controlled equipment it is possible for a number of machines to be interconnected to a master computer to form an integrated automated production unit.

This type of machinery requires high capital investment, but gives the benefits of maintained high output with a consistent level of product quality. Built-in or *in-process* measuring systems constantly monitor the size, shape or other parameters of the product at each critical control point and provide feedback to a comparison unit which provides correction automatically when there is a drift towards the pre-determined, specified limits.

There are a number of advantages in using CNC:

- The program can be stored on disk or tape for re-use when repeat batches are required.
- The writing and loading of a program is much quicker than manufacturing control cams of complex design.
- There is total management control over performance, quality and costs.
- The machine control system can be easily linked with other computer-controlled equipment and systems.

Although an engineering example has been shown in Figs. 3.9 and 3.10, CNC is equally applicable to joinery (woodworking) machines, weaving looms and knitting machines and many other types of machines used in the manufacturing industries. The application of computing to the automation of manufacturing will be discussed further in Chapter 4.

3.4.3 The working environment

The first task to be considered in the preparation for manufacture is the space around the equipment and machinery. This can be determined before the machinery or equipment is installed by the use of the scale models that are available for most standard machine types and sizes. Whole workshops can be planned, three-dimensionally, in this way.

Space

Raw materials and part-finished components must be delivered to the machine and its operator in a steady flow. If too much is delivered, the space around the machine becomes dangerously cluttered. If too little is delivered the machine and its operator will be kept waiting and production targets will not be met. Similarly processed materials must be removed from the work zone on a regular basis to prevent congestion and to prevent the next operation in the production sequence from being kept waiting. There must be adequate space around the machine for the free flow of the work being processed.

Floor

The floor space surrounding the machinery and equipment should be clean, even and uncluttered. Substances such as cutting fluid, oil, grease or fat could, if left on the floor, cause an operator to slip and sustain an injury.

Lighting

It is essential that operators can clearly see all the machine's controls, screens, dials and gauges and on manual machines the item undergoing processing. As well as good overall lighting, the machine or equipment should have its own lighting system that can be positioned by the operator to illuminate the actual point of processing.

Cleanliness

Ensuring that machinery and equipment is clean and free from dirt is a fundamental part of preparation for processing, and is an essential action in the manufacture of food products in the interests of hygiene. Clothing and upholstery manufacture also requires the utmost cleanliness if the products are to be kept clean, unmarked and fit for sale. Before any cleaning takes place make sure that the machine is switched off at the mains and if necessary post a warning notice on the switchbox. Do not rely on interlock switches or machine controls. Always follow the manufacturer's instructions on cleaning the machinery.

Safety

Ensure that all safety devices are in place and available. In particular, ensure that all guards are in position, secured, and in a serviceable condition. Never operate machinery or equipment without the guards in position. If there is any doubt about the condition and/or function of safety equipment and guards, notify your supervisor immediately. Do not operate the machine or equipment until the fault has been corrected. Guards must only be set and/or adjusted by a suitably qualified person and must not be tampered with by the operator.

Tooling

If it is required, make sure that the correct tooling is available for the operation to be carried out and that it is in a serviceable condition. Should the tooling be damaged or blunt it must be returned to the stores for replacement or refurbishment. The use of tools that are incorrectly selected or in poor condition will result in loss of production and poor quality products. Further, the use of such tools can be dangerous.

Settings

All controls and instruments must be set correctly for the process to be carried out. Settings will include speeds, feeds, depth of cut, approach and runout distances, cycle times, processing temperatures, and thread tensions in sewing machines. Let's now have a look at some machine settings.

Test your knowledge 3.5

1. Briefly compare the advantages and limitations of automated and manually operated machine tools.
2. Describe the requirements for a safe and efficient working environment.

3.5 The preparation of tools and equipment

Tools fall into three general categories:

- Cutting
- Shaping
- Assembly and fixing.

3.5.1 Cutting tools

Cutting tools are used to remove surplus material during manufacture. Some trimming tools may also fall into the cutting tool category. The materials from which cutting tools are made must have the properties of hardness in order to maintain a keen cutting edge, but with sufficient toughness to prevent chipping. They must be easily resharpened or disposable. Starting with the simplest cutting tools, Fig. 3.11 shows a selection of shears, scissors and knives used for cutting out and trimming.

Scissors

Scissors are used to cut textiles, plastics, paper, card, metal foil, woven glass fibre, glass fibre mat and food, the correct type being selected for the process to be carried out. They should be checked for sharpness and cleanliness especially those with inserted blade sections. Scissors range in size from the very small as used for needlework, up to the very large size (shears) used in industries incorporating textiles in their products. While scissors would be used for cutting out hand-made garments, for example bespoke tailoring, power-driven rotary shears and even CNC laser-cutting machines are used

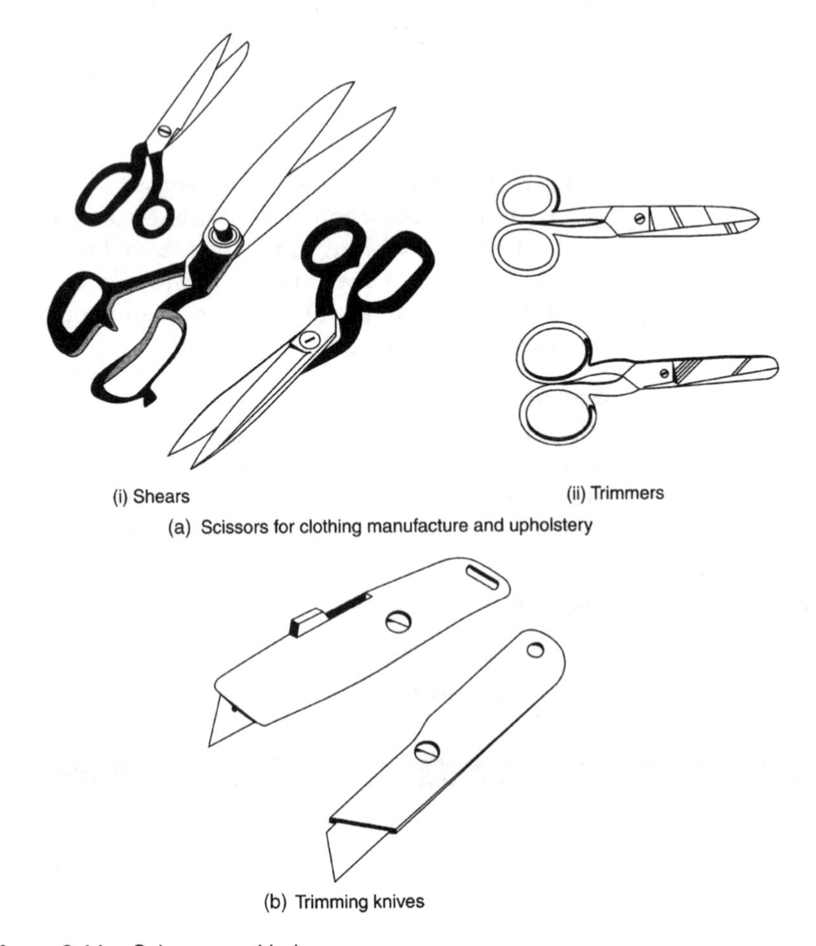

Figure 3.11 Scissors and knives.

for mass-produced clothing with several thicknesses of cloth being cut at the same time.

Knives

Knives are used for cutting and for trimming such materials as food, plastics, wood, card, glass fibre roving and leather. The correct type of knife should be selected for the process to be carried out. Knife blades should be kept clean and sharp and when not in use should be covered or sheathed to prevent damage to the cutting edge and for safety. Knives may have blades that are fixed, adjustable, retractable or disposable.

Shears

Shears or 'snips' are used for cutting out thin sheet metal such as tinplate. They are available with straight and curved blades and come in a variety of sizes. For thicker materials bench shears are used and, for the thickest sheet and thin plate, power driven guillotine shears are used.

Guillotines shears

Guillotine shears are used to cut sheet materials such as metal, plastics, paper, card and composite material and may be treadle operated or power driven. Examples of shearing equipment are shown in Fig. 3.12. Before using this or any machine, the operator should be fully conversant with the controls, ensure that all the guards are in position and know the emergency stop procedure. On some power driven machines a photo-electric cell is positioned in front of the blade to prevent operation if any object breaks the beam. For efficient use, the blades should be kept sharp and the clearance between the blades should be adjusted to suit the type and thickness of the material being cut.

Hacksaws

Hacksaws are used to cut metal, plastic, composite materials, frozen meat and bone, and may be hand or power operated. The blades used in hacksaws can be changed to suit the job in hand. It is important to select the correct type and size of blade for the task to be carried out. Hacksaw blades are classified by the pitch of their teeth and length, and are selected according to the thickness of the material to be cut. They range in size from 150 mm long for a 'junior' hacksaw blade up to 600 mm for a power hacksaw blade, the most common size being 300 mm for a hand hacksaw.

Carpenter's saws

Saws are used to cut wood, wood products such as chip and block board and some composite materials. The correct type and size must be selected for the task to be carried out. Saws are classified by size and type, and may be selected according to the pitch of the teeth to suit the thickness or width of the material to be cut. Typical saws are shown in Fig. 3.13.

Files

Files are mainly used on metals to shape, deburr and finish surfaces. They can also be used on other materials such as plastics, composites and

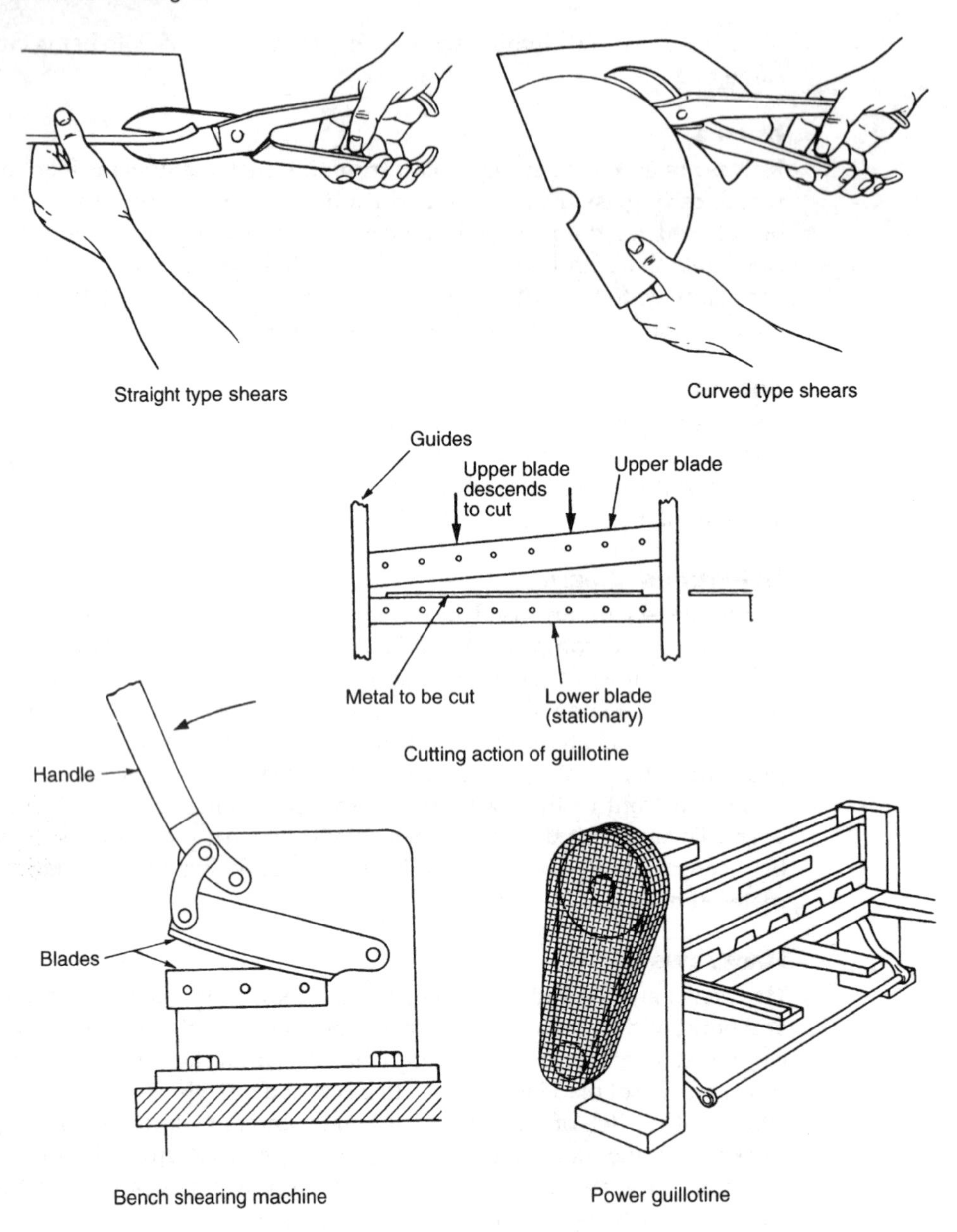

Figure 3.12 Sheet metal cutting tools.

hardwoods. Files should be selected for the process to be carried out according to their shape, grade, size and length. They are classified by shape of their cross-section, e.g. flat, hand, round, half round, square and three-square (triangular) in the following grades of cut: coarse, bastard cut, second cut, smooth and dead smooth. For very fine work, needle files and riffler files are available.

Before use, files should be checked for serviceability and cleanliness. The handle should be properly fitted, free from cracks, and the ferrule should

Figure 3.13 Typical saws.

be in place. A properly fitted handle prevents the pointed tang of the file piercing your hand or wrist with serious consequences. A cracked or split handle should be immediately replaced. Any metal or dirt clogging the teeth should be removed with a file card (a special type of wire brush). Clogging can be greatly reduced by rubbing chalk into the cutting faces. Also, due to their extreme hardness and consequent brittleness, files should not be tapped on hard surfaces to remove swarf (particles of cut metal) as they may break or fracture. Some typical files are shown in Fig. 3.14.

Scrapers

Scrapers, as used in cabinet making and the leather goods industry, are used to remove rough grain, burrs and sharp edges. Fitters in the engineering industries use scrapers to produce flat surfaces, or make shafts and bearings fit each other. Engineer's scrapers can be flat, half round and three square (triangular) and should be selected to suit the task in hand. Scrapers should be kept clean, sharp, and free from grease. When not in use the blade should be cased to protect the cutting edge. Cracked or split handles should be immediately replaced. Various types of engineer's scrapers and typical applications are shown in Fig. 3.15.

Drills

Drills are used to cut holes in materials such as metals, wood, plastics, brick, masonry, glass and composites. They are made of high carbon or high speed steel. Drills used to cut holes in brick, masonry and glass have cutting edges

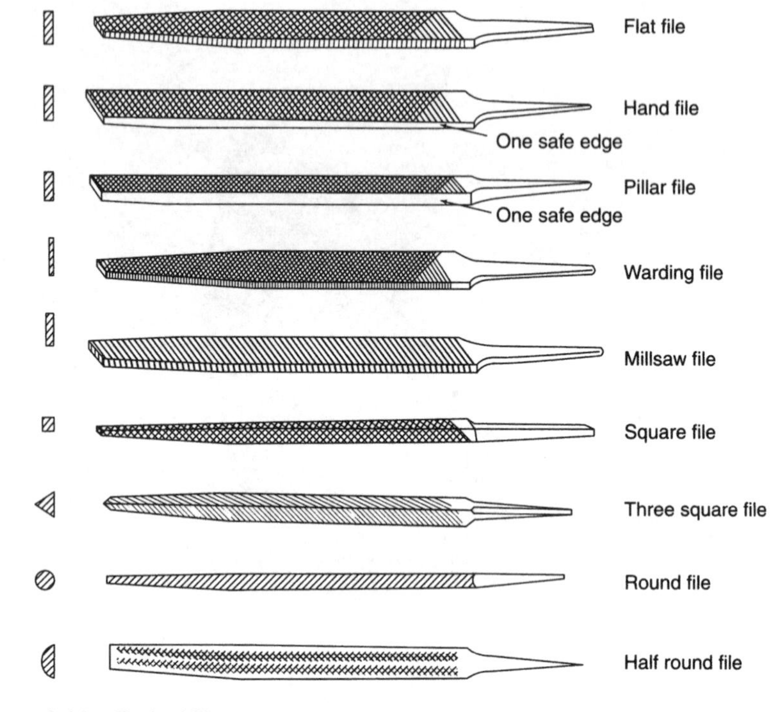

Figure 3.14 Typical files.

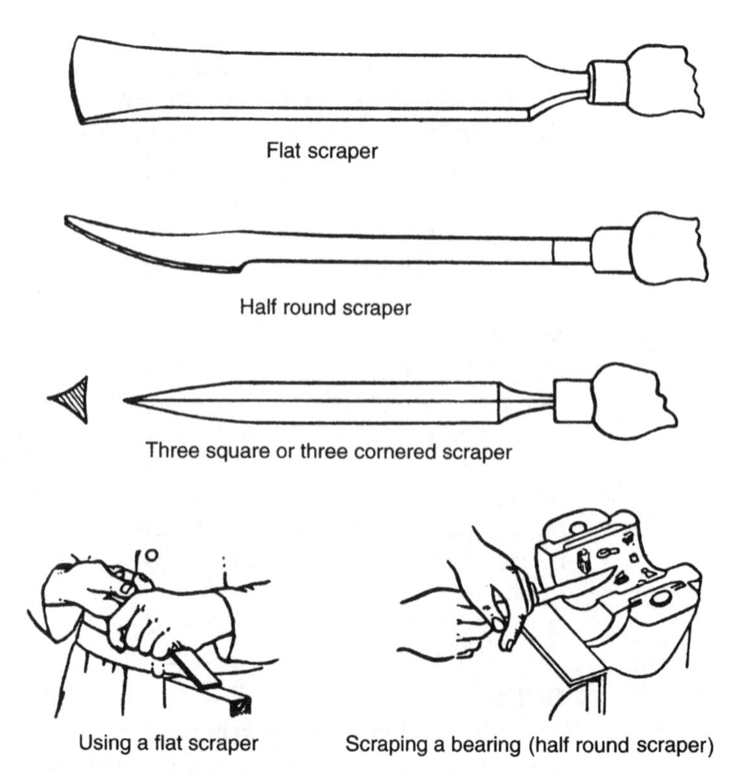

Figure 3.15 Engineer's scrapers.

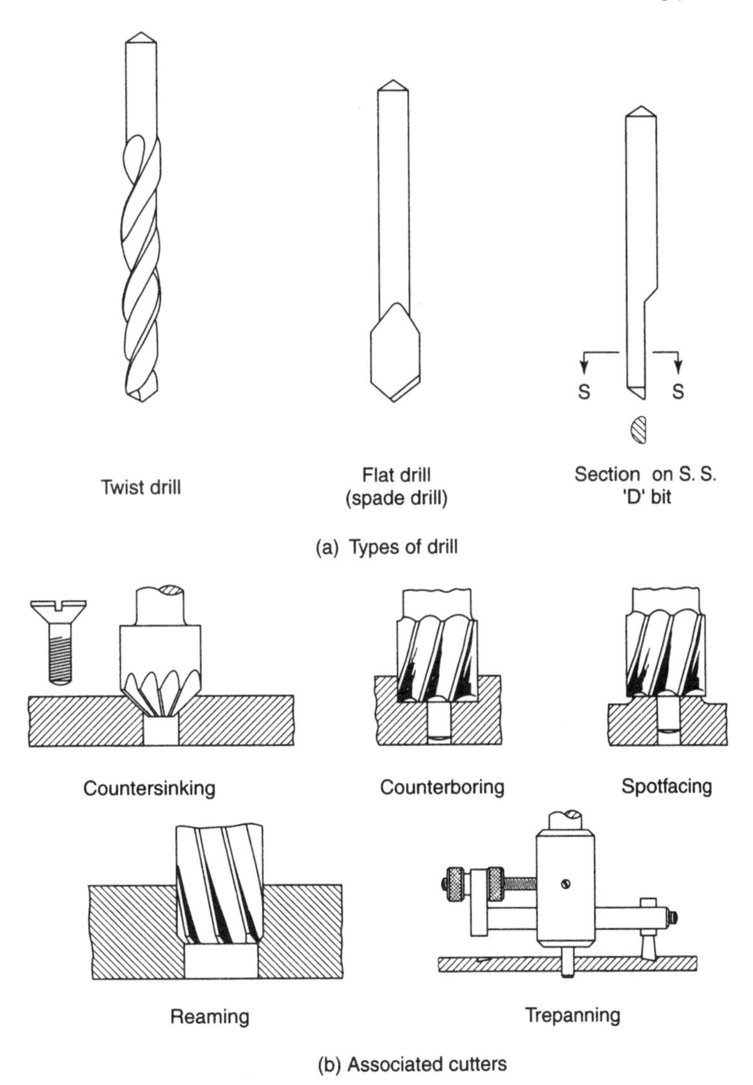

Figure 3.16 Drills and associated cutters.

in the form of a tip of sintered carbide to resist the highly abrasive effects of these materials when cut. The correct size and type of drill should be selected for the hole to be drilled, and should be sharp and clean. The drills are rotated by hand or, more usually, by electrically or pneumatically powered drilling machines. These may be fixed or portable. Always ensure that electrically powered drilling machines are checked for electrical safety before use, for example, ensure that the plug is secure, and that the cable is not cut, frayed or twisted. Eye protection in the form of safety spectacles or goggles should always be worn when drilling. Chuck guards are not required on portable drilling machines but must be fitted to floor or bench mounted drilling machines. Some typical drills and associated cutters are shown in Fig. 3.16.

Abrasive paper and cloth

Although not strictly a cutting tool, abrasive papers and cloths have been included in this section since they remove small amounts of material by a cutting action. The cutting action is provided by particles of hard abrasive material such as 'emery' (a form or aluminium oxide) bonded to paper or cloth. Abrasive papers and cloths are used to produce a smooth finish on wood, plastics, metals, glass, stone, painted surfaces and composite materials.

The correct type and grade should be selected for the process to be carried out, for example, glass paper should be used on wood and plastics. Emery papers and cloths are used on metals. The types of abrasive papers and cloth available include glass-paper that can be supplied in nine grades with abrasive grain sizes ranging from 180 (finest) to 50. Waterproof paper sheets (wet and dry) can be supplied in 17 grades with abrasive grain sizes ranging from 1200 (finest) to 60. Emery cloth sheets and strip supplied in 10 grades with abrasive grain sizes from 320 (finest) to 40.

3.5.2 Shaping (forming) tools

These are tools used to change the shape of materials without cutting them, for example forging dies, extrusion dies and plastic moulds. The tablet forming devices used in the pharmaceutical industry can also be considered as a type of shaping tool. Some typical sheet metal forming tools are shown in Fig. 3.17.

3.5.3 Fixing and assembly tools

Fixing and assembly tools are used on components and products to join, assemble or secure in some way during the manufacturing process. Some typical examples are shown in Fig. 3.18.

Screwdrivers

Screwdrivers, as their name implies, are used to drive screws into materials such as wood, plastics and metal. The correct type and size of screwdriver should be chosen to suit the head of the screw being driven. For example, the types of screwdriver available include flat blade, 'Phillips', 'Posidrive', socket and various other shapes introduced by the computer industry. Screwdrivers should be regularly checked for damaged blades and the condition and security of the handle.

Spanners

Spanners are used to turn items such as nuts and bolts when assembling and securing components and products made of wood, metal and plastics. The correct size and type of spanner must be used to prevent bolt heads and nuts being damaged and to prevent possible injury to the user. The types of spanner available include open-ended, combination, ring, box, adjustable and socket. Socket spanners must be used with a wrench that may be of the ratchet type or fixed.

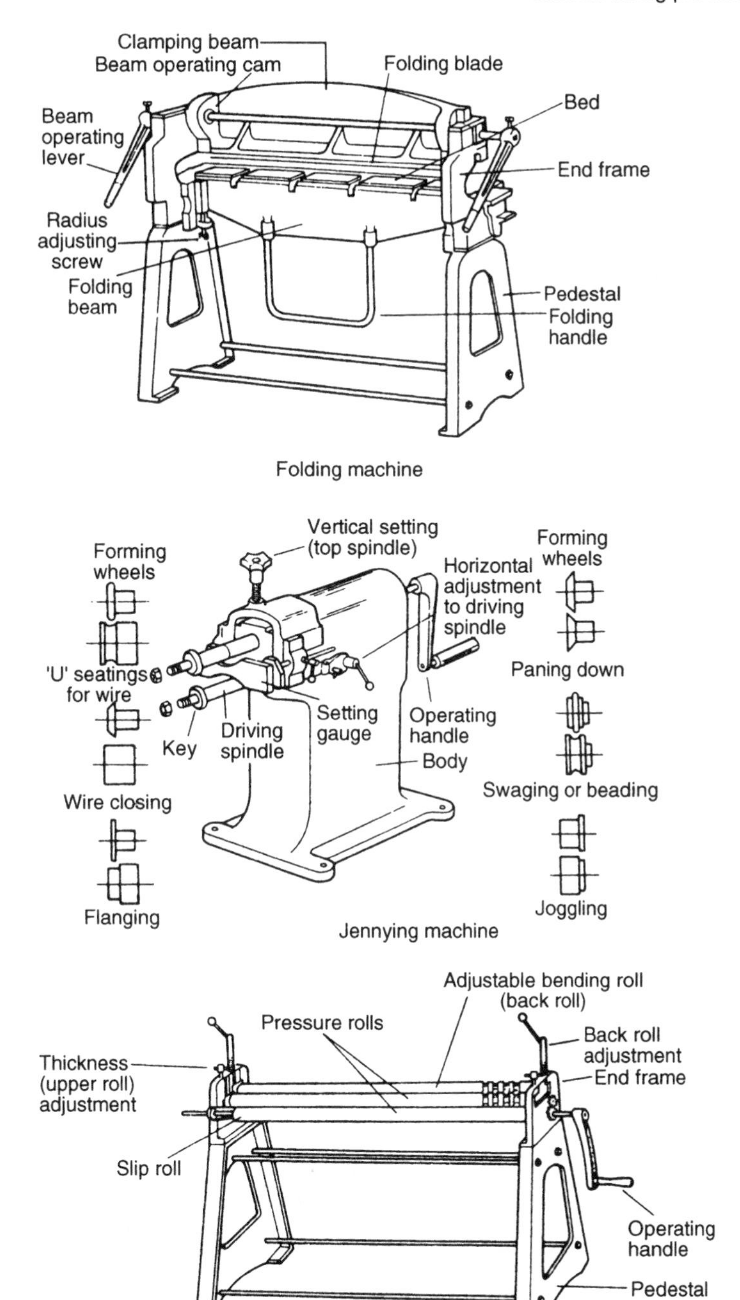

(a) Sheet metal working tools

Figure 3.17 Typical sheet metal forming tools and operations. From *Basic Engineering Craft Course Workshop Theory (Mechanical)* by R. L. Timings, Longman. Reprinted by permission of Pearson Education Ltd.

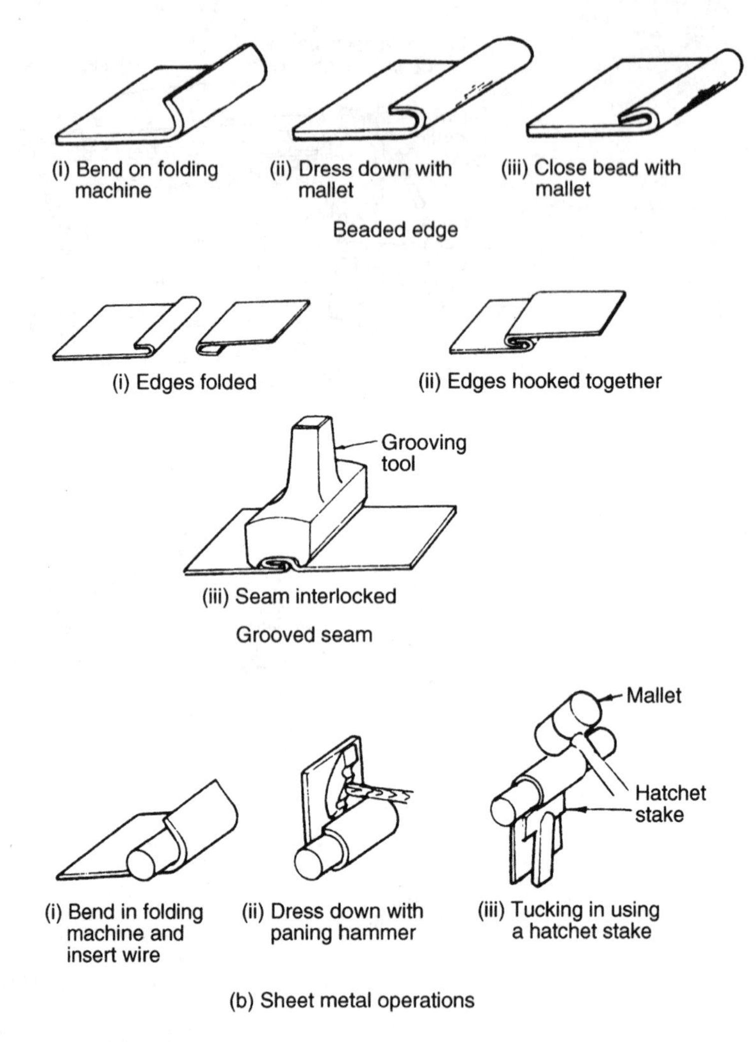

Figure 3.17 (*Continued*)

Hammers and mallets

Hammers and mallets are mainly used on metals and wood. The correct type should be selected for the material and process to be carried out. Hammers are classified by the shape of the head or head feature and to a lesser extent weight and include, ball pein, cross pein, straight pein, claw, planishing, creasing, stretching and blocking. Mallets are classified by their face material and size and include boxwood, hide, lead, aluminium, zinc, nylon, copper, rubber, polyurethane and polypropylene. Before use hammers and mallets should always be checked to ensure that the head is secure and the handle clean, dry, free from grease, and in good condition.

Soldering irons

Soldering irons may be of the traditional gas-heated type which are used mainly on sheet metal and allied products, or electrically heated that are used

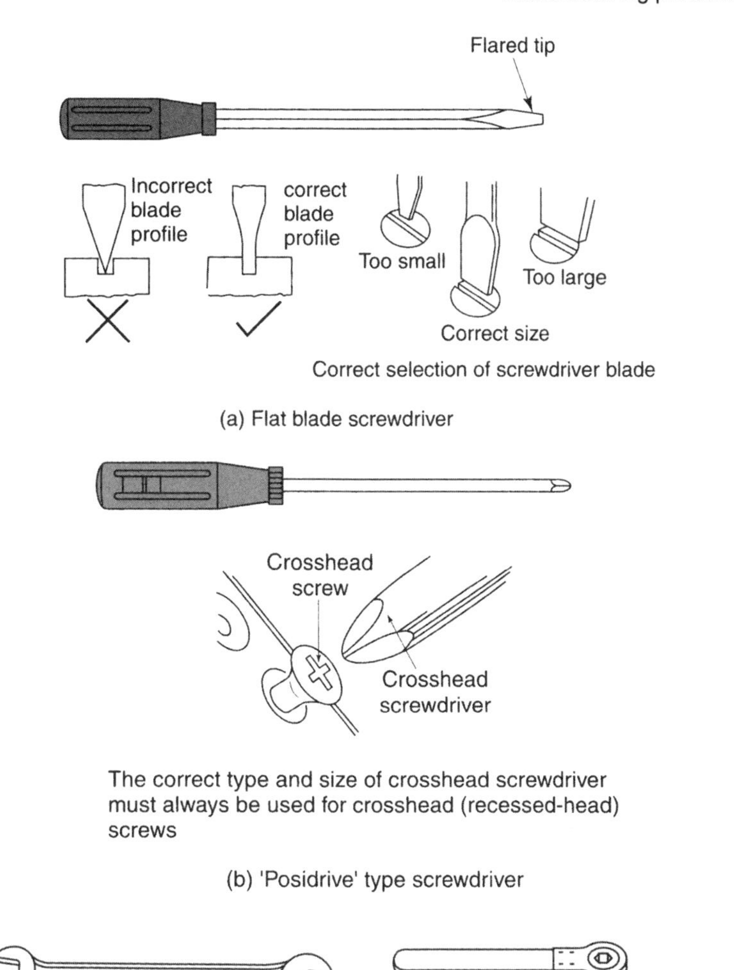

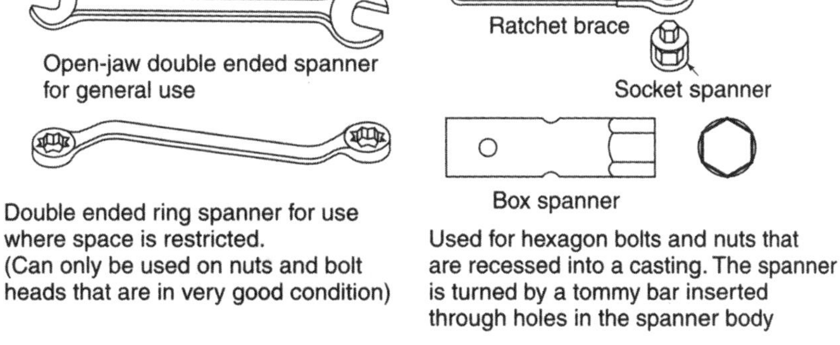

(c) Spanners

Figure 3.18 Fixing and assembly tools.

mainly on electric and electronic components and assemblies. The gas-heated type requires a separate heat source (soldering stove). For each type of soldering iron the correct size and or power rating should be chosen for the task to be carried out. Both types should be inspected for serviceability and safety

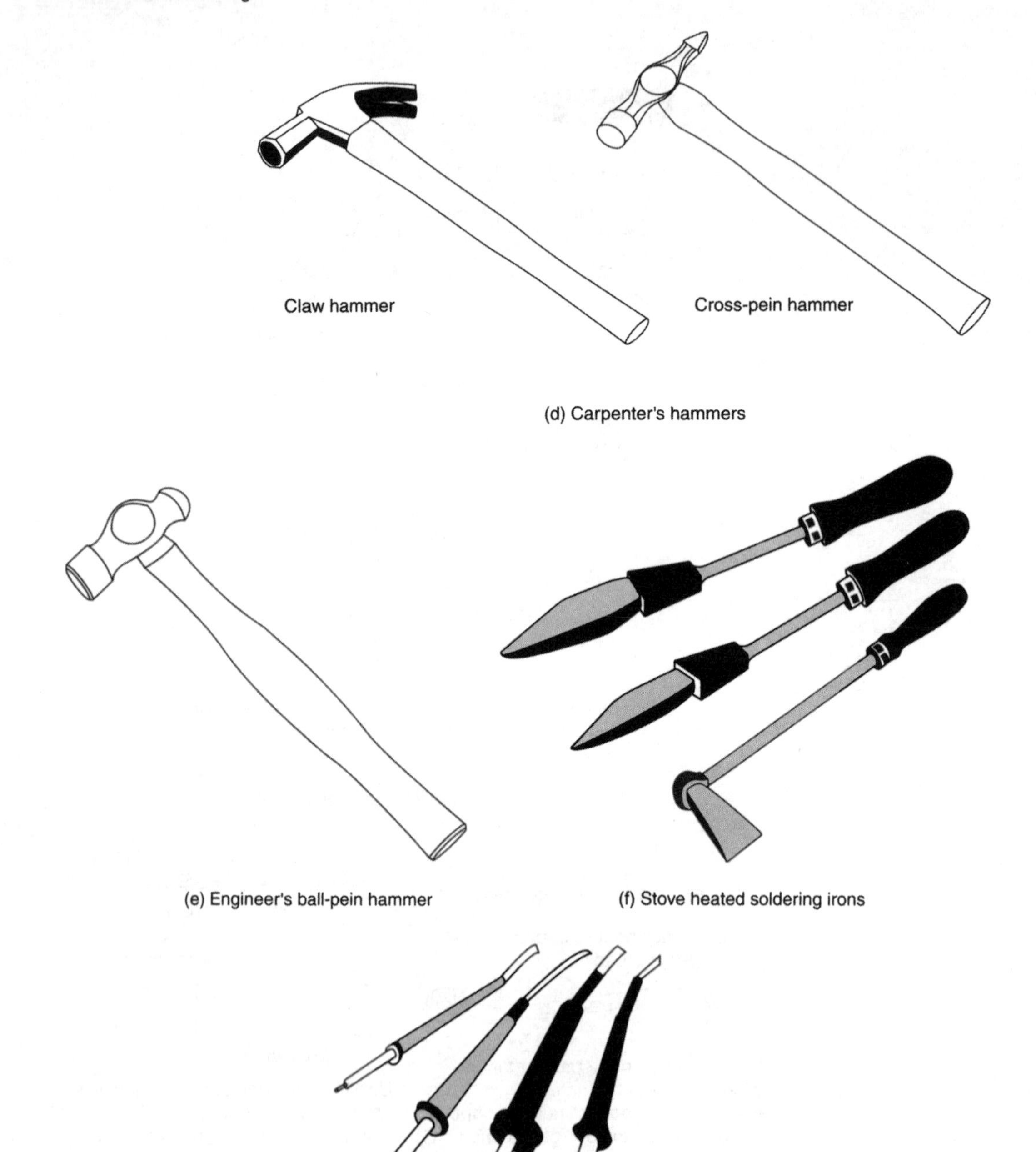

(g) Electrically heated soldering irons

Figure 3.18 (*Continued*)

prior to use, with the bit being clean and secure, as should be the handle. The wiring and connections should also be checked.

Riveting tools and guns

Rivets are used to join sheet metal, plastic, composite materials and components by means of hand or powered riveting gun. Hand riveting can be carried out with a ball-pein hammer and riveting snaps. These are special tools shaped to form the rivet head. Hollow or pop rivets can be fitted by hand using a special tool. This pulls a headed mandrel into the plain side of the rivet to form a head and hence secure it, the protruding part of the mandrel is then broken off by the riveting 'gun' and the joint is made. These riveting techniques are considered in Section 3.8.2. Some POP® riveting guns have a magazine of rivets to save the operator having to load and place the rivets one at a time.

Riveting guns are usually pneumatically powered and may be used for cold riveting small steel rivets and non-ferrous rivets of copper or aluminium. Hot riveting is mainly used on the large rivets required in structural steel work, bridge building and ship building where it is largely giving way to welding. In this process the rivets are of steel. They are raised to red heat before being placed in the hole ready for forming the head (closing the rivet) in order to secure the rivet in place.

Needles

Needles are used for hand or machine stitching textiles, leather, and canvas as used in the clothing, bedding and upholstery industries. The correct type and size of needle should be selected for the task to be carried out. Needles for hand-stitching are available in 26 sizes from number 1 which is long and fat to number 26 which is short and thin.

Needles used for industrial sewing-machines are classified by the shape of their points and their cross-sections. Point shapes for piercing needles may be sharp, rounded or ballpoint, the body having a circular cross-section. Point shapes for cutting needles are mostly spear point, and the cross-section shape may be lozenge, triangular, or flattened oval, the body having a circular cross-section.

The major difference between domestic hand and machine sewing needles and those used on industrial machines, is that the industrial needles have the eye at the point end and the body has a groove running along it to the eye.

Test your knowledge 3.6

1. Select suitable tools and equipment for making single prototypes of the following articles:
   (a) A tinplate tray 250 × 100 × 25 mm deep with soldered corner joints or the link shown in Fig. 3.19
   (b) A garment of your own choice
   (c) A garden bird table or a nesting box.
2. Describe how you would check and prepare the tools used in (1) above.
3. Explain briefly why it is bad practice and possibly dangerous:
   (a) To use a spanner that is a poor fit on the nut
   (b) Extend a spanner to obtain more leverage.

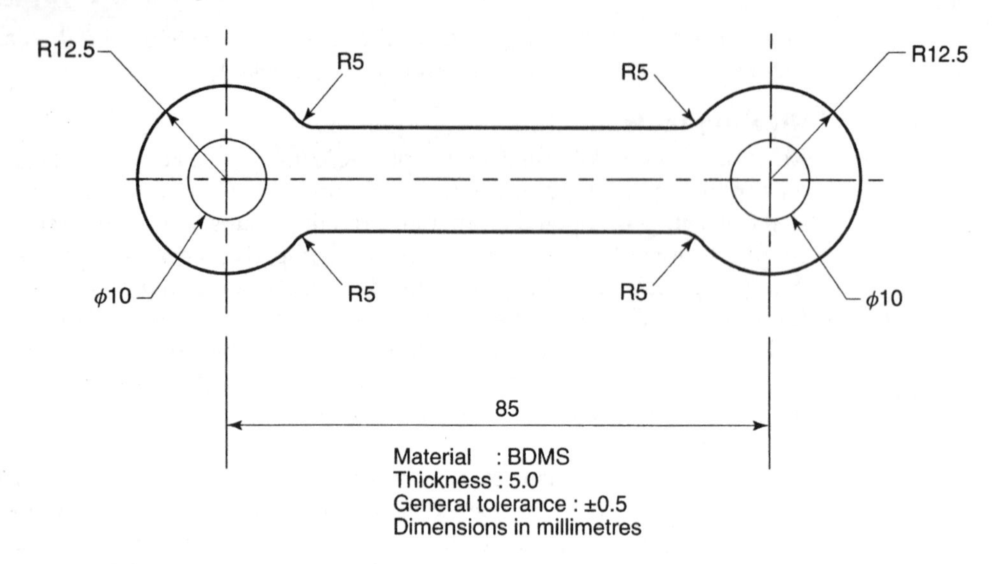

Figure 3.19 Link.

3.6 Processing methods

There are a wide variety of processes that can be used to manipulate materials and products during manufacture. The processing method selected will depend upon:

- The product to be made
- The materials used
- The size, shape and mass of the finished product
- The number to be produced
- The cost and time constraints imposed by the customer, the market and competitors.

Figure 3.20 shows a flow chart indicating the route that is taken in selecting the method of processing.

The control and adjustment of machinery and equipment must be carried out in the correct manner in order to process materials and components to specification and required quality. Quality standards are specified by the customer and will include consideration of such items as materials and tolerances in order to meet specification. Let's consider some of the elements that control the overall quality of a product.

3.6.1 Control and adjustment of manual equipment and machinery

Manually operated machines need continuous setter or operator attention. The controls have to be adjusted in order to process materials to the desired quality specification. For work where very precise measurements are not necessary, as in clothing manufacture, sheet metalwork and woodwork, patterns and templates may be used for guiding the initial cutting out operations. Engineering operations are generally of a more precise nature and the operator may work to accurately scribed lines for one off, prototype components. Where greater accuracy is required, the scribed lines are only a guide and

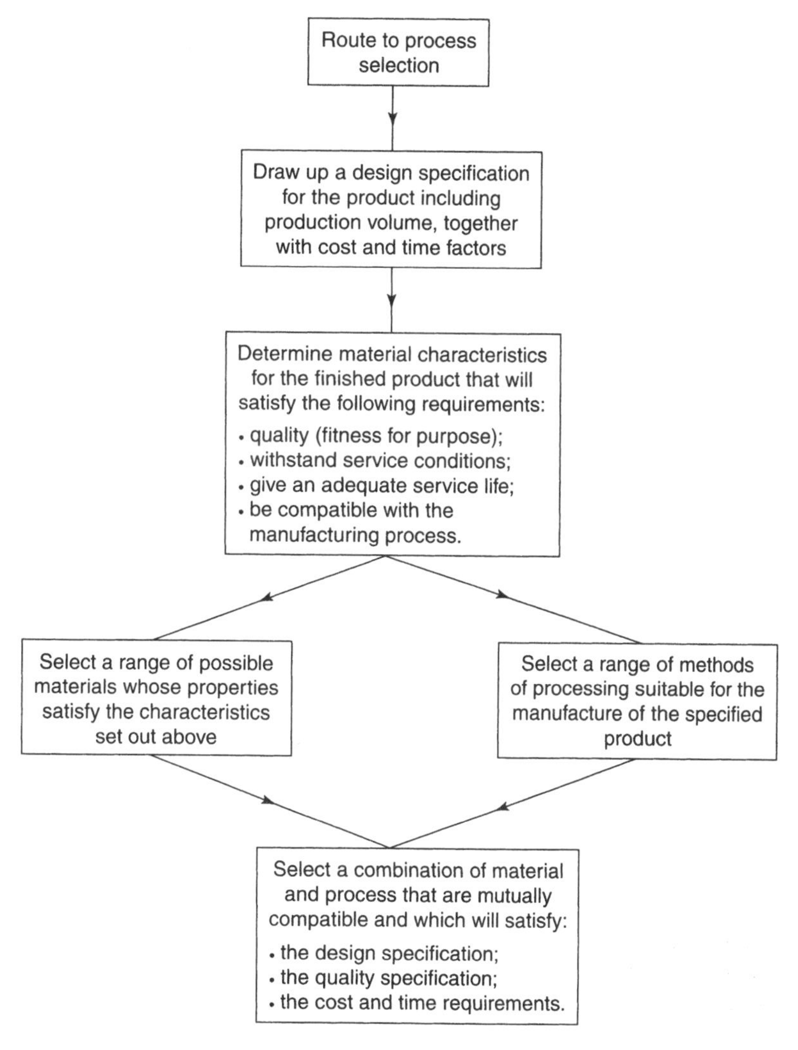

Figure 3.20 Route to process selection.

dimensional control relies upon the micrometer dials of the machine controls together with precision measuring instruments such as micrometer and vernier calipers. For batch production, jigs and fixtures may be used. These may:

- Position one component relative to another for assembly purposes (welding jig);
- Position a component relative to a machine cutter (milling fixture);
- Position a component relative to a machine cutter and guide the cutter (drilling jig).

Control and adjustment is not constant, it is only required when the machine produces components that are approaching the dimensional limits or are in danger of drifting outside the desired specification and quality standard. Some machines can be adjusted whilst they are running (adjusted 'in process'). They may also be adjusted after a pilot run. The final adjustments being

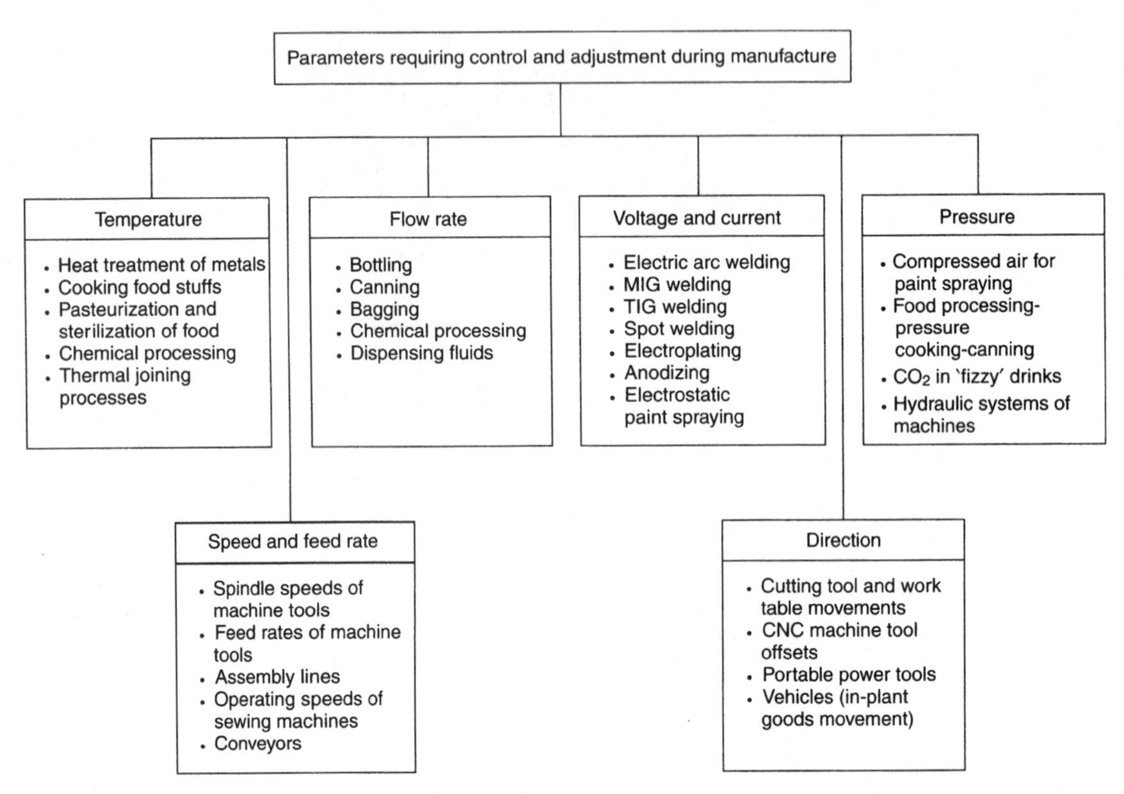

Figure 3.21 Examples of parameters requiring control and adjustment during manufacture.

made after the product has been inspected and the machine allowed to settle down. Controls that need to be adjusted whilst the machine or equipment is in operation, are positioned so that there is no need to remove guards and no risk to the operator. Such controls are sometimes referred to as 'in-process' controls. During machining or operation a tool may wear and have to be changed and some minor adjustment made as a result. This is a normal part of manual machine or equipment operation.

Typical examples of the parameters that may require adjustment during the operation of manufacturing plant and machinery are shown in Fig. 3.21. An example of control and adjustment of flow rate is the filling of bottles in a bottling plant. The bottles pass the filling station on a conveyor at a set speed. Therefore the *flow rate* of liquid into the bottles determines the level to which the bottles are filled. If the flow rate is too fast the bottles will contain too much liquid, and if too slow the bottles will not contain enough liquid. Therefore the flow rate is controlled and adjusted as required during the process.

3.6.2 Automated equipment and machinery

Automated machines are pre-programmed to operate within specified parameters such as speed, temperature, feed rate, and flow rate. The operator

Seq No	Code Programme	Explanation
%	GØØ G71 G75 G9Ø	Default line
N1Ø	X – 5Ø.Ø Y – 5Ø.Ø S1000 T1	
N2Ø	MØ6	TC posn spindle tool/offset (06 mm)
N3Ø	X25.Ø Y75.Ø Z1.Ø	Rapid to posn (1)
N4Ø	GØ1 Z – 3.Ø F1ØØ	Feed to depth
N5Ø	GØ1 X75.Ø F35Ø	Feed to posn (2)
N6Ø	GØØ Z1.Ø	Rapid tool up
N7Ø	X – 5Ø.Ø Y – 5Ø.Ø S8ØØ T2	
	MØ6	TC posn
N8Ø	X – 5Ø.Ø Y1Ø.Ø Z1.Ø	Rapid to posn (3)
N9Ø	GØ1 Z – 6.Ø F1ØØ	Feed to depth
N1ØØ	GØ1 Y5Ø.Ø F35Ø	Feed to posn (4)
N11Ø	GØØ Z1Ø	Rapid tool up
N12Ø	X – 5Ø.Ø Y – 5Ø.Ø S11ØØ T3	
	MØ6	

Figure 3.22 Portion of a part program for a CNC milling machine (BOSS6 software) (Note: BOSS6 = Bridgeport's own software system, version 6, is the software language for which the program segment shown is written).

would need to observe and check that the machine is safe and to specification whilst carrying out any control and adjustments that might be necessary. Automation can be mechanical where the machine movements are controlled by cams, or pneumatic/hydraulic where the machine movements are controlled by sequencing valves. CNC is widely used these days. The setter operator programmes the machine from the data on the component drawing or may receive the programme already prepared on tape or disk. Having loaded the programme into the machine controller, the operator then has to adjust the 'off-sets' to allow for such variables as cutter length and diameter, lathe tool nose radius, etc. The machine will then produce identical components until the batch is complete or the cutters need to be changed. When the cutters are replaced the 'off-sets' will need to be reset but the programme remains unchanged. For your interest a portion of a programme for a CNC milling machine is shown in Fig. 3.22.

Figure 3.23 shows some examples of flow charts for processing typical materials and processes in the manufacturing industries.

It is no good starting up a production line unless the flow of materials and part-finished goods are available to maintain a steady rate of production. If a large batch or continuous production is involved, it is also unlikely that there will be sufficient room to take delivery of all the materials in one delivery. Therefore it will be necessary to work out the flow rate per hour and arrive at the amount of material that needs to be available to supply a complete shift without interruption. An example is shown in Fig. 3.24.

Whenever an employee or trainee has any doubts as to the correct and safe operation of any tool, piece of equipment or machinery, always enquire who is the competent person to give you advice, and ask that person. Always remember 'IF IN DOUBT ASK'.

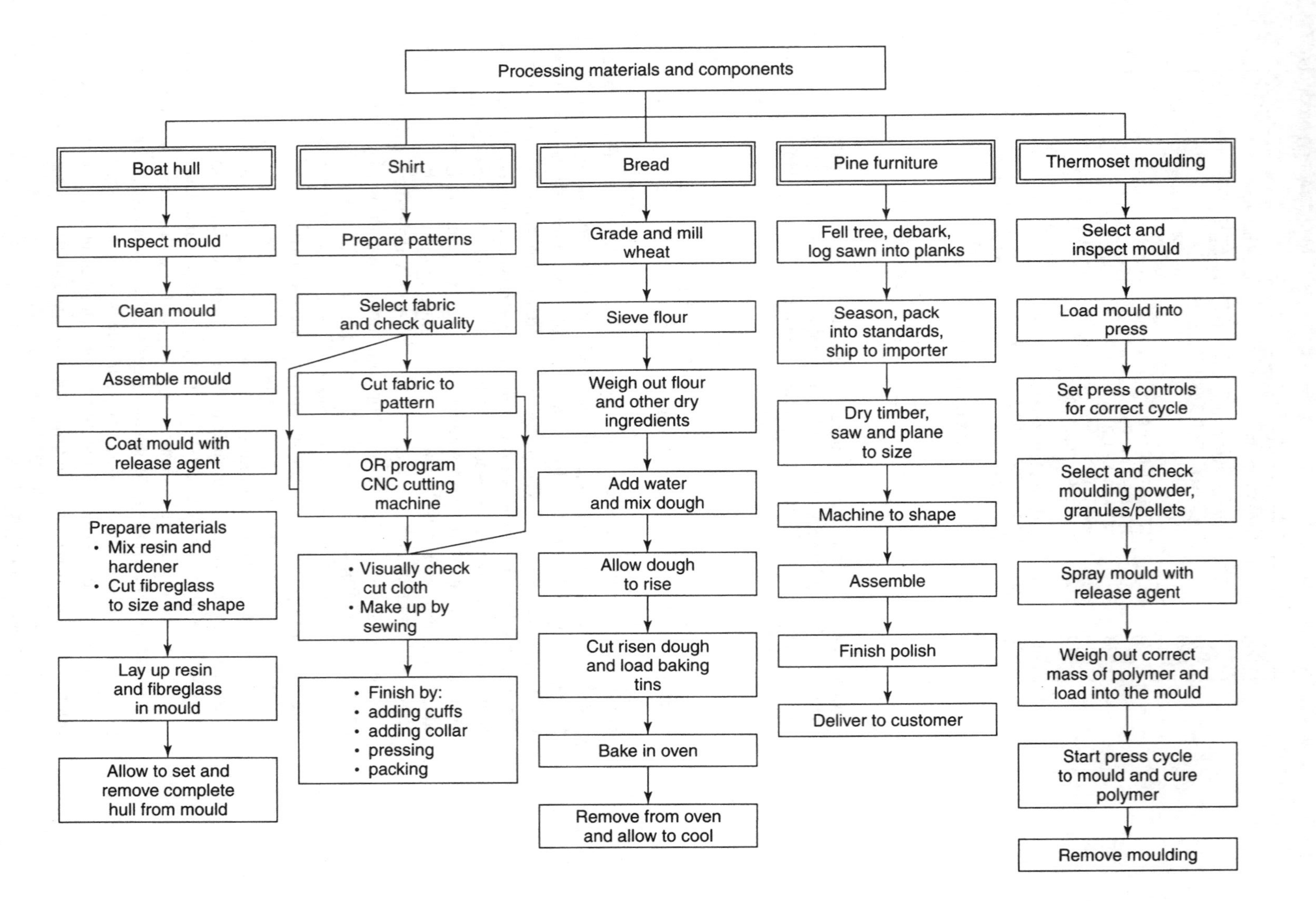

Figure 3.23 Processing materials.

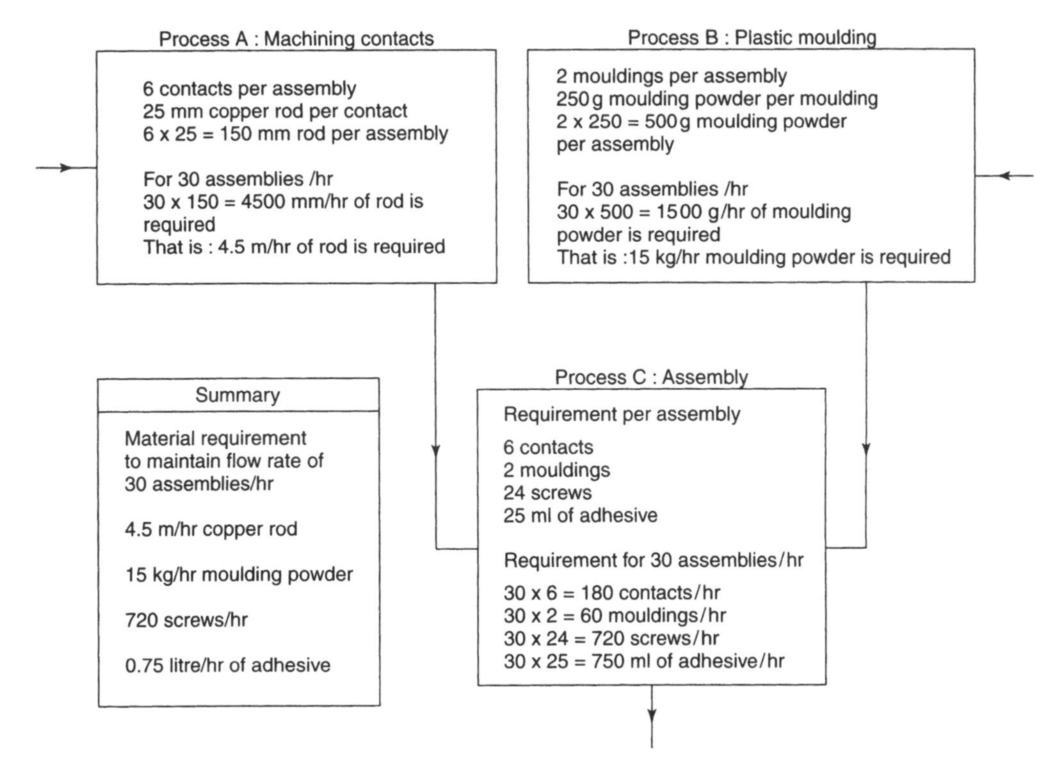

Figure 3.24 Material flow rate to maintain production of 30 assemblies per hour.

Test your knowledge 3.7

1. State the action you should take if you are inadvertently directed to operate a machine that is unfamiliar to you.
2. A high precision component is dimensioned as being 50.15 ± 0.01 mm diameter. Explain why checking this dimension alone will not necessarily be sufficient to ensure that the component conforms to its quality specification.
3. Describe how the duties of an operator using a manually operated machine differ from the duties of an operator using an automatic machine.
4. Draw up a flow chart to show the processing stages for a product of your own choosing.
5. Given the following data, calculate the flow rate of materials that is required to maintain uninterrupted manufacture.

Key notes 3.2

- Do not operate any machine or piece of equipment without prior training and, if necessary, appropriate certification of competency. Do not operate any machine without permission from your supervisor.
- Do not operate any machine or piece of equipment that you consider unsafe or unfit for service until it has been checked and made safe by a competent person.
- If you are directed to operate a machine that is unfamiliar to you, point this fact out to your supervisor and ask for instruction before you attempt to use it.
- IF IN DOUBT ASK.
- Manually operated machinery and equipment is controlled continuously by the operator all the time it is in operation.
- Automated machines are set up by the operator or setter and will then run unsupervised except for periodic checks to ensure all is well.

- Automation systems may be mechanical (cam operated), pneumatic/hydraulic using sequence valves, or CNC.
- Material flow is necessary to maintain an even rate of production without interrruption.

3.6.3 Processing methods (heat treatments)

Annealing

Annealing is a process carried out on metals and glass in order to soften the materials, render them less brittle, and remove the effects of previous working, thereby increasing their ductility and plasticity. The process consists of heating the material to a given temperature for a set period of time followed by slow cooling, usually in the furnace.

Normalizing

Normalizing is a process mainly carried out on metals to eliminate the stress effects caused by casting, hot and cold working, and welding. The process consists of heating the material to a given temperature for a set period of time followed by slow cooling in still air. This slightly more rapid cooling prevents the excessive grain growth of annealing. The metal will be less ductile but will machine to a better finish. Normalizing is often carried out after rough machining castings and forgings to remove any residual stresses before finish machining. This prevents the finished components from warping.

Solution treatment

Some aluminium alloys such as 'duralumin', containing copper, harden naturally under normal atmospheric conditions and have to be softened immediately before use. This softening process is called *solution treatment* since it causes the copper and aluminium to form a solid solution. Solution treatment consists of heating the aluminium alloy to 500°C and allowing it to soak at this temperature for a short while. This renders the alloy soft and ductile and it will still be in this condition when it cools to room temperature.

Hardening

Hardening is a process carried out on metals to improve strength, wear and indentation resistance. The principal methods of hardening are:

- *Cold working* by rolling, drawing or pressing during which the grain structure is distorted and this imparts a degree of hardness. This is the only way of hardening many non-ferrous metals and alloys.
- *Precipitation hardening* This is a natural process that occurs in certain aluminium alloys, such as 'Duralumin', which contain copper. We have just seen that this type of aluminium alloy can be softened by a process called solution treatment. However, in this condition the alloy is unstable and at room temperature the aluminium and copper will slowly combine together to form a *hard intermetallic* compound of copper and aluminium particles that will precipitate out of solution. This is called *precipitation age hardening* and the metal will lose much of its ductility and become harder and more rigid. This process of natural ageing can be speeded up by again heating the metal. Alternatively it can be delayed by refrigeration.

- *Quench hardening* in which medium carbon, high carbon and alloy steels are heated to a high pre-determined temperature (red-heat) followed by rapid cooling (quenching) in brine, water or oil to give a high degree of hardness. The degree of hardness depends upon the composition of the metal and the rapidity of the cooling. Once the hardening temperature has been reached, there is no benefit in increasing the temperature further.
- *Tempering* is the name of a process carried out after quench-hardening to remove the brittleness of the metal and increase its toughness. There will be some small loss of hardness. The metal is reheated to the required temperature and again quenched. Typical tempering temperatures for plain carbon steels are listed in Table 3.5. The temperature can be judged by the colour of the metal surface after polishing it with emery cloth. Other materials such as glass and some polymers also benefit from suitable tempering processes to reduce their brittleness.

Steam pressing

Steam pressing is a process carried out in the clothing industries for pressing clothing. Steaming is also used in the millinery industry for forming or blocking hats.

Pasteurization

Pasteurization is a process discovered by Louis Pasteur whereby certain food and drink products (particularly milk and milk products) are heated to a high enough temperature to kill any pathogenic organisms and hence make it safe for human consumption.

Ultra-heat treatment (UHT)

Ultra-heat treatment (UHT) is the process of producing UHT milk by using higher temperatures than pasteurization, thereby killing all pathogenic

Table 3.5 Tempering temperatures

Colour*	Equivalent temperature (°C)	Application
Very light straw	220	Scrapers; lathe tools for brass
Light straw	225	Turning tools; steel engraving tools
Pale straw	230	Hammer faces; light lathe tools
Straw	235	Razors; paper cutters; steel plane blades
Dark straw	240	Milling cutters; drills; wood engraving tools
Dark yellow	245	Boring cutters; reamers; steel cutting chisels
Very dark yellow	250	Taps; screw cutting dies; rock drills
Yellow-brown	255	Chasers; penknives; hardwood cutting tools
Yellowish brown	260	Punches and dies; shear blades; snaps
Reddish brown	265	Wood boring tools; stone cutting tools
Brown-purple	270	Twist drills
Light purple	275	Axes; hot setts; surgical instruments
Full purple	280	Cold chisels and setts
Dark purple	285	Cold chisels for cast iron
Very dark purple	290	Cold chisels for iron; needles
Full blue	295	Circular and bandsaws for metals; screwdrivers
Dark blue	300	Spiral springs; wood saws

* Appearance of the oxide film that forms on a polished surface of the material as it is heated

organisms. The process gives the milk a longer shelf life but, unfortunately, it alters the flavour and some people find this unpalatable.

Canning

Canning is a process used in the food processing industry that relies on high temperatures to destroy enzymes and micro-organisms in order to maintain the food in a good and safe condition. The food is processed after it has been sealed in the can to prevent oxidation and re-contamination. Canning can also be considered as a packaging process.

Test your knowledge 3.8

1. Choose suitable heat treatment processes for the following purposes:
   (a) Softening a quench hardened steel;
   (b) Softening a cold-rolled brass sheet;
   (c) Stress relieving an iron casting after it has been rough machined;
   (d) Softening aluminium alloy rivets before use. Also state how they can be kept soft;
   (e) Hardening and tempering a high-carbon steel cutting tool.
2. List the main advantages and disadvantages of the following heat treatment processes as applied to milk: pasteurization, sterilization, ultra-heat treatment (UHT).

3.6.4 Processing methods (shaping)

Moulding and forming

Moulding is a shaping process commonly carried out on polymers and to a lesser degree on glass and clays. Note that polymer materials do not have a defined melting temperature or temperature range like metals. They become increasingly soft and 'plastic' until their viscosity is sufficiently low that they will flow into moulds under pressure. The following techniques being the most widely used:

Compression moulding

Compression moulding is the process of simultaneously compressing and heating a polymer powder, loose or as a compressed and heated pre-form, in a mould. This process is one of the major methods of shaping *thermosetting* plastics. These are polymer materials that undergo a chemical change during moulding. This change is called *curing* and the plastic material can never again be softened by heating. Mouldings made from *thermosetting* plastics (thermosets) tend to be more rigid and brittle than those made from *thermo-plastics*. Compression moulding is suitable for high volume production, and may involve the use of positive, semi-positive (flash) or transfer moulds. Sections through typical moulds are shown in Fig. 3.25.

Injection moulding

Injection moulding consists of the injection, under pressure, of a charge of *thermoplastic* material that has been heated to soften it sufficiently to flow, under pressure, into a water-cooled mould. Unlike the thermosetting plastics used in compression moulding, thermoplastics soften every time they are heated.

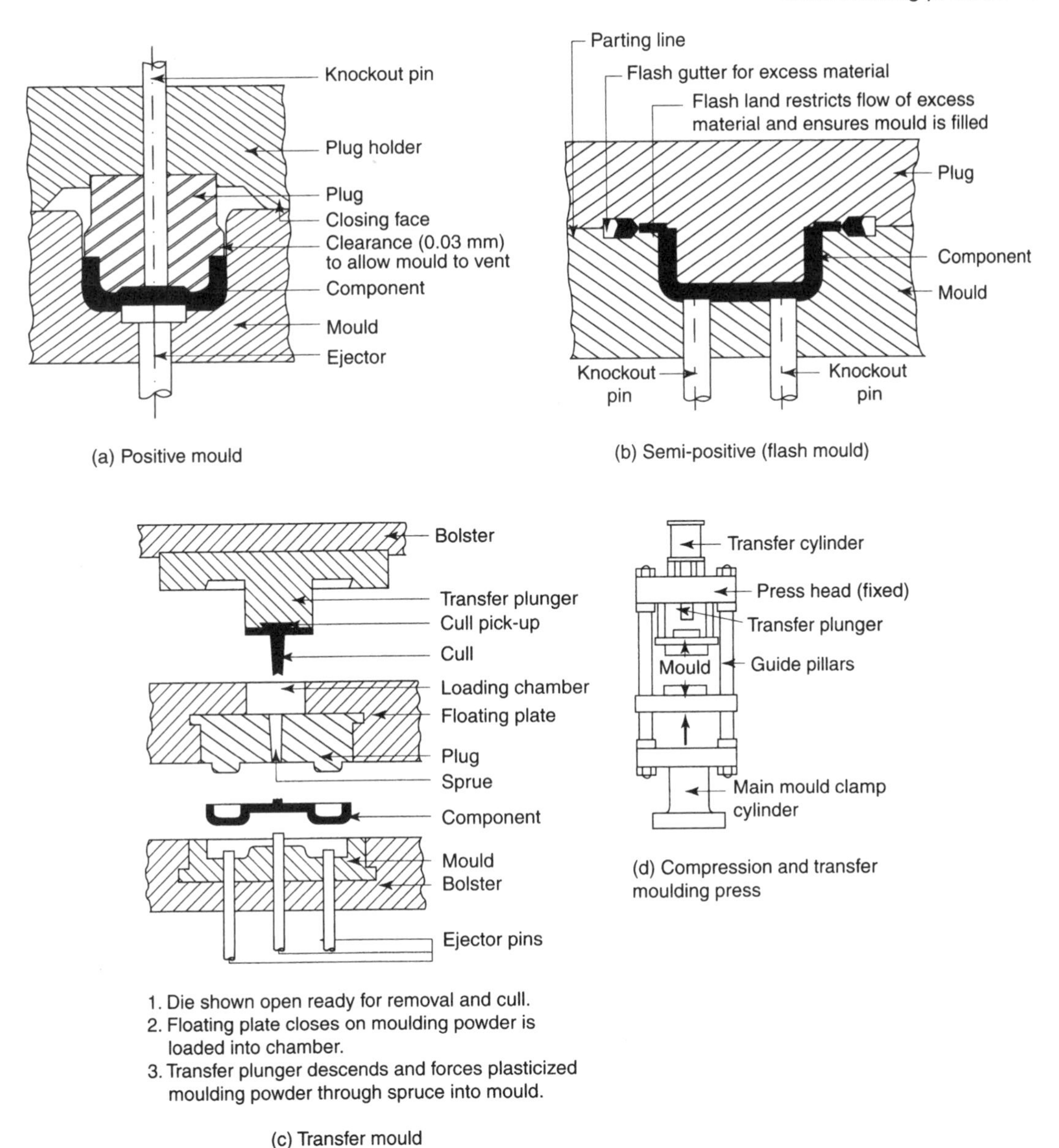

Figure 3.25 Compression moulding.

Care has to be taken not to overheat them or the plastic will become degraded. The plastic material takes the shape of the mould impression when solidified. Injection moulding is used mainly for the manufacture of thermoplastic products in large quantities. The principle of the process is shown in Fig. 3.26.

Extrusion moulding

Extrusion moulding consists of the continuous ejection, under pressure, of heated thermoplastic material through a die orifice to produce long lengths

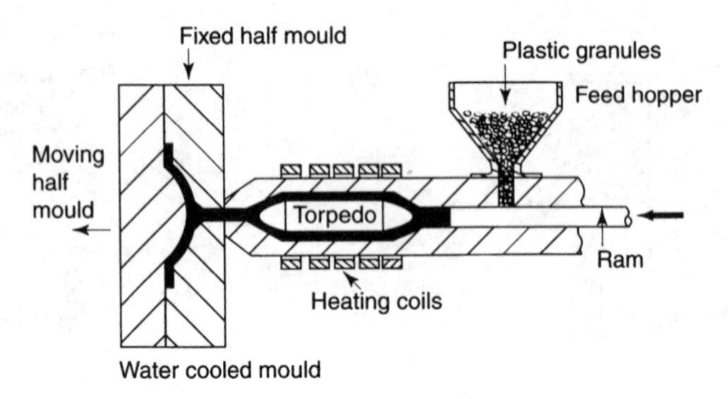

(a) Ram injection moulding machine

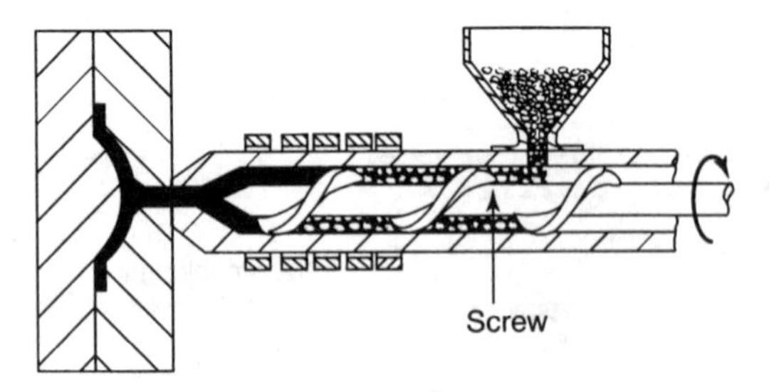

(b) Screw plasticizer injection moulding machine

Figure 3.26 Injection moulding.

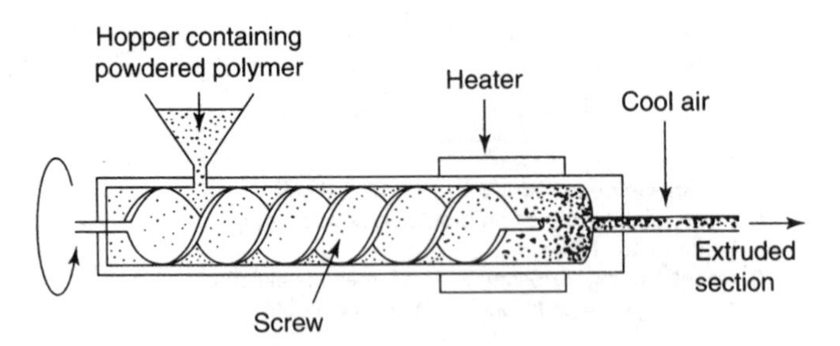

Figure 3.27 Extrusion (plastics). Reprinted by permission of W. Bolton.

of formed plastic strip, for example, curtain rail strip. The principle of the process is shown in Fig. 3.27.

Blow moulding

Blow moulding consists of blowing compressed air into a tube of heat-softened thermoplastic material clamped between the two halves of a mould, whereupon the plastic is forced outwards into the mould impression. The blow moulding process is used to manufacture products such as plastic soft drink and beer bottles, chemical containers and barrel shapes in large quantities. The principle of this process is shown in Fig. 3.28.

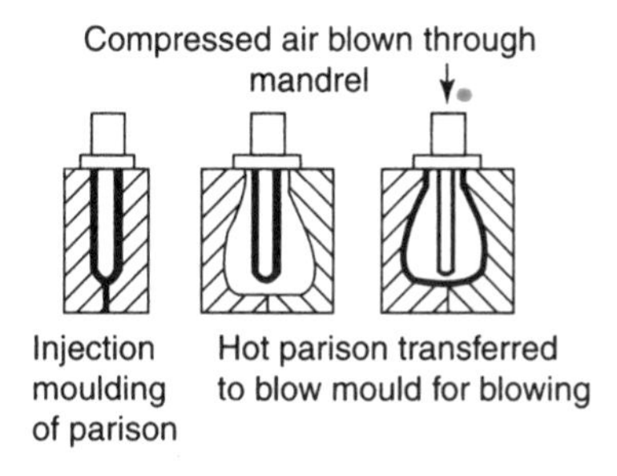

Figure 3.28 Blow moulding.

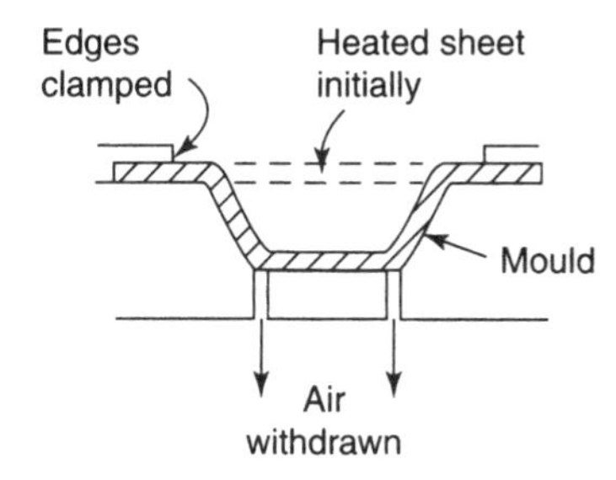

Figure 3.29 Vacuum forming. Reprinted by permission of W. Bolton.

Vacuum forming

In vacuum forming, a preheated sheet of thermoplastic material is laid across a mould. The air is pumped out of the mould and the atmospheric pressure above the sheet forces the softened sheet into the shape of the mould. Products such as plastic washing-up bowls are made in this manner. The principle of this process is shown in Fig. 3.29.

Casting

Casting consists of shaping components (castings) by pouring or injecting molten metal into a mould of the required shape and then allowing it to solidify by cooling. Various casting techniques are used depending upon the type of metal, the size and shape of the finished casting and the number of castings required.

Sand casting consists of making a mould by ramming moulding sand around a wooden pattern. The pattern is made the same shape as the finished product, but slightly larger so as to allow for shrinkage of the metal as it cools. The pattern is removed and the molten metal is poured into the mould cavity until it is full. When the metal has cooled and solidified, the sand is broken away to expose the casting. The mould can only be used once. A section through a simple sand mould is shown in Fig. 3.30.

Where a large number of castings of the same component are required alternative casting techniques are used, for example as in the manufacture of such products and components as bench vice bodies, machine tool beds, motor vehicle engine blocks and cylinder heads.

Shell moulding is where the mould is made from a mixture of sand and a resin bond around a heated former. This technique produces a rigid shell

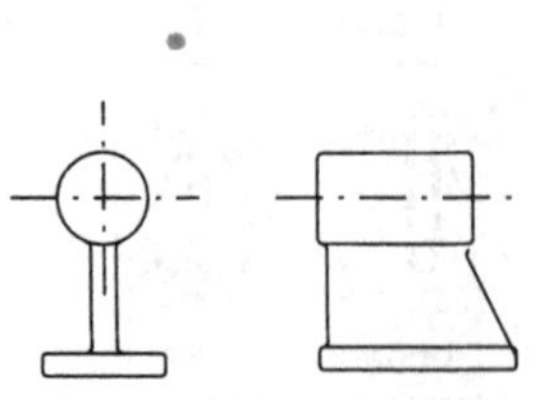

(a) Component to be cast

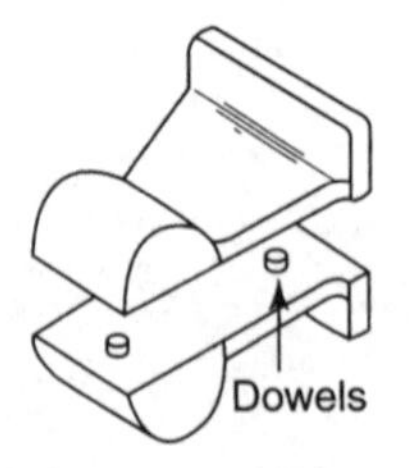

(b) Split pattern for casting

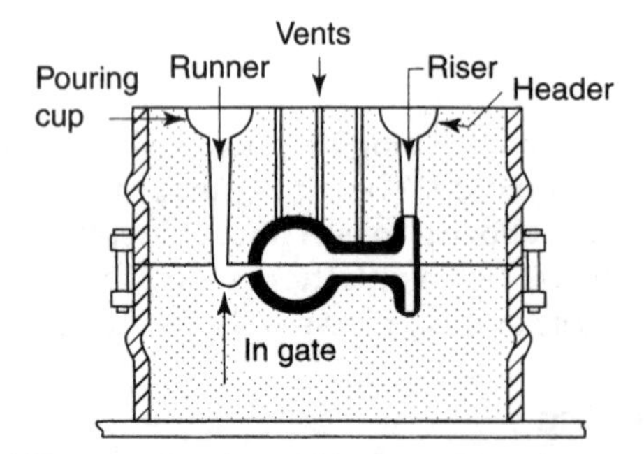

(c) Lower half of mould (drag) complete

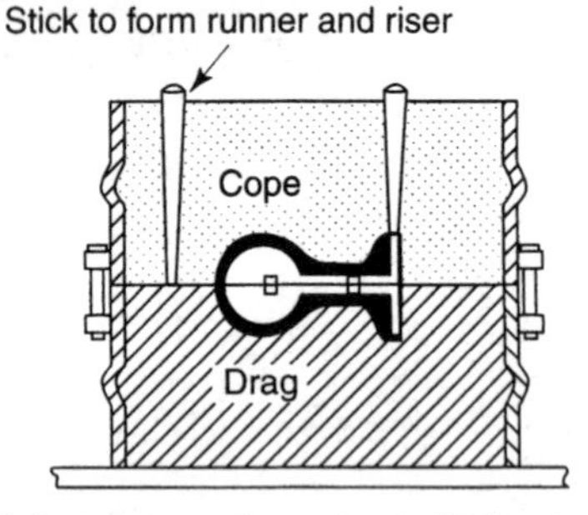

(d) Drag is turned over, top half of pattern is added and the cope is made on top of the drag

(e) The completed mould is opened and the pattern is removed. The sticks forming the runners and risers are also removed. The in gate is cut in the drag and vents and pouring cups are cut in the cope. The cope and drag are then reassembled ready for pouring the molten metal

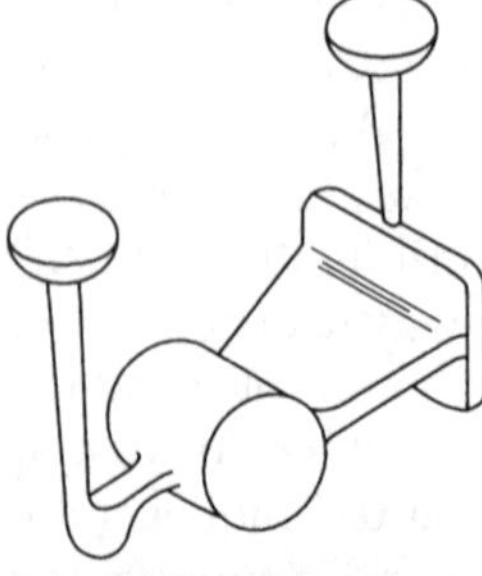

(f) The casting as removed from the mould ready for the runner and riser to be cut off and any blemishes trimmed off (fettling)

Figure 3.30 Sand casting.

that, after baking in a kiln, can be handled without damage. The shell moulds can be mass-produced and stored until required.

Gravity die-casting uses metal dies in place of the sand moulds of the previous processes. The dies have to be in two parts so that the casting can be removed when it has cooled. These dies are expensive to manufacture but can be used over and over again. The number of components that can be made before the dies have to be replaced ranges from hundreds to thousands depending upon the melting temperature of the metal being cast and the alloy steel used for the dies. This process is used for making castings from metals and alloys having a low melting point.

Pressure die-casting consists of injecting a charge of molten metal into metal dies under very high pressure. Again split metal dies are used. The castings are usually made from metals with a low melting temperature such as aluminium and its alloys and zinc based alloys such as 'Mazak'. This process is widely used for the manufacture of car door handles, fuel pump bodies, badges and other small components.

Extrusion

Like plastics, metals can also be extruded. However, the extrusion of metals involves the use of very much higher temperatures and very much greater forces. The extrusion process consists of forcing a heated billet of metal through a die of the desired profile. The extruded section can be in the form of products of constant cross-section such as tube, rod and complex profiles, such as double glazing window frame sections.

Forging

Forging is the shaping of very hot metal, by force, in a die. The materials used are carbon and alloy steels. The process consists of heating the metal in the form of a billet to a specified temperature such that it is soft and plastic (but not molten). The heated billet is placed in the bottom half of the die whereupon the top half of the die is brought down at high speed and pressure forcing the metal to fill the die cavity. Forging is used to produce very tough, shock resistant products and components such as crankshafts, cam shafts and connecting rods for the automobile industry; and hand tools such as spanners and wrenches. Forging tools and their uses are shown in Fig. 3.31. Forging temperature ranges for various materials are shown in Fig. 3.32.

Sheet metal pressing

Pressing is the shaping of sheet metal, in a die of the desired shape, under force whilst cold. The material must be in the annealed (soft) condition prior to pressing. Pressing is used to manufacture products such as automobile body parts, domestic appliance parts and cooking utensils. A typical press and a pressing are shown in Fig. 3.33.

Test your knowledge 3.9

1. Name a suitable process for manufacturing the following products giving the reason for your choice:

   (a) Plastic mouldings for a scale model aeroplane kit
   (b) A stainless steel sink unit

(c) A car engine connecting rod
(d) A car engine cylinder block
(e) A plastic bucket
(f) A fizzy drink bottle
(g) Plastic hosepipe
(h) A plastic case for an electricity meter
(i) Aluminium alloy pistons for a motorcycle engine
(j) Lengths of aluminium section for a greenhouse.

2. Describe the essential difference between a thermosetting plastic material and a thermoplastic material.

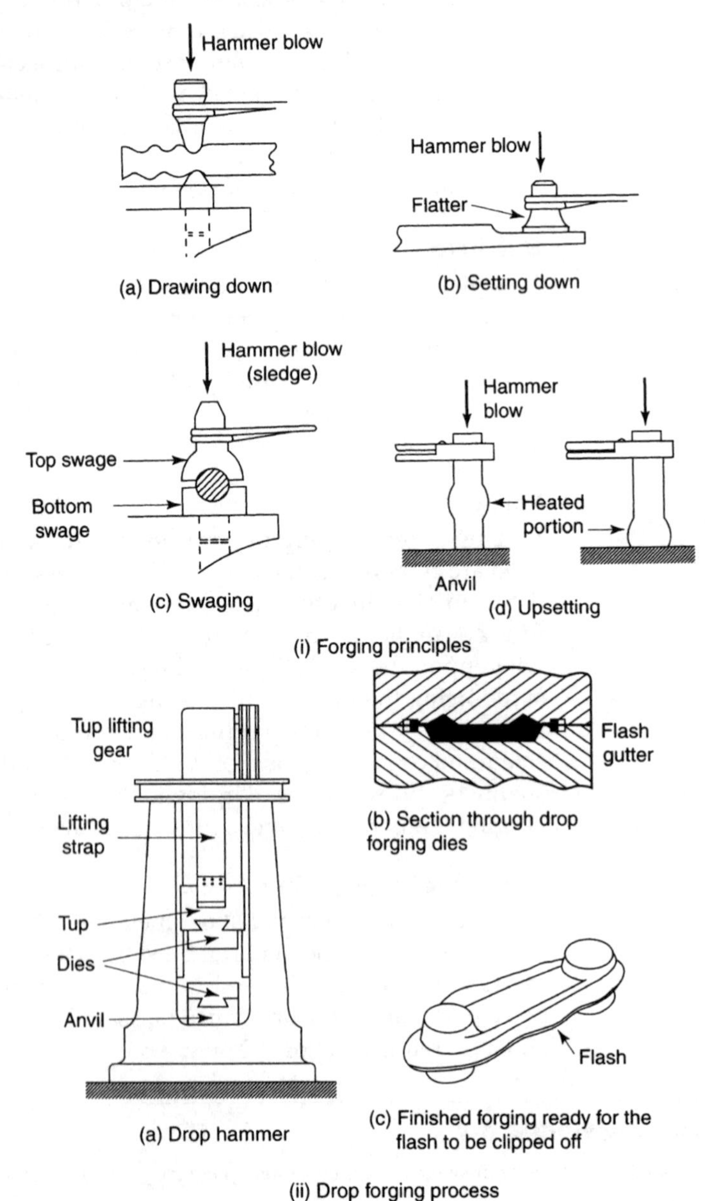

Figure 3.31 Forging.

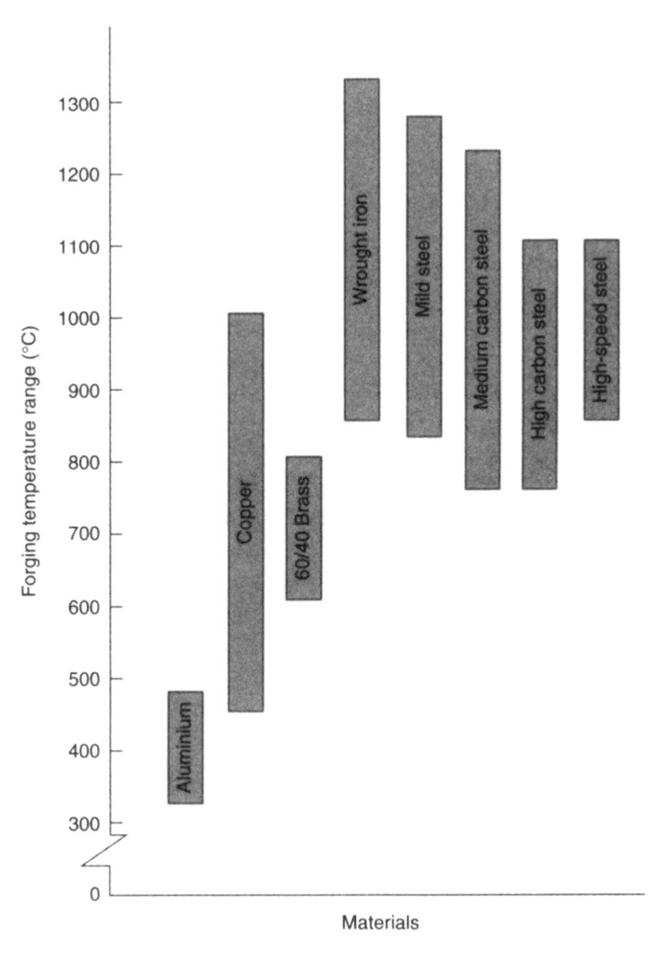

Figure 3.32 Forging temperatures.

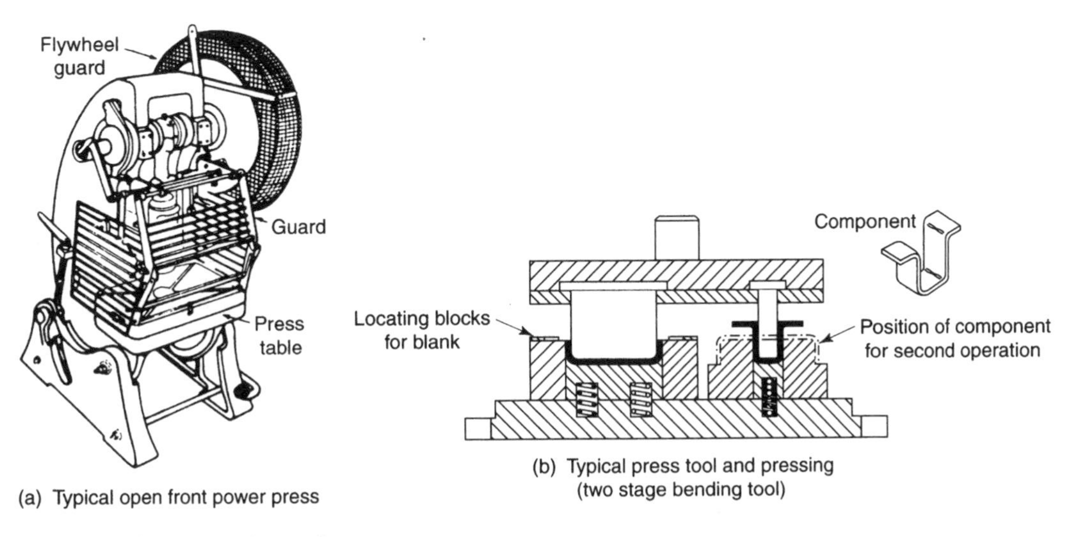

Figure 3.33 Sheet metal pressing.

3.7 Handling and storage of materials

In general the location of storage accommodation and areas should be:

- *situated close to the work area* to avoid time wasting and labour intensive handling and transportation;
- *easily accessible* so that the stores personnel do not continually have to use steps and ladders or move one lot of goods to get to another;
- *secure* against unauthorized entry and theft;
- *dry* to avoid damage due to corrosion, staining or decomposition;
- *well ventilated* to provide a pleasant working environment and to avoid any possible build up of flammable gases;
- *well lit* for easy product identification and avoidance of accidents.

It is very important that materials, components and products are correctly stored and handled to ensure that they are in good condition for processing and in perfect condition for delivery. The major factors that affect handling and storage are:

- Radioactivity
- Toxicity
- Oxidation
- Flammability
- Perishability
- Contamination
- Hygiene
- Sharpness
- Discolouration.

Storage areas should not be overcrowded. There should be ample room for handling the products they contain and the storage facilities should be accessible without having to climb about. Appropriate *safety notices* should be displayed where they can be clearly seen. Mechanical handling aids should be provided for lifting and stacking heavy objects and the stores staff should be properly trained in the use of such equipment. Stored goods must not obstruct emergency exits.

3.7.1 Radioactivity

Radioactive isotopes are used for taking X-ray photographs of welded joints on pressure vessels and high-pressure pipe runs. These isotopes emit highly dangerous gamma rays. They must only be used and handled by suitably qualified persons and must be stored in accordance with the appropriate guidelines and legislation. They must be securely stored against theft and misuse and kept separate from all other materials.

3.7.2 Toxicity

Toxicity is the property of a material or substance to have a poisonous effect on the body by touching, swallowing or inhaling. Toxic materials may be liquid, solid, powder or gaseous in form, and should be stored in a cool, dry and well ventilated and drained place. Toxic materials can affect our bodies in two ways:

- *Irritants* affect our flesh both externally and internally. Externally irritants can cause skin complaints such as sensitization and industrial dermatitis. More serious is the internal irritation caused by poisons that can affect the major organs of the body. Such irritants can cause inflammation, ulceration, poisoning and the formation of cancerous tumours.
- *Systemics* affect the fundamental organs and nervous systems of our bodies. They affect the brain, liver, kidneys, lungs and bone marrow. Systemics can lead to chronic and disabling illness and early death.

Narcotics, even in small concentrations, cause drowsiness, disorientation, giddiness and headaches. Such effects seriously affect the judgement of a worker and his or her ability to carry out assigned tasks correctly or to control machines and equipment. In larger concentrations narcotics can produce unconsciousness and death.

Toxic and narcotic substances include degreasing chemicals, solvents for paints and adhesives, liquid petroleum gases, process chemicals used in surface treatments and heavy metals such as lead, cadmium and mercury. They also include the pesticides, herbicides and organophosphates used in the agrochemical manufacturing industries. The storage procedures and regulations laid down for each substance must be rigidly adhered to, these may include storage on pallets, temperature control and limits on the number of containers of a material that can be stacked on top of each other.

Toxic and narcotic liquids should be stored off the ground to aid speedy leak detection and improve the ventilation and dispersion of any dangerous vapours. The appropriate regulations relating to the stored substances should be strictly enforced and inspections carried out regularly and each inspection and its findings logged.

All toxic and narcotic substances should be handled with extreme caution and suitable protective clothing worn as necessary. Containers and packaging should be secure at all times, and clearly indicate their contents. They must be handled with care and be kept in an undamaged condition so that no spillages or leaks occur.

3.7.3 Oxidation

Oxidation is a chemical reaction with the gas oxygen. Oxygen gas is an important constituent of the air we breathe. Metals, plastics and foodstuffs are all subject to deterioration by oxidation on exposure to the atmosphere. On some metals, especially in the presence of heat and moisture, oxygen will cause corrosion, i.e. the surface of the metal becomes oxidized. Plastics may degrade due to the effects of oxygen and sunlight, especially outdoors, unless they are stabilized in some way. The environmental stress cracking (ESC) of some polymers is liable to occur when in the presence of moisture or immersed in liquids.

Food will spoil due to increasing bacteriological action in the presence of oxygen at room temperature over a period of time. It is therefore very important that in all the above cases storage must be by such means as to reduce the effects of oxygen to a minimum.

3.7.4 Flammability

Flammability is the ability to combust or burn. Burning is a very rapid oxidation accompanied by the generation of heat. Flammable materials should be stored in cool, dry, well-ventilated and well-drained accommodation in which all statutory regulations concerning such materials are rigidly enforced. Flammable materials include the following:

- gases
- paints
- fuels
- solvents
- chemicals
- plastics
- timber
- fabrics
- paper
- explosives
- lubricants.

All these substances must be stored and handled with great care ensuring that there must be no risk of sparks or naked flames in their vicinity. It is mandatory that they are all kept in non-smoking areas away from direct sunlight. Many of these substances are subject to local authority regulations and also they come within the jurisdiction of the local fire prevention officer. Special insurance requirements may also be enforced.

3.7.5 Perishability

Perishability is the property of a material to perish, spoil or rot over a period of time and this applies mainly to foodstuffs and rubber goods such as tyres, gloves and aprons. Foodstuffs can be kept fresh and deterioration retarded by using a combination of the correct storage temperature and packaging materials. The packaging materials will protect the product from physical damage and keep within its own micro-environment. The correct storage temperature will maintain freshness over the designed shelf life (the period of time for which the product can be displayed before being purchased). Table 3.6 gives the industry recommended temperatures for the various types of food storage area.

3.7.6 Contamination

Contamination is the inadvertent mixing due to contact, of one substance with another. Raw foodstuffs are particularly prone to contamination from each other. This contamination may simply be strong smelling and strong flavoured foodstuffs affecting the taste and smell of more delicately flavoured foodstuffs by contact or close storage, for example, meat, fish, cheese, cut garlic and onions. More seriously is the risk of infection when foodstuffs such as raw meats are in contact with cooked meats and other foods not requiring cooking.

Table 3.6 Storage temperatures

Environment	Temperature range
Dry food store	10°C to 15°C
Cold stores	3°C to 4°C
Refrigerators	−1°C to 4°C
Freezers	−20°C to −18°C

Liquids, powders and granules are also prone to contamination and must be stored and handled with great care prior to packaging. Detailed attention must be paid to their method of containment for sale. Compression moulding powders are often pressed into a pre-form or pellet prior to moulding to avoid the risk of contamination, to aid handling and to ensure correct filling of the mould.

3.7.7 Hygiene

Hygiene is the maintenance of cleanliness of persons and equipment and is of prime importance when handling and storing food, medicines, and pharmaceuticals. In all instances personal hygiene is particularly important. Food processing machinery must be capable of being cleaned and serviced relatively easily. Suitable lighting and ventilation is also a factor in maintaining hygiene.

Food processing operatives and handlers must keep every part of their person that comes into contact with food, clean. Cuts and sores must be covered with a waterproof dressing or non-permeable gloves, in the case of hands. A more responsible approach is to redeploy a person with a cut, sore or infection until it has healed. Clean and regularly laundered industrial clothing suitable for the task should also be worn.

Any infectious or contagious illness contracted by persons involved in the preparation and production of processed foods will necessitate their being taken off production until a doctor has certified that they have the necessary clearance to return.

3.7.8 Sharpness

Sharp or rough edged and abrasive materials should be handled with caution and suitable protective clothing or equipment must be worn when handling them. Materials in this category include:

- paper
- cardboard
- timber
- aggregates
- stone
- laminates
- glass
- metals
- gravel
- plastics
- bone.

A selection of suitable gloves and 'palms' for hand protection are considered in Section 3.11.4.

3.7.9 Discolouration

Discolouration is a change in colour or pigmentation due to the effects of natural light, in particular that of the ultraviolet (UV) and infra-red (IR) wavelengths. It can also be caused by long term immersion in liquids such as detergents, water, brine solutions and seawater. Reaction with atmospheric oxygen can also cause discolouration. Materials subject to fading or discolouration such as plastics products, stained or painted surfaces, fabrics and timber products should not be stored outside where they would be subject to weathering and direct sunlight. Such conditions cause loss of colour due to changes in the pigmentation (fading).

Products such as wall coverings, carpets and furniture (indoor products) will also fade and discolour when exposed to sunlight. This can be avoided by covering any windows with a thin film of tinted plastic sheet to reduce the intensity of light and filter out the harmful rays. Most commercial and industrial stock is stored in artificially lit storerooms. Some products such as earthenware pipes for rainwater (stormwater) and sewage are stored outdoors, in the open and are not subject to fading or discolouration because these products have become light stabilized during manufacture. They are, however, susceptible to frost damage.

3.7.10 Shelf life

Some products can be affected by one or more of the factors previously discussed. This makes it necessary for a range of precautions to be taken. In addition to controlled storage conditions, the length of time in storage is also important. Many products, and particularly foodstuffs, may slowly deteriorate however carefully they are handled and stored. Therefore their storage time must be limited to ensure they reach the customer in prime condition. This time is called the product's *shelf life*. The 'sell by date' should be clearly marked on the product and/or its packaging. This does not necessarily mean the products have become unsafe. In the case of preserved foodstuffs it may only mean that the colour, texture and flavour are below standard. Biscuits and crisps may start to become soft after their 'sell by date' has passed. Products such as batteries may have a shortened service life if they are old stock.

Test your knowledge 3.10

1. List the main precautions that should be taken to ensure the safe and effective storage and handling of materials.
2. List the main precautions that should be taken to ensure the safe handling of materials.

3.8 Combining and assembling components

Assembly is the bringing and joining together of the necessary parts and subassemblies to make a finished product. Sometimes the parts and subassemblies will have been finished before assembly and sometimes after assembly. For instance the body-shell, doors, boot lid and bonnet of a car will have been painted before final assembly. On the other hand, the sheet metal parts and rivets that make up a farm feeding trough will be hot dip galvanized after assembly in order to seal the joints and make them fluid tight as well as sealing the raw edges of the metal against corrosion. There are three major categories of joining processes used in assembly:

- Mechanical
- Thermal
- Chemical.

3.8.1 Semi-permanent fastenings

Also called *temporary fastenings*, these included nuts, bolts, screws, locking washers, retaining rings and circlips, and are amongst the most common

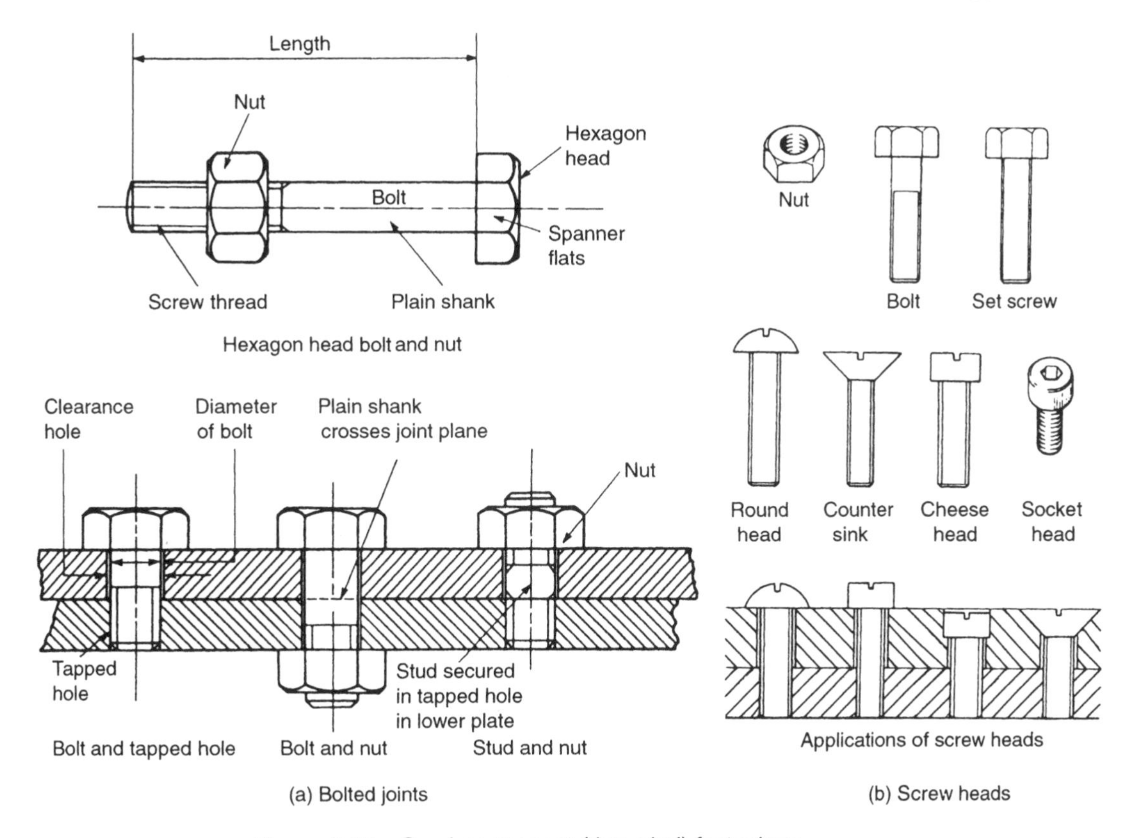

Figure 3.34 Semi-permanent (threaded) fastenings.

method of joining materials. They are used where a joint has to be dismantled and remain from time to time, for example an inspection cover. When one side of the work is blind, that is inaccessible, or when dissimilar materials are to be joined, self-tapping screws are often used. This type of screw is used to secure such materials as sheet metal, plastics, fabrics, composite materials and leather. Examples of semi-permanent (threaded) fastening are shown in Fig. 3.34.

Since semi-permanent (temporary) fastenings can be taken apart, they are liable to work loose in service. For critical assemblies such as the controls of vehicles and aircraft this could be disastrous. Therefore locking devices are used to prevent this from happening. Some examples are shown in Fig. 3.35.

3.8.2 Permanent fastenings (mechanical)

Permanent fastenings are those in which one or more of the components involved have to be destroyed to separate the joint. For example, the ring gear shrunk on to the flywheel of a motor car engine has to be split before it can be removed from the flywheel when it needs replacing. At the other end of the scale the cotton or thread used to sew fabrics together has to be destroyed when a seam is unpicked to make alterations or repairs to a garment.

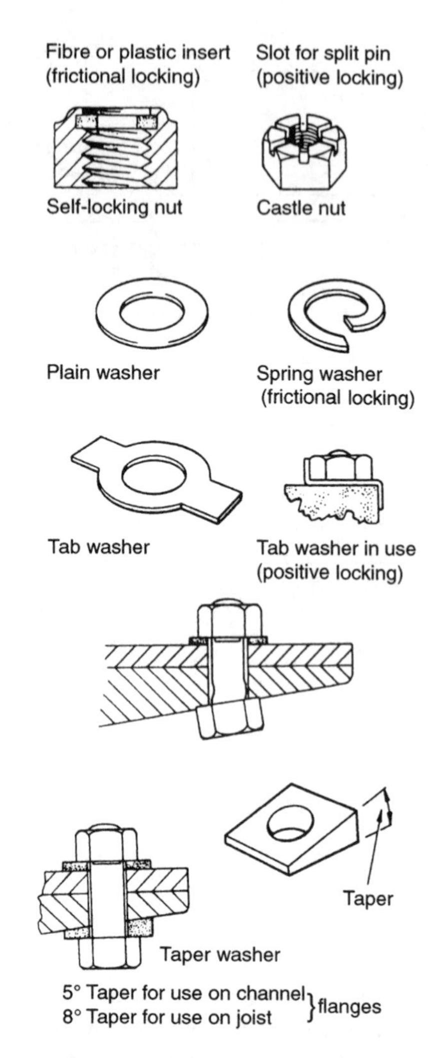

Figure 3.35 Washers and locking devices for threaded fasteners.

Riveting

Rivets are made from malleable materials and depend on deformation to fasten and hold components in place. Rivets and riveted joints have already been introduced in Section 3.5.3. They may be closed either in the red-hot or cold condition depending on the materials to be joined, their thickness and the rivet material. Some techniques for making riveted joints are shown in Fig. 3.36, and some examples of riveted joints are shown in Fig. 3.37.

Compression joints

Compression joints rely upon the elastic properties of the materials being joined. No additional fastenings, such as screws or rivets are required. In principle, a mechanical compression joint, as shown in Fig. 3.38(a), consists of a slightly oversize peg being forced into a component with a slightly undersize hole. This results in the peg being compressed and the metal surrounding

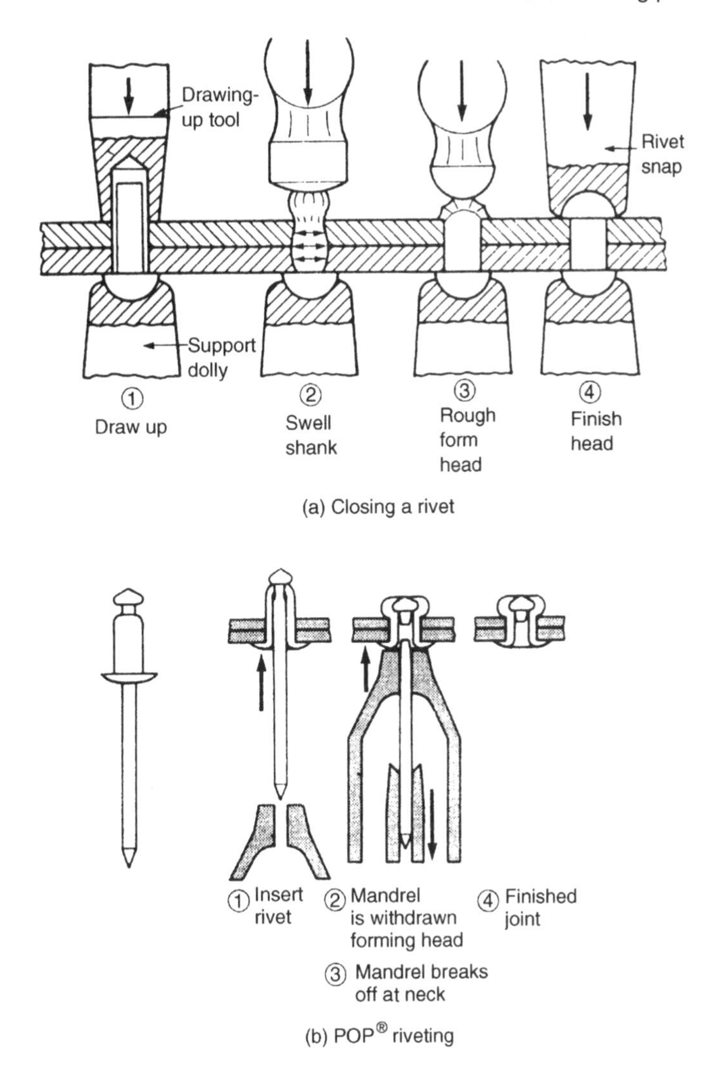

Figure 3.36 Riveting processes.

the hole being stretched. Since the material is elastic, the spring back in the material will grip the peg tightly in the hole where it will be held by friction.

There is a limit to the degree of compression that can be obtained purely by mechanical insertion of the peg into the hole. Where an even greater degree of grip is required two other techniques can be used. In a *hot-shrunk compression joint* the outer component is heated up so that it expands. This enables a much greater difference in size to be accommodated. On cooling, the degree of interference between the components causes one to bite into the other resulting in very considerable grip. The principle of this process is shown in Fig. 3.38(b).

Unfortunately, when the outer component is heated sufficiently for the required degree of expansion, the temperature may be high enough to alter the properties of the material. The alternative is to use a *cold expansion joint* in which the inner component is refrigerated in liquid nitrogen to cause

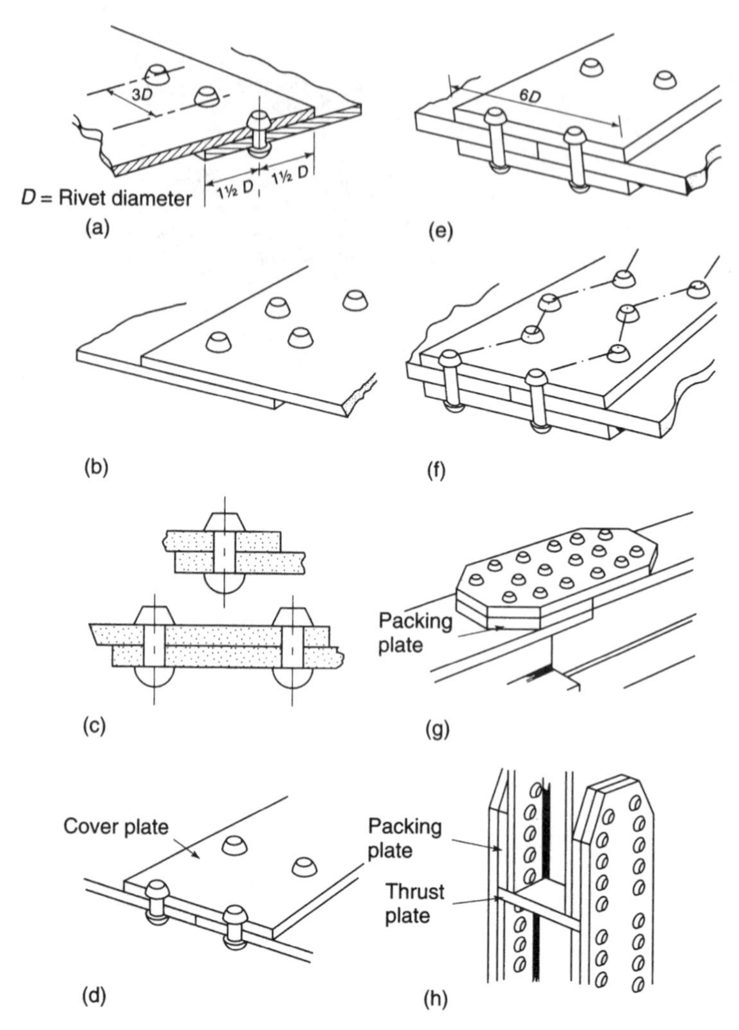

Figure 3.37 Typical riveted joints: (a) single-riveted lap joint, (b) double-riveted lap joint, (c) assembly of lap joints, (d) single-cover-plate butt joint, (e) double-cover-plate butt joint, (f) double-riveted, double-cover-plate butt joint, zigzag formation, (g) splice joint (horizontal), (h) splice joint (vertical).

it to shrink. On warming to room temperature after assembly, the inner component expands and a compression (expansion) joint is again formed as shown in Fig. 3.38(c). Liquid nitrogen can be difficult and dangerous to store and use, so this latter process is only suitable for factory controlled manufacturing processes. Since this process does not alter the properties of the component material it is used to fit the starter ring gear of a motor vehicle engine on to the flywheel.

Test your knowledge 3.11

1. State the essential difference between permanent and temporary (semi-permanent) mechanical joints and name an example of each type of joint.
2. Compare the main advantages and disadvantages of mechanical, hot, and cold compression joints.

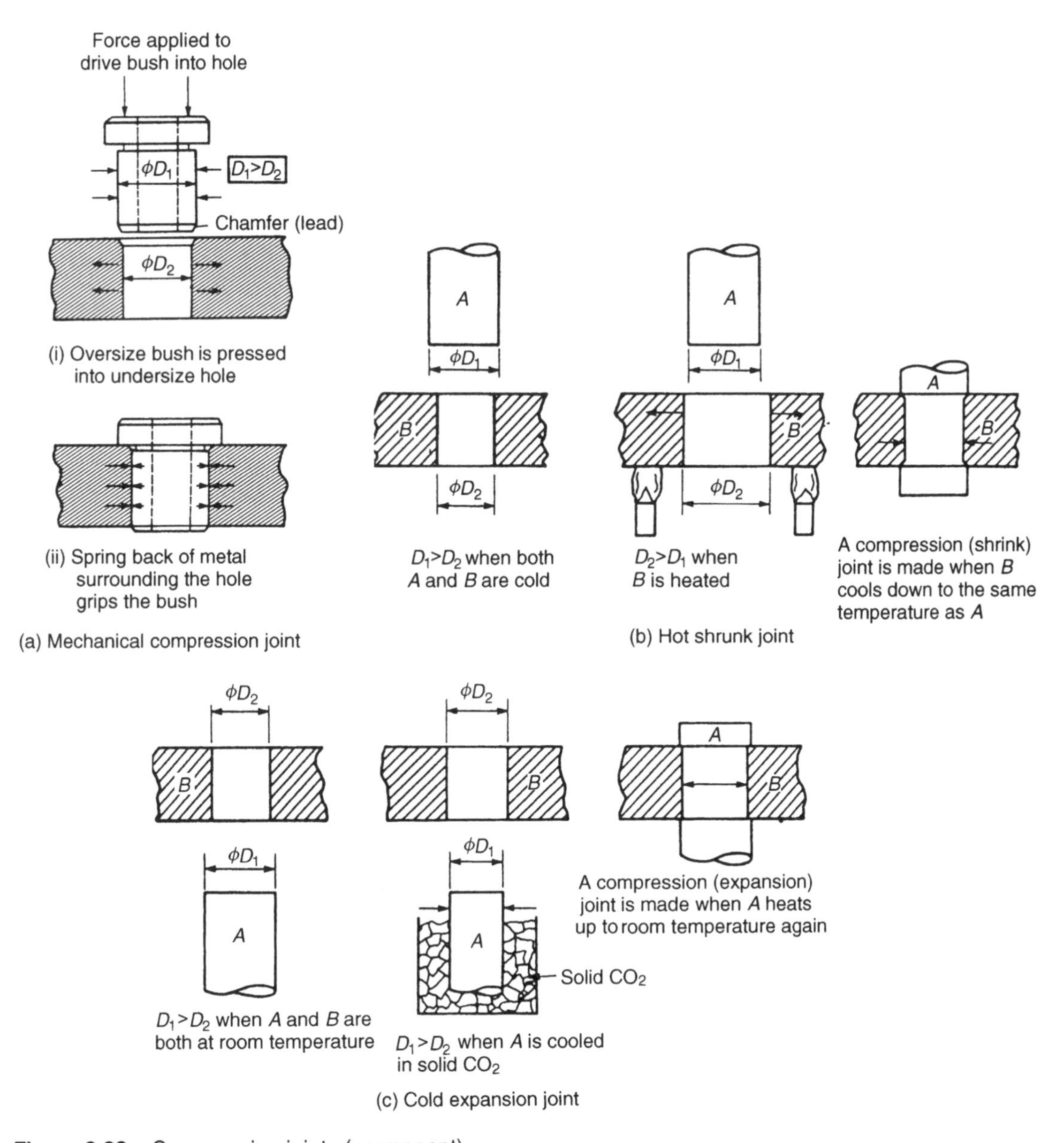

Figure 3.38 Compression joints (permanent).

3. One component of a compression joint is made from cast iron, the other is made from steel. State which material would be used for the inner component and which material would be used for the outer component. Give reasons for your choice.
4. Explain briefly why locking devices are sometimes used in conjunction with threaded fasteners and give an example of a typical application.

3.8.3 Permanent fastenings (thermal)

The main methods of thermal joining are shown in Fig. 3.39.

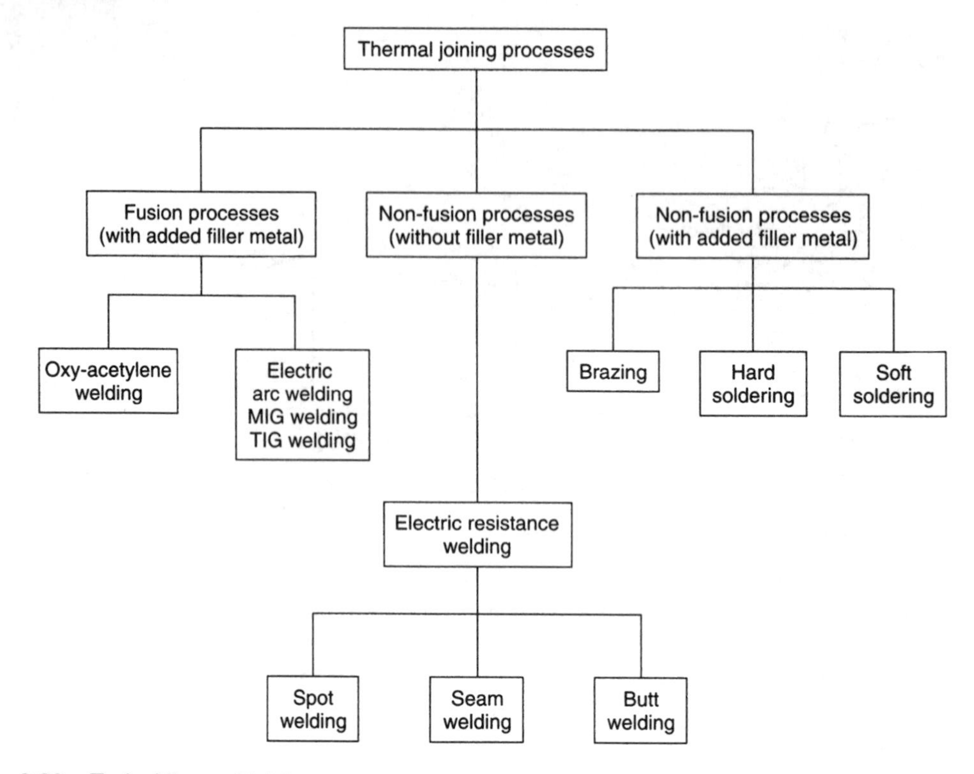

Figure 3.39 Typical thermal joining processes.

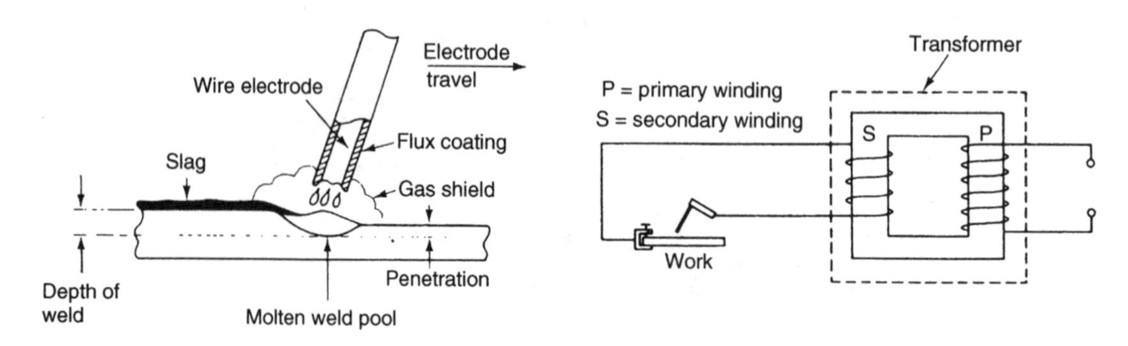

Figure 3.40 Electric arc welding.

Electric arc welding

Electric arc welding is one of the fusion welding processes. As in all fusion welding processes the filler rod (gas welding) and the electrode (arc welding) together with the edges of the parent metal being joined are all *melted and fuse together* to complete the joint.

The principle of the *electric arc-welding process* is shown in Fig. 3.40. In this process the heat required to melt the electrode is generated by an electric arc struck between a flux coated consumable electrode, which also provides the filler metal, and the workpiece (the parent metal). High temperatures in the range of 4000 to 6000°C are generated by the electric arc. An electric arc is and elongated and continuous spark. At such high temperatures the metal melts almost instantly in the weld pool. The flux coating of the electrode also melts and some vaporizes. The combination of molten flux and flux vapour protects the weld from atmospheric contamination. On cooling the flux forms a coating over the weld. Arc-welding is mainly used to weld carbon and alloy steels in the fabrication, civil engineering and shipbuilding industries. A heavy current at a low voltage is used for welding. This is supplied by a special welding transformer that has an adjustable output to suit the material and thickness of the workpiece.

Oxy-acetylene (gas) welding

Oxy-acetylene welding is a process in which the heat for welding is produced by burning approximately equal amounts of oxygen and acetylene. The gas is stored in cylinders and is fed via gas regulating valves, to a blowtorch. The welding flame produces a temperature of approximately 3500°C and this is sufficient to melt the filler rod, and the edges of the metals to be joined. Oxy-acetylene welding is a manual process widely used in the engineering, fabrication and automobile industries on metals such as carbon and alloy steels, cast iron, aluminium and aluminium alloys, brasses and bronzes. The gas is very expensive compared with electricity and the heat energy available is limited. Therefore gas welding is limited to sheet metal and thin plate work. The equipment and principle of oxy-acetylene welding is shown in Fig. 3.41. A flux is rarely needed when oxy-acetylene welding since *the products of combustion* of the oxy-acetylene flame form a blanket over the weld pool and prevent atmospheric oxidation.

Brazing

Brazing, together with hard (silver) and soft soldering is a non-fusion thermal joining processes. That is, only the filler material becomes molten. The materials being joined remain solid but must be raised to a temperature where a reaction takes place bonding the filler material to them.

In the brazing process the filler material is a non-ferrous alloy such as brass (hence the name of the process). Brass filler materials are often referred to as *spelter*. The heat source is a gas torch burning a mixture of oxygen and a fuel gas such as acetylene or propane. The melting temperature of the filler material is above 700°C, but less than the metals to be joined. In all brazing and soldering processes cleanliness of the joint materials is essential. A chemical flux is used, the type being dependent upon the materials to be joined. The entire joint area must be fluxed prior to the application of heat. Once the desired temperature has been reached, the filler rod melts and flows, being drawn between the metals to be joined by capillary attraction to form a strong, durable joint. One advantage of this process is that dissimilar metal may be joined, for example steel tubes to malleable iron fittings when making cycle frames. The process is widely used on a wide and varied range of work, on metals such as carbon steel, copper, cast iron and aluminium.

Soldering (hard)

Hard soldering is similar to brazing except that the filler material contains the precious metal silver. For this reason the process is often referred to as

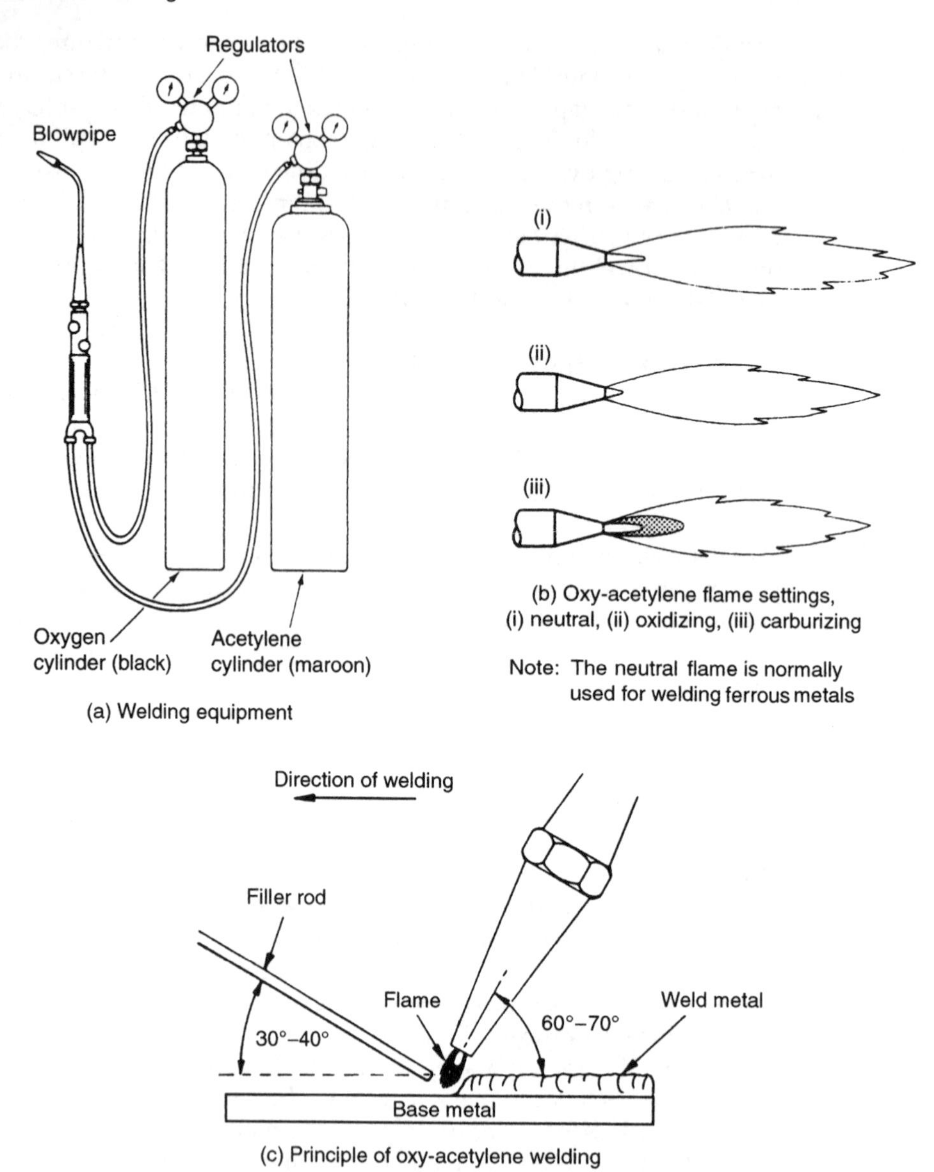

Figure 3.41 Oxy-acetylene welding (gas welding).

silver soldering. Hard soldering is carried out at temperatures above 600°C, the heat for the process being provided by a welding torch or blowtorch depending on the size and intricacy of the component to be joined. Since the solder contains a high percentage of the precious metal silver, it is relatively expensive so that it can be only used economically for the finest work. It produces a neat, strong and ductile joint. A flux has to be used, and its composition depends upon the composition of the solder being used. A suitable flux is supplied by the manufacturer of the solder. The materials that can be joined are steel, copper, tin bronze and brass, and some precious metal alloys in similar or dissimilar combinations on light fabrications and jewellery.

The solder can be supplied with melting points in 5°C steps to allow a number of components to be sequence soldered on to the same assembly.

Soldering (soft)

The soft soldering process is carried out at temperatures below 250°C. The solder is essentially an alloy of tin and lead. The heat for the process is provided by a gas or electrically heated soldering iron in the manual process. The solder is supplied in the form of a stick or a wire. For production and automated welding some form of hot plate, resistance heating, oven or induction heating is used, the solder being in the form of a paste, foil or fine wire preform. For electrical and electronic wiring, the solder is often hollow and has a flux core.

As for all brazing and soldering processes, the area to be joined must be physically and chemically clean. This is why the joint must be fluxed prior to soldering, and the iron cleaned, fluxed and tinned before use. Soldering is used to join light sheet metal fabrications, electrical wiring and decorative goods. Fluxes may be *active*, in which case they chemically clean the joint but leave an acid residue that must be washed off after processing. This is unsuitable for electronic wiring or for food canning where a *passive* flux must be used. This will only prevent oxidation and has no chemical cleaning action. In the case of food canning the flux must also be non-toxic. Figure 3.42 shows a typical manual soft-soldering operation.

Friction welding

Friction welding is the process of joining metals or plastics materials by rotating one part against the other thereby generating heat by friction until the welding temperature is reached. When the welding temperature has been reached the two parts to be joined cease to rotate and are pressed together until the joint area cools and the joint is made. In addition to similar metals the dissimilar metal combinations that can be joined include steel to aluminium, steel to copper and steel to cast iron. PVC pipes can be, and frequently are, joined by friction welding. PVC is the only plastic material that can be welded by this method.

Resistance (spot and seam) welding

This is not a fusion welding process. The welding temperature is slightly below the melting point of the metal and the weld is completed by the application of pressure. It is similar in principle to a blacksmith's forge weld.

In resistance welding processes, an electric current is passed between two pieces of sheet metal at the point where they are clamped together by the two electrodes. The resistance of the metal to the flow of current causes the metal to heat up locally to its pressure welding temperature. At this point the electric current is turned off and the joint is squeezed tightly between the two electrodes and a spot weld is formed. The cycle of operations is controlled automatically.

Spot welding, as the name implies, makes a weld at the point or spot of contact of the electrode with the sheet metal. This is shown in Fig. 3.43(a). Spot welding is widely used in the manufacture of sheet metal fabrications in the automobile industry.

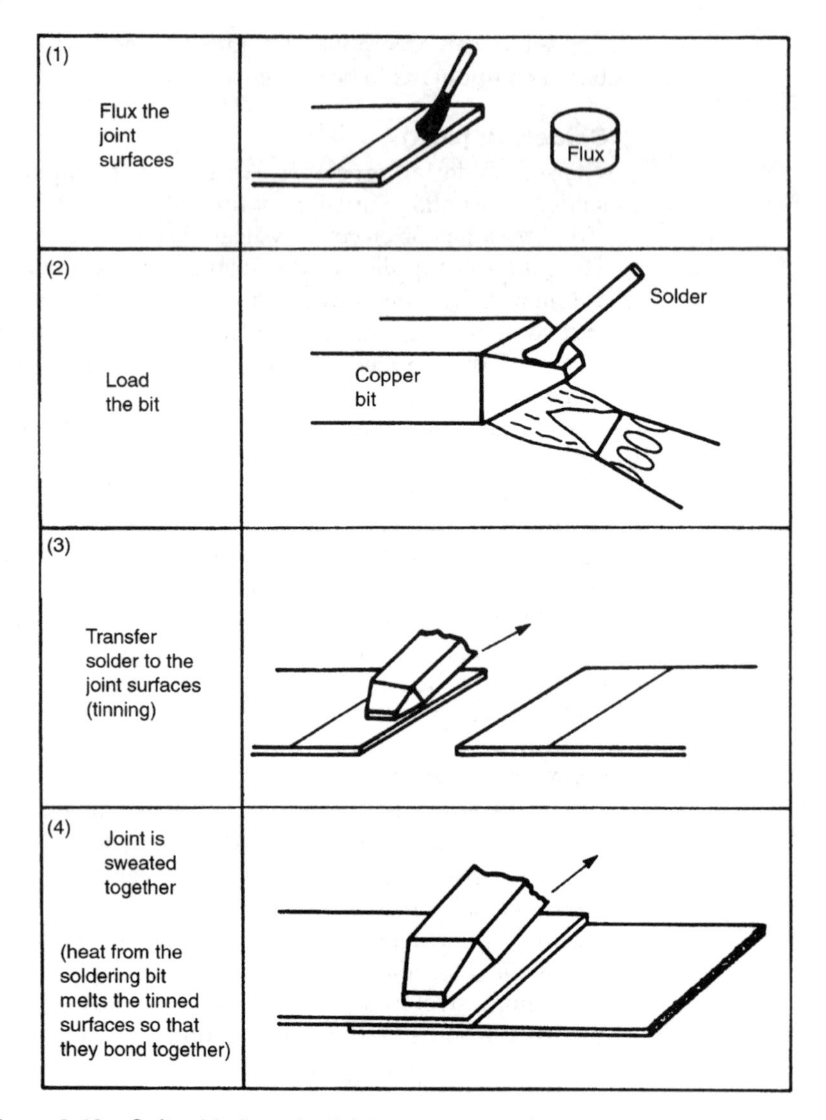

Figure 3.42 Soft soldering a lap joint.

Resistance seam welding employs two power driven wheels as the electrodes, between which the sheet metal to be joined is passed. The wheels press down hard on the components being joined. A pulsed electric current is passed through the wheels as they rotate. This heats the metal, and the pressure of the wheel load makes the weld. In fact the joint consists of a series of overlapping spot welds as shown in Fig. 3.43(b). Resistance seam welding is widely used to make fluid tight joints in the canning industry, in the production of aerosol cans and the production of road vehicle fuel tanks.

Ultrasonic welding

The ultrasonic sound waves used for this process are at too high a pitch for the human ear to hear. When focused, the high-speed vibrations they produce

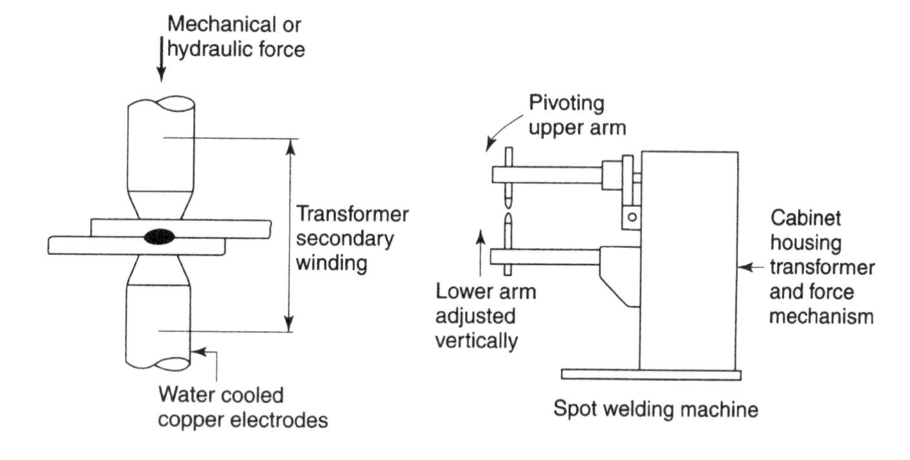

(a) Spot welding

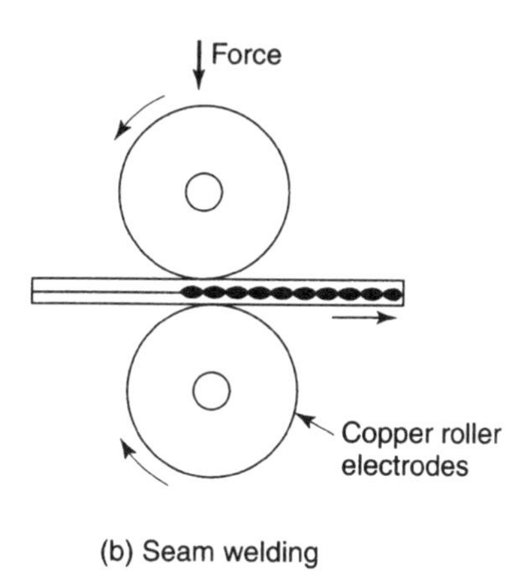

(b) Seam welding

Figure 3.43 Resistance welding processes.

in the materials being joined produce a build-up of heat energy enabling a weld to be made when pressure is applied. The process is similar to spot welding but can be used on plastic materials that will not conduct electricity. Instead of electrodes the work is gripped between the ultrasound resonators.

Test your knowledge 3.12

1. Name the process you would use for the following applications, giving reasons for your choice:

   (a) Welding 25 mm thick steel plate
   (b) Connecting electronic components to a circuit board
   (c) Building a copper boiler for a model steam engine
   (d) Assembling small sheet steel components together
   (e) Welding sheet metal ducting on site.

2. Explain why ultrasonic welding is used in preference to spot weld when joining plastic materials.

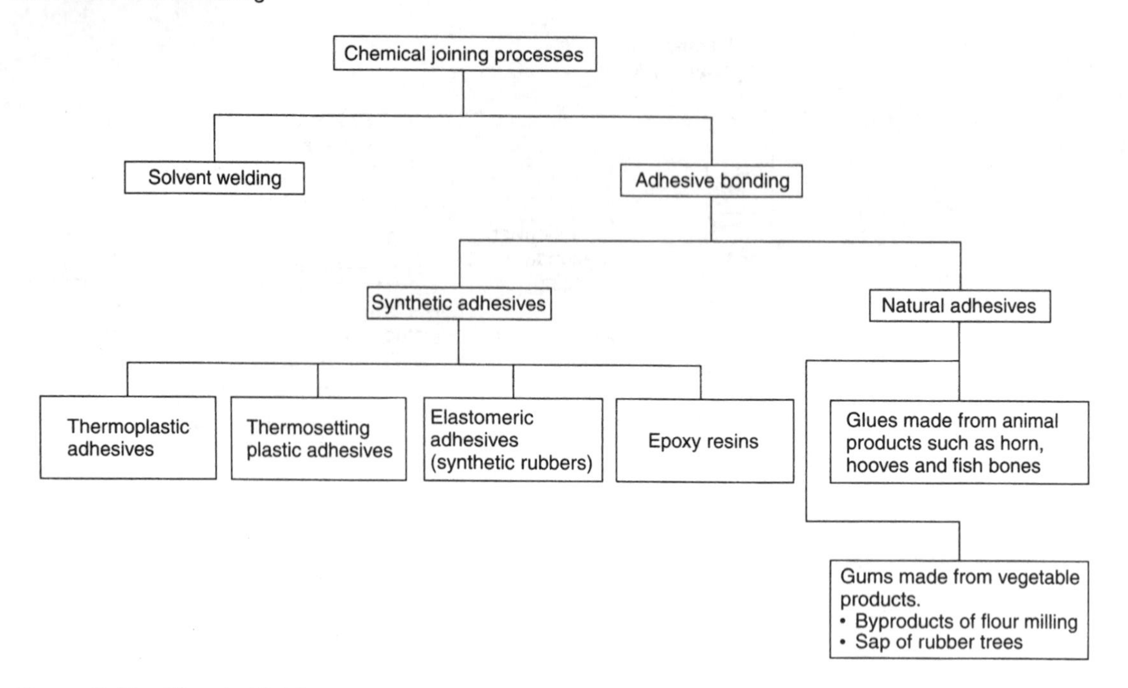

Figure 3.44 Chemical joining processes.

3.8.4 Chemical joining processes

The relationship between some joining processes depending upon chemical reactions is shown in Fig. 3.44.

Solvent welding

Solvent welding is the process of joining certain thermoplastics by the application of a solvent to the surfaces to be joined. Polystyrene and polymethylmethacrylate (acrylic) are the two most widely used materials in this process. There are many solvents that can be used for solvent welding and also numerous proprietary solvent adhesives available. The solvent is applied by means of a brush or pad, the two faces being brought together immediately after application and allowed to dry for a sufficient period of time to ensure adequate joint strength. Solvent welding cements are in this group. For example the balsa cement used in the building of model aeroplanes. This cement consists of cellulose material dissolved in an acetone solvent. It has 'gap-filling' properties and sets by the evaporation of the solvent.

Adhesive (chemical) jointing

Adhesive jointing is used extensively in industry today. There are many different types of adhesives available and they come in many different forms. It is essential to choose the correct adhesive for each application and the manufacturers' literature should be consulted. No matter what the application or the type of adhesive used it is essential that the surfaces being joined are clean and that the humidity and temperature of the workshop is correct.

Adhesives may be applied by brush, tube, gun and dipping. The main types of adhesives available are:

- Natural
- Thermoplastic
- Thermosetting plastic
- Elastomeric (rubber based).

Natural adhesives

Natural adhesives are of vegetable or animal origin and the simplest in common use. Gums are derived from vegetable matter and include starches, dextrin and latex products. Their use is generally limited to paper, card, and foils. Being non-toxic they are safe for use in the food industry for labelling and packaging. Being non-toxic they are suitable for products where the adhesive is licked, e.g. envelope flaps and stamps. Latex adhesives are used for self-sealing envelopes. Animal glues that are produced from hide, bone, horn and hooves have higher joint strength and were traditionally mainly used in woodworking. They had to be softened by heating over boiling water and set again on cooling (a natural form of thermoplastic). Natural glues have been almost entirely superseded by synthetic glues that are stronger and less affected by their service environment.

Thermoplastic adhesives

These adhesives soften when heated and are subject to creep when under stress, but have good resistance to moisture and biodeterioration. They are used for low loaded and stressed assemblies of wood, metal and plastics.

Thermosetting plastic adhesives

Thermosetting adhesives cure through the action of heat and/or chemical reaction. Unlike thermoplastic adhesives they cannot be softened. These 'two-pack' adhesives consist of the resin and the hardener that have to be kept separated until they are mixed immediately before use. Once mixed the 'curing' reaction commences and there is only a limited time during which the adhesive can be worked. Such adhesives provide high strength joints that are creep resistant, have a high-peel strength. However, they tend to be brittle and have low impact strength. Therefore they are unsuitable for joining flexible materials. Epoxy adhesives are in this group.

Elastomeric adhesives

These types of adhesive are based on natural and synthetic rubbers. In general these have relatively low strengths but high flexibility and are used to bond paper, rubber and fabrics. Neoprene-based adhesives are superior to other rubber adhesives in terms of rapid bonding, strength and heat resistance. Impact or contact adhesives are in this group.

Test your knowledge 3.13

1. Name the type of adhesive you would use for the following applications giving the reasons for your choice:

   (a) The application of 'formica' sheet to the chipboard top of kitchen units
   (b) Assembling the polystyrene components of a model aeroplane kit

(c) Labelling food products
(d) Assembling a coffee table
(e) Resoling a shoe.

2. Describe the essential difference between solvent welding and using an adhesive.

3.9 Assembly of components and subassemblies to specification and quality standards

Components and subassemblies must be assembled to the given specification and quality standards to ensure customer satisfaction and confidence in the product. For a simple assembly the general arrangement (GA) drawing is adequate. It shows the relationship between all the component parts, the methods of joining them together and the number of parts required. It does not show the order in which the parts should be assembled.

For more complex assemblies and when moving parts or mechanisms are involved, more detailed and comprehensive assembly drawings are required in addition to the GA drawing. Assembly drawings often incorporate exploded views that indicate the relative positions of the various components and the order in which they should be assembled. In addition to such drawings a schedule of the type and quantities of the components required would also be issued together with a schedule of any assembly jigs and/or special tools that are required.

The process of assembly is the fitting of component parts together to produce either a complete product or a subassembly of a larger and more complex assembly as shown in Fig. 3.45. A typical example of this process is the production of a motor car in which the subframe, engine and transmission, steering mechanism, body-shell, doors, and seats are all subassemblies that, on final assembly, result in the finished car.

The type of assembly process will depend upon the type of product and the number being produced. For example motor cars are built in very large

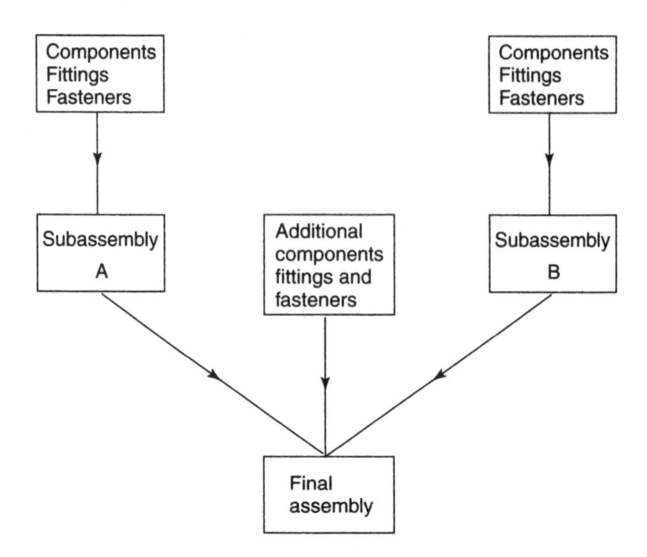

Figure 3.45 Assembly of components and subassemblies.

numbers on production lines involving the use of industrial robots for some assembly processes and assembly line operatives where only human skills can produce the required result. On the other hand ships and bridges are built as one-off projects almost entirely by hand, except for the use of mechanical lifting and handling aids because of the size and weight of the parts.

A bottling plant or a food canning plant can also use automated conveyor line assembly techniques because of the volume of products made and the repetitious nature of the process. A clothing factory is more likely to work on a batch basis. The ready cut cloth is stacked by the machinists who make up the garments and let them fall into baskets as they are completed ready to be wheeled away to the next workstation.

No matter what method of assembly is used, the quality standards set out in the specification must be achieved. For example, this can involve random sampling in the case of canned foods or bottled drinks. In the case of assemblies with moving parts such as machine tools each machine will not only be rigorously inspected for dimensional accuracy, it will be operated over its full range of feeds and speeds to test for smooth running. Most likely a test piece will be machined to ensure all is well before delivery to the customer.

3.9.1 Finishing methods

Finishing is the final process in the manufacture of a product and as such is very important particularly with regard to product quality. Finishing methods may range from lacquering a polished metal product to wrapping and labelling a food product, and from pressing a shirt prior to packaging to sharpening a knife. The reasons for applying some sort of finish to a product include providing:

- Environmental protection against corrosion and degradation
- Resistance to surface damage by wear, erosion, and mistreatment
- Surface decoration.

In general, product finishes should have the following properties:

- They should be colourfast
- They should provide an even covering of uniform colour and texture
- They should not run, blister or peel
- They should be able to resist all environmental and operating conditions
- They should be resistant to physical damage (e.g. erosion and scratching)
- They should be resistant to chemical and electrochemical damage (e.g. corrosion)
- They should be resistant to staining
- They should be easy to keep clean.

Different finishes are applied to different products to suit their application and market, for example a metal product may be electroplated to improve its corrosion resistance and/or appearance. On the other hand, a fabric may require a finishing treatment to render it waterproof or crease resistant. In some cases it may be necessary to apply more than one finishing process to a product. Furnishing fabrics used for upholstery may need to be finished to render them both stain resistant and flame resistant.

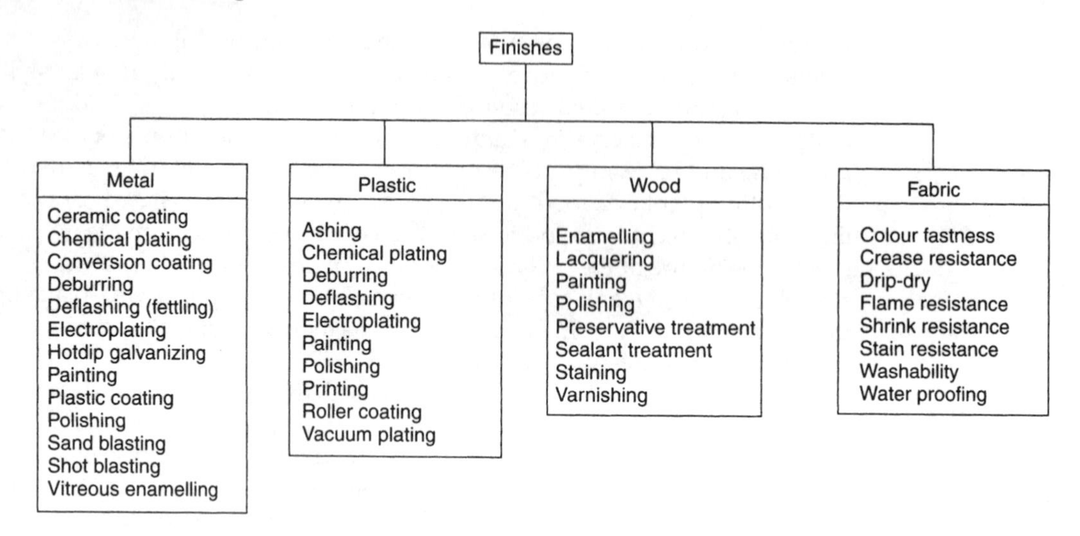

Figure 3.46 Finishes.

The range of finishes used in manufacturing is very wide and as we have already seen the finishes vary according to the product and the materials used. One way to consider finishes is to group them according to the group of materials to which they will be applied. Some examples are shown in Fig. 3.46.

Another way to group finishes is as follows:

- Applied liquids (organic)
- Applied liquids (inorganic)
- Coatings
- Heat treatment processes.

Let's now look at each of these categories in turn.

Applied liquid finishes (organic)

Organic liquid finishes include paints, enamels, lacquers and varnishes, etc. Paint consists of pigments (colourants) suspended in a liquid base that can set hard once it has been applied. Traditionally, linseed oil was used because this natural oil 'cures' and sets hard in the presence of atmospheric oxygen. Nowadays synthetic liquids are used that set quicker and harder, for example polyurethane. A complete paint system usually consists of three coats, a priming coat to aid adhesion and prevent rotting or corrosion, an undercoat to build up the colour and a top coat containing a varnish to seal the system and retard or prevent the absorption of moisture. Organic coatings form a decorative, durable and protective coating on the metal or wood to which they are applied. Paints, lacquers and varnishes may be applied in a variety of ways. Some of the more common used in manufacturing industries are:

- Brushing
- Dipping and stoving
- Spraying
- Roller coating
- Silk screen printing.

Applied liquid finishes (inorganic)

Compared with organic finishes, inorganic finishes are harder, have increased rigidity, and are more resistant to the high temperatures used in cooking. In manufacturing, two of the most widely used groups of inorganic finishes are porcelain enamels and ceramic coatings.

Porcelain and vitreous enamels

These are based on ceramic materials and metallic oxide, metallic sulphide, and metallic carbonate pigments. The metal products are coated with a slurry of these materials and then fired in an kiln. The resulting finish is hard, abrasion resistant and suitable for ovenware. The high gloss surface is easy to keep clean and hygienic, hence its popularity for cooking utensils. Unfortunately these finishes are very brittle compared with ordinary paints and varnishes and are easily chipped.

Ceramic coatings

These are hard refractory coating materials, used to protect metal components from the effects of heat corrosion and particle erosion. They are used in such products as:

- Gas turbine blades
- Rocket motor nozzles
- Chemical processing plant
- Metal processing plant
- Textile processing plant
- Thermal printers for data-processing.

Metallic coating

Metallic coating is the plating of a metal product with another metal that has better corrosion resistant and/or decorative properties, thus providing improved:

- Surface protection
- Wear resistance
- Decoration
- Dimensional control (gauges are often hard-chrome plated to make them more wear resistant. Hard chrome plating is used to build up worn gauges so that they can be reground to size).

Metallic coatings may be applied by a number of processes such as galvanizing and electroplating.

Hot dip galvanizing

In this process, steel components and assemblies are cleaned physically and chemically and immersed in molten zinc to give them a protective coating. A small amount of aluminium is usually added to the melt in order to give the finish an attractive 'bright' appearance. Although the zinc coating is much more resistant to corrosion than steel, nevertheless it will very slowly corrode away in the presence of acid rain and in marine environments. Since in protecting the steel, the zinc slowly corrodes away it is said to be *sacrificial.* To retard this corrosion even further, galvanized goods may also be painted to protect the zinc finish.

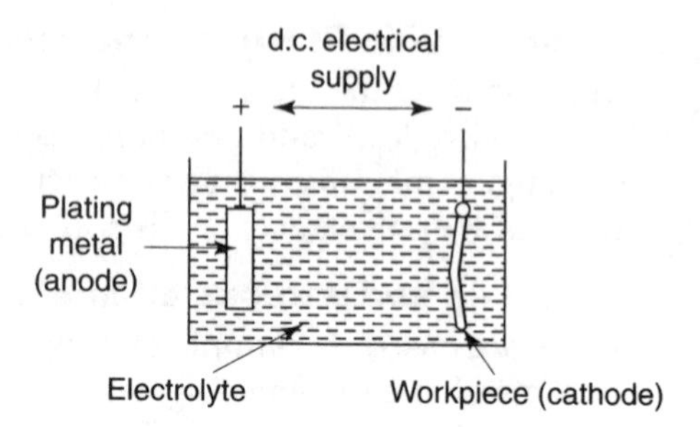

Figure 3.47 Electroplating. Reprinted by permission of W. Bolton.

Electroplating

A thin film of a metal is deposited on the surface of the component to be protected by an electrochemical process as shown in Fig. 3.47. The metal film deposited may be intended to prevent corrosion or to impart a more attractive finish or both. The bright parts of a car body trim or a motorcycle are often nickel plated to provide corrosion resistance and then given a further film of chromium to improve the finish.

Anodizing

Although aluminium and its alloys are corrosion resistant to most environments, components made from these metals become dull and grey in time. To prevent this happening or to give them an attractive colour, aluminium and aluminium components are often *anodized*. In this process, a transparent film of aluminium oxide is deposited on the surface of aluminium or aluminium components. The finish may be self-colour or dyed after treatment. Unlike electroplating where the components to be treated are connected to the negative pole (cathode) of an electrolytic cell, in this process the components to be treated are connected to the positive pole (anode) of an electrolytic cell, hence the name of the finish. The composition of the electrolyte, which in anodizing is a solution of acids, will be formulated according to the finish required.

Plastic coating

Metals may be coated with plastic materials to provide decorative and/or protective finishes. Bathroom towel rails are sometimes finished in this way. Various processes are available but the most widely used are:

- Fluidized bed dipping where the heated components are lowered into plastic powder. The powder is supported on a column of air blown through the porous base of the dipping chamber as shown in Fig. 3.48. This enables a uniform coating to be built up.
- Plastisol dipping, where the work is immersed in a liquid plastisol. No solvents are used and the process provides no toxic hazards to the workers involved.

Conversion coatings

The components to be treated are cleaned physically and chemically and immersed in hot chemical solutions. Complex compounds of metal phosphates,

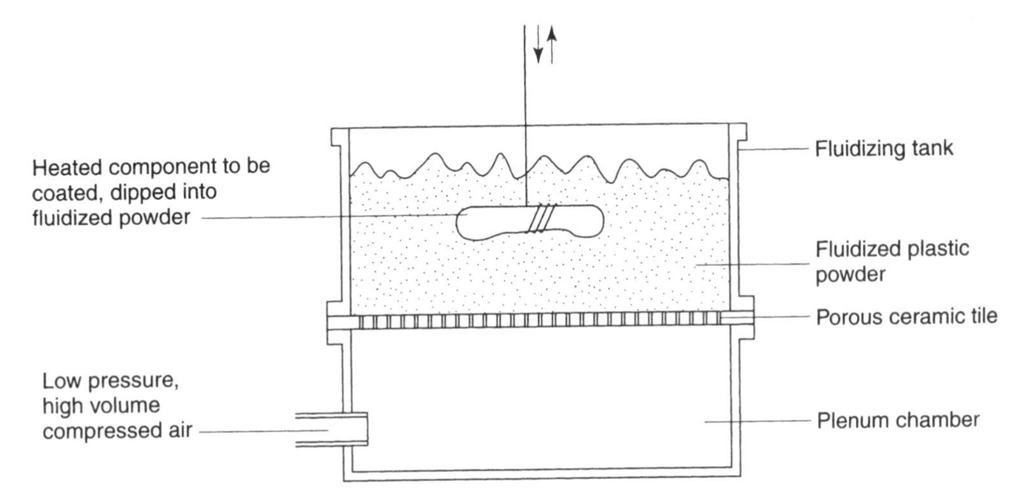

Figure 3.48 Fluidized bed dipping (plastic coating of metal parts).

chromates or oxides are formed by the chemical conversion on the surface of the metal depending upon the solution used. These surface films not only protect the metal from corrosion directly, they also provide an excellent 'key' for any subsequent painting process. One or other of these processes is frequently used in treating car body panels before painting.

Glazing

The term glazing means, literally, coating with glass. In fact, the vitreous finish given to ceramic products such as bone china and pottery is exactly this. The ceramic products are given their first firing and this leaves them with a dull porous finish in self-colour. They are then dipped into a slurry of glazing powders suspended in water. It looks rather like a thin mud. The second firing melts the powders and causes certain chemical reactions to take place. The surface is coated in a glass-like vitreous finish. This *vitreous* finish imparts a colourful, attractive, shiny and impervious finish to these products. Metal products such as roasting tins may also be finished with a *vitreous enamel* to withstand oven temperatures and maintain a smooth, hygienic surface.

The term is also used in many other contexts where a shiny finish is imparted. For example, pastry may be given an attractive brown and shiny finish by applying a beaten egg with a little milk to the surface before cooking. Textiles, papers and cards may also be given a smooth and lustrous finish by treating them with various chemicals.

Laminating

Kitchen unit work-tops are often finished by a process called 'laminating'. The wooden, chipboard panels are finished by bonding a thin plastic sheet to them. The plastic sheet not only gives a pleasing colour to the work-top but the plastic is chosen to be heat resistant, as well as being smooth and easy to keep clean and hygienic.

Buffing (polishing)

Many metal and leather products are finished by polishing them using a buffing process. Rapidly rotating cloth polishing 'mops' or leather 'basils' are impregnated with a polishing compound and the goods to be finished are pressed against them. Metal components are usually polished in this way before electroplating. They are sometimes given an additional polish after plating as well.

Colouration

Dyes and pigments are used to either colour products directly as in the dyeing of fabrics, or are used in surface treatment processes such as painting, lacquering and enamelling. Paints and enamels use pigments and lacquers use dyes.

Dyes are transparent colourants that are soluble. They are derived from natural sources such as insects, berries and wood or from synthetic organic chemicals that are a byproduct of the petrochemical industry. Cloth to be dyed can be treated by immersing it in vats containing solutions of the dyes. The dyed fabrics are often 'fixed' in a second solution to render them colourfast and prevent fading by exposure to sunlight and washing. Alternatively, the fabrics are woven from yarns that have already been dyed before weaving. Knitwear is usually produced from coloured yarns.

Pigments are opaque colourants and, with the exception of carbon black, are derived from natural or synthetic inorganic sources such as metal oxides, metal sulphides or metal carbonates ground to a fine powder. These powdered pigments may be incorporated into plastic materials to provide 'self-colour' mouldings or incorporated into paints and enamels for surface application.

Test your knowledge 3.14

1. For a product of your own choosing, describe the final inspection it should receive to ensure it has met its quality specification.
2. Name the finishing process you would use, giving the reason for your choice, for:
   (a) Rust-proofing a low-carbon sheet steel feeding trough for farm use
   (b) Finishing the exhaust system of a motorcycle
   (c) Colouring a woven cloth
   (d) Providing a china dinner service with a shiny, impervious and colourful finish.
3. Explain why the colourant used in a paint will be a pigment and not a dye
4. Explain the essential difference between electro-plating and anodizing.

Key notes 3.3

- The purpose of assembly is to combine together all the components and subassemblies necessary to make the finished product.
- There are three groups of joining processes used: mechanical, thermal and chemical.
- Mechanical joining techniques include threaded fasteners, keys, rivets and compression joints.
- Thermal joining techniques include soldering, brazing, fusion welding, friction welding, resistance welding and ultrasonic welding.
- In fusion welding the edges of the metals being joined are melted and allowed to run (fuse) together.
- Oxy-acetylene welding uses these gases burnt in a welding torch as the heat source. A separate filler rod is used to add metal to the joint.
- Electric arc-welding uses the heat of an electric arc (continuous, elongated spark) struck between an electrode and the work as the heat source. The temperature is much higher and the amount of heat energy

available is much greater than for gas welding, therefore thicker metals can be joined. The electrode is also the filler material and is surrounded by a flux coating.
- In soldering and brazing only the filler material (solder or brazing spelter) becomes molten. The materials being joined are not melted. A flux is required.
- Chemical joining techniques include chemical welding and adhesive bonding. The adhesive used has to suit the materials being joined and the service conditions.
- Assembly processes are just as much subject to quality specifications as manufacturing processes. However the quality assessment of assemblies is more concerned with finish and performance.
- Finishing processes treat the materials used during manufacture in order to provide decoration, retard corrosion and degradation, and protect the surfaces from wear.
- Fabrics may be self-colour or they may be dyed and printed to improve their appearance. They may be treated to make them waterproof, crease-resistant, flame-resistant, make them drip-dry, colour-fast, and stain resistant.
- Plastic products may be self-colour, painted, vacuum-metallized, electroplated, and polished.
- Wood products may be painted, varnished, lacquered, sealed and creosoted to improve the finish and render them rot resistant.
- Foodstuffs may be finished with icing, icing sugar, bun-wash, glaze, and edible decorations.
- For higher temperature finishes such as for ovenwear, metal products can be vitreous enamelled.
- Metal components such as gas turbine blades are often ceramic coated to make them more resistant to high temperature erosion.
- Metal components may be given protective and decorative coatings by such processes as electroplating, anodizing, galvanizing, etc.
- Metal components may be protected by conversion coatings, for example chromating and phosphating processes, prior to painting.
- Metals, glass and fabrics may be finished by a variety of heat treatment process after manufacture. These may harden, soften or stress relieve the products. In the case of fabrics they may be hot pressed to remove creases and to make pleats more permanent.
- Safety is as important to assembly and finishing processes as it is to manufacturing processes. This is particularly the case when welding is used.
- Welding equipment must never be used except by persons closely supervised whilst under instruction and persons who are fully trained in its use.
- When welding, the appropriate protective clothing must be worn and the appropriate protective equipment used. Gas welding goggles are unsuitable when electric arc-welding and arc-welding masks are unsuitable when gas welding.
- Oxygen and acetylene gases are highly dangerous and must be used and stored with the greatest care and in accordance with the practices recommended by the suppliers and the local authority fire officer.
- Electric arc-welding equipment must be inspected regularly by a qualified electrician.

3.10 Quality and production control

Product quality has been introduced from time to time during the preceding elements. This element brings together many of the factors involved in manufacturing a product that is 'fit for purpose'. Quality assurance is achieved by adopting proven systems and procedures of inspection and record keeping. Typically, a company that is approved to BS EN 9000/1 will provide goods and services whose quality can be assured.

3.10.1 Quality indicators

Quality indicators have already been introduced, and we saw that they could be *variables* or *attributes*. An indication of quality can be obtained by objective measurement (the component is or is not the correct size), or subjective assessment (the taste or feel is satisfactory in the opinion of the inspector). We saw that quality indicators are applied at critical control points during production. These quality indicators essential in attaining the quality specification agreed with the customer.

3.10.2 Frequency of analysis

For 'one-off', prototype and small batch production every product is inspected. This is 100 per cent sampling and inspection. It is also used for key components where a fault could cause a major environmental disaster or the loss of human life. Under these conditions, the high cost of 100 per cent inspection would be small compared with the cost of failure in service. For the large volume production of many items, 100 per cent inspection is too time-consuming and expensive if carried out manually. Either fully automated 100 per cent inspection is built into the production line or statistical sampling is used.

In statistical sampling, which we discussed in Section 2.6.6, samples are taken from each batch and the quality of the batch is determined by analysis of the sample. If the sample satisfies the criteria of the quality specification, then the whole batch is accepted. If the batch does not satisfy the quality specification the following lines of action may be taken:

- The batch is re-inspected to confirm the original findings
- The batch is sold at a reduced price to a less demanding market
- The batch is reworked to make it acceptable
- The batch is scrapped.

Figure 3.49 shows the inspection criteria for our wooden carrying box which was first introduced in Fig. 3.2.

Inspection is a comparative process. The quality indicator is compared with a known standard of acceptable quality.

- Dimensional measurements are compared with dimensional standards such as rulers, micrometers, calipers and gauges. These, in turn, are checked by comparing them with even more precise standards such as slip gauges.
- Paint and dye finishes are compared visually with colour charts.
- Food and drink products are assessed for quality by highly skilled and experienced inspectors. These experts can compare the flavour of the current batch with their own memory.

3.10.3 Defects and their causes

During manufacture defects will have many causes. Let's now look at some of the more common ones.

Operator errors

There are two main categories of operator error:

- Inadvertent errors are caused by carelessness, fatigue and inattention.
- Skill errors occur as the result of poor technique caused by inadequate training and supervision.

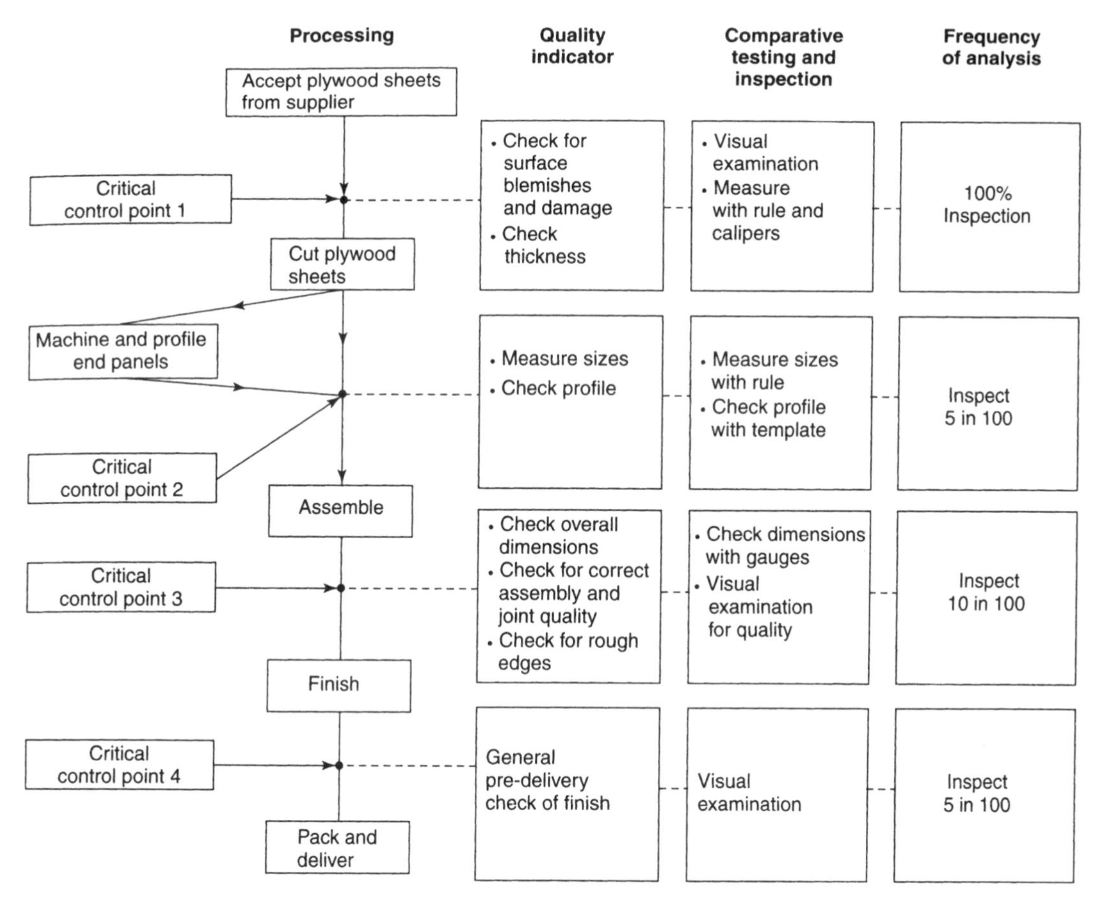

Figure 3.49 Critical control points and inspection criteria for wooden carrying box (Fig. 3.2).

Inattention resulting in inadvertent error can occur at any time and is due to the operator not paying attention to the job in hand. The situation can be improved by 'foolproofing' the process. That is designing the product or the process so that the parts can only be loaded into a machine or assembled the correct way round. This reduces the dependence on human attention. Job rotation within a work team, scheduled rest periods and designing the job to make it more interesting can all be used to improve operator attention and motivation.

Lack of skill or technique errors can occur when an operator lacks the skill or knowledge to perform a task correctly. This can not only result in manufacture of products that do not meet the quality specification, but also can put the operator at risk of injury through the incorrect operation of equipment. This problem can be remedied by giving the operator extra training followed up by regular training reviews.

Process condition changes

Process conditions can change for a variety of reasons. For instance, a machine may be unable to hold its settings because of faulty maintenance. Precision measuring equipment can become inaccurate due to expansion and distortion if the temperature of the inspection area rises too much on a hot summer's day. Automated equipment is often extremely complex and a component failure or a virus in the control computer can cause a malfunction without necessarily causing the equipment to fail or shut down. An example of such a situation is when a soft drinks company has its bottle filling operation automatically controlled. A component failure or a computer virus will cause the process conditions to change and the control mechanism underfills the bottles such that the volume of liquid falls outside the acceptable tolerance limits.

Composition proportion changes

A common cause of the reduced quality of a product is brought about by changes in the composition of a product. For example precast concrete beams are made to a specification in which the proportions of cement, aggregate and water are closely controlled. The proportions may change for a variety of reasons. For example the metering and control equipment may get out of adjustment due to poor maintenance. Alternatively, an unscrupulous manufacturer might deliberately reduce the amount of cement and increase the amount of aggregate beyond the limits laid down to increase the profitability of the process. Such changes would reduce the quality of the product.

Changes of process conditions or changes in composition proportions resulting from the malfunction of equipment must be rectified immediately. Planned, preventative maintenance should be organized to prevent such occurrences happening.

Material substitution

Material substitution in a product can affect product safety. Designers and materials engineers must carry out stringent and extensive testing before any change of material specification is permitted. Material substitution is usually made in an attempt to reduce costs or to tap available sources more easily. If this results in an unauthorized substitution of inferior or substandard materials, this will affect the quality of the product.

Test your knowledge 3.15

1. A batch of 5000 cans are each to be filled with 750 ml ± 2 ml of a soft drink. Table 3.7 gives the results of testing a sample of 50 cans for the volume of the contents.

   (a) A frequency distribution table
   (b) A histogram for the results in the frequency distribution table.

2. Classify the following defects as: critical, major, minor or incidental.

   (a) A blocked jet in a lawn sprinkler
   (b) A knot in the wood of a window frame
   (c) A fatigue crack in a gas turbine blade
   (d) A slight run in the colour of a tea towel.

3. Give examples of your own choosing where the quality of a product would be reduced by:

   (a) A change in the process conditions
   (b) A change in composition proportions
   (c) The substitution of material.

Table 3.7

Can no.	Volume (ml)	Can no.	Volume (ml)	Can no.	Volume (ml)	Can no.	Volume (ml)	Can no.	Volume (ml)
1	749	11	750	21	751	31	751	41	750
2	749	12	749	22	749	32	748	42	749
3	747	13	751	23	751	33	750	43	751
4	750	14	748	24	749	34	752	44	750
5	748	15	750	25	750	35	746	45	750
6	750	16	749	26	752	36	750	46	753
7	750	17	751	27	748	37	752	47	747
8	750	18	747	28	748	38	749	48	754
9	749	19	751	29	750	39	750	49	749
10	751	20	750	30	752	40	753	50	748

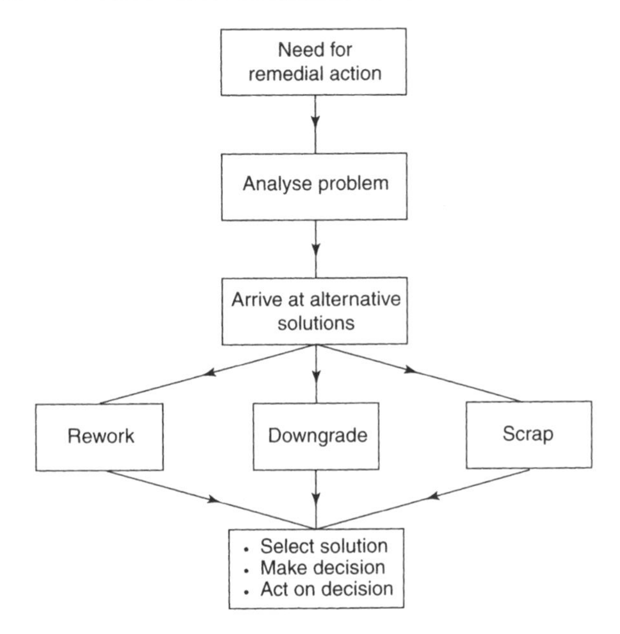

Figure 3.50 Planning remedial action.

3.10.4 Dealing with defective products and components

Defective products are those which are unacceptable because they do not conform to either the product specification or the quality specification. For instance the door seals on a batch of washing machines may be defective causing flooding of the customer's kitchen. The remedial action that needs to be carried out on defective products depends upon the type of defect, the severity of the defect, and the number of defective components involved. Let's now look at the procedures for dealing with defective products and components. Figure 3.50 shows a simple flow chart for planning remedial action.

Investigate cause

A quality deficiency in a component or product should be identified by the quality control system by inspection at the critical control points. If statistical process control is employed any deficiency will show up on the control chart, which will then be analysed by a quality control engineer in order to identify the defect and its probable cause. Once the defect and its cause have been identified as originating in a particular area of production or machine, a production engineer or a machine setter will be given the responsibility for taking remedial action to eliminate the cause of the defect.

The most likely causes of defects in manufactured goods could result from one or more of the following:

- Faulty or unsuitable tools and equipment
- Unsuitable materials
- Untrained or badly trained operators
- Unsuitable production methods
- Inadequate quality control system
- Poor product design.

Recommend solutions

The solutions recommended to eliminate the cause of the defect in the product will, of course, depend upon the product, the nature of the defect, and the sector of the manufacturing industry concerned. It is easier and cheaper to correct a defect in a wheelbarrow than in a high performance military aircraft. Let's now look at some of the solutions available:

- Improve the tools and equipment being used.
- Introduce alternative materials or, possibly, materials of improved quality.
- Procure materials and components from more reliable sources.
- Improve training schedules and methods and retrain underskilled operators.
- Analyse and improve the production methods used so that quality standards can be met without increased costs.
- Introduce efficient quality control systems, or improve the efficiency of existing systems.
- Improve the product design, but avoid overengineering so that it becomes too costly.

Rework

Reworking is the rectification of defects on a product to meet the required specification. This is not always possible. Oversize products can be cut down in size but undersize products cannot always be rectified. If the wrong material has been used no rectification is possible (see Downgrade). Reworking is often expensive and the cost of reworking must be considered carefully against the other alternatives.

Downgrade

A defective product is downgraded if it is usable but does not meet the required product specification, and is sold at a reduced price as 'substandard' or 'seconds'. This is often done with clothing and pottery, where 'seconds' are sold off to the public in the factory shop.

Scrap

Scrap is a defective product that is so far outside the specification that it cannot be reworked, used, or sold, and is only suitable for recycling. The scrap value of the material can be recovered and offset against the loss made on the batch.

Test your knowledge 3.16

1. Decide whether you would rework, downgrade, or scrap the following defective products giving reasons for your decisions.
   (a) A batch of M16 bolts that are tight in M16 nuts
   (b) A batch of wing fixing bolts for a microlight aircraft made from defective material
   (c) A china dinner plate with a fault in the glaze
   (d) A batch of canned food that may have been contaminated
   (e) A batch of neckties that are slightly short
   (f) Curtain material that has faded whilst in store
   (g) A batch of wooden window frames that are slightly oversize.

Key notes 3.4

- Quality indicators are applied at critical control points and may be variables or attributes.
- Frequency of analysis is the sample size taken for inspection.
- Inspection is a comparative process where a feature of the product being manufactured is compared with a known standard that satisfies the quality specification.
- Formats for recording data may be manual, computer generated, tabular, or graphical.
- The recorded inspection data may be presented as a frequency distribution table, a frequency distribution histogram, a cumulative frequency distribution table, or a cumulative frequency curve (an *ogive* curve).
- A *critical defect* can result in a catastrophic failure of a component or assembly resulting in loss of life or a major environmental disaster.
- A *major defect* will prevent the product operating correctly but will not involve injury or damage.
- A *minor defect* will have little or no effect on the performance of the product but will have lowered its quality so that it has to be disposed of as second grade at a reduced price.
- An *incidental defect* is so trivial that it will have no effect on the performance, appearance, or selling price of the product.
- Operator errors occur due to fatigue, boredom, or lack of training.
- Condition changes occur when equipment is unable to maintain its settings due to wear, inadequate maintenance, or computer viruses.
- Composition proportions occur when materials are being mixed or blended. The proportions may be changed accidentally or deliberately to increase the profitability of the product. In both cases the quality of the product is likely to suffer.
- Substitution of cheaper materials can be made to advantage. This should not be done without the consent of the designer, the materials engineer and the customer, as it may lead to a reduction in quality.
- Defective products may be *reworked* if this is economical or can be done without lowering the quality of the product. Defective clothing may be cut down and reworked as a size smaller providing the cost does not exceed the value of the cloth saved.
- Defective products may be *downgraded* if there is a ready market for the lower quality product. Second grade pottery and china is often sold off to the public in factory shops.
- *Scrap* is the name given to defective products that can neither be reworked nor downgraded. It is also the name given to the waste material left over when a shaped blank is cut from a standard rectangular sheet of the material.

3.11 Safety equipment, health and safety procedures and systems

It is said that safety is 'everyone's business'. This is particularly true in all fields of manufacturing. It is most important that safety equipment, procedures, and systems are inspected and checked regularly to ensure that they are in place and functioning whenever and wherever a manufacturing process takes place.

3.11.1 Risk assessment

Whenever a new process is introduced or a new material is to be used a *risk assessment* must be carried out. For example if a new range of shirts is to be manufactured then, if they are to be cut out using a power driven rotary shear this would be classified as a high-risk operation. On the other hand, sewing on the buttons or packing the individual products would be classified as a low-risk operation. The substitution of a synthetic adhesive that is potentially toxic for natural glue in the manufacture of furniture products increases the level of risk.

Whenever a high-risk operation or a high-risk material is introduced or whenever the level of risk is increased in any way, adequate safety precautions must be taken to safeguard the workforce and the customers who will use the product. The workforce must be fully trained in the safe use of the new equipment or materials, the equipment itself must be fully guarded and undergo regular safety inspections. The ultimate aim must always be to phase out high-risk equipment and operations for ones that are of lower risk wherever possible.

3.11.2 Emergency equipment (fire)

Emergency equipment may include fire extinguishers, hose reels, axes, fire blankets, and sprinklers. Examples of these devices are shown in Fig. 3.51 and will be described in greater detail below. It is essential that such equipment is checked regularly and kept in good condition. Such checks must be logged in a register. If an extinguisher is used, it must be recharged without delay as soon as the fire has been put out. A fire certificate may also be required for premises depending on the type of business and the number of people employed.

For combustion to commence and to continue three things are necessary: *heat*, *oxygen* (air) and *fuel*. If any one of these is removed the fire will go out. The purpose of fire extinguishers is to prevent air getting to the burning materials (fuel) and also to cool them down. They may be adequate for putting out small, localized fires or containing fires until the professional brigade can arrive. Remember that human lives are more important than property. If you are using an extinguisher always make sure you have an escape route in case the fire spreads and gets out of control.

Fire extinguishers are colour coded to indicate their contents and the class of fire on which they are to be used as listed in Table 3.8. The classes of fire are as follows:

- Class A. Fires involving *solid materials*, usually of organic nature, in which combustion normally takes place with the formation of glowing embers, e.g. wood, paper and textiles, etc
- Class B. Fires involving *liquids* such as oil, fat, paint, etc., and liquifiable solids

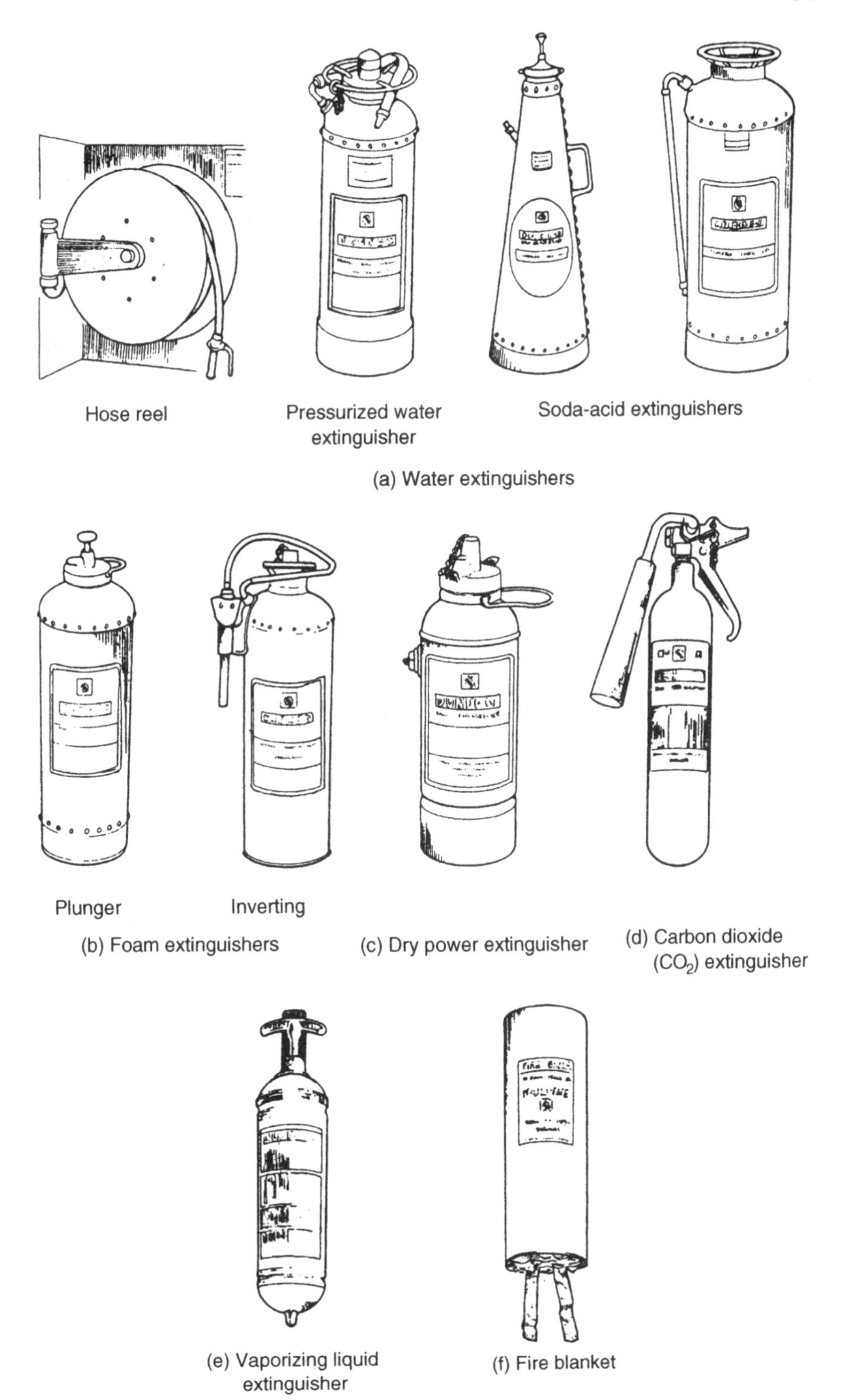

Figure 3.51 Typical fire-fighting equipment.

Table 3.8 Fire extinguishers

Type of extinguisher	Colour code	Class of fire
Water	Red	A
Foam	Light cream	B
Dry chemical powder	Blue	D
Carbon dioxide (CO_2)	Black	Electrical fires
Vaporizing liquid	Green	A, B, C and electrical fires

- Class C. Fires involving *gases*
- Class D. Fires involving *metals*, magnesium, sodium, titanium, zirconium.

3.11.3 Emergency equipment (first aid)

Minor and major injuries are an ever-present occurrence in the manufacturing industries due to negligence, carelessness and the flouting of safety rules, no matter how stringently the requirements of the HSE are implemented. Therefore we need to consider the first aid equipment that must always be available in a manufacturing plant. The basic equipment consists of:

- First aid kit
- Stretcher
- Blankets
- Protective clothing.

The Health and Safety Regulations (First Aid) 1981 specify the contents of first aid kits and boxes, and that these must be checked at regular intervals and replenished as necessary. Any out-of-date contents must be discarded. A number of people, sufficient for the number of employees in a company, factory site or workshop must be trained and hold an appropriate first aid qualification approved by the Health and Safety Executive (HSE). The contents of the first aid box is specified according to the number of people in a particular work area.

The approved code of practice issued with the regulations stipulates that an employer should provide a suitably equipped and staffed first aid room where 400 or more employees are at work or in cases where there are special hazards. In cases where employees are using potentially dangerous machinery or equipment, portable first aid kits must be provided. Every accident that occurs in the workplace no matter whether it is a minor or a major incident must, by law, be entered in the accident book.

3.11.4 Personal safety clothing and equipment

There is a wide range of personal safety and hygiene clothing and equipment available for personnel in the manufacturing industries.

- Workers in the engineering and chemical industries require protection from the materials, machines and equipment they use as shown in Fig. 3.52.
- Workers in the food, drinks and pharmaceutical manufacturing industries wear protective overalls and equipment such as hats, gloves, etc., to prevent contamination of the products they are handling.

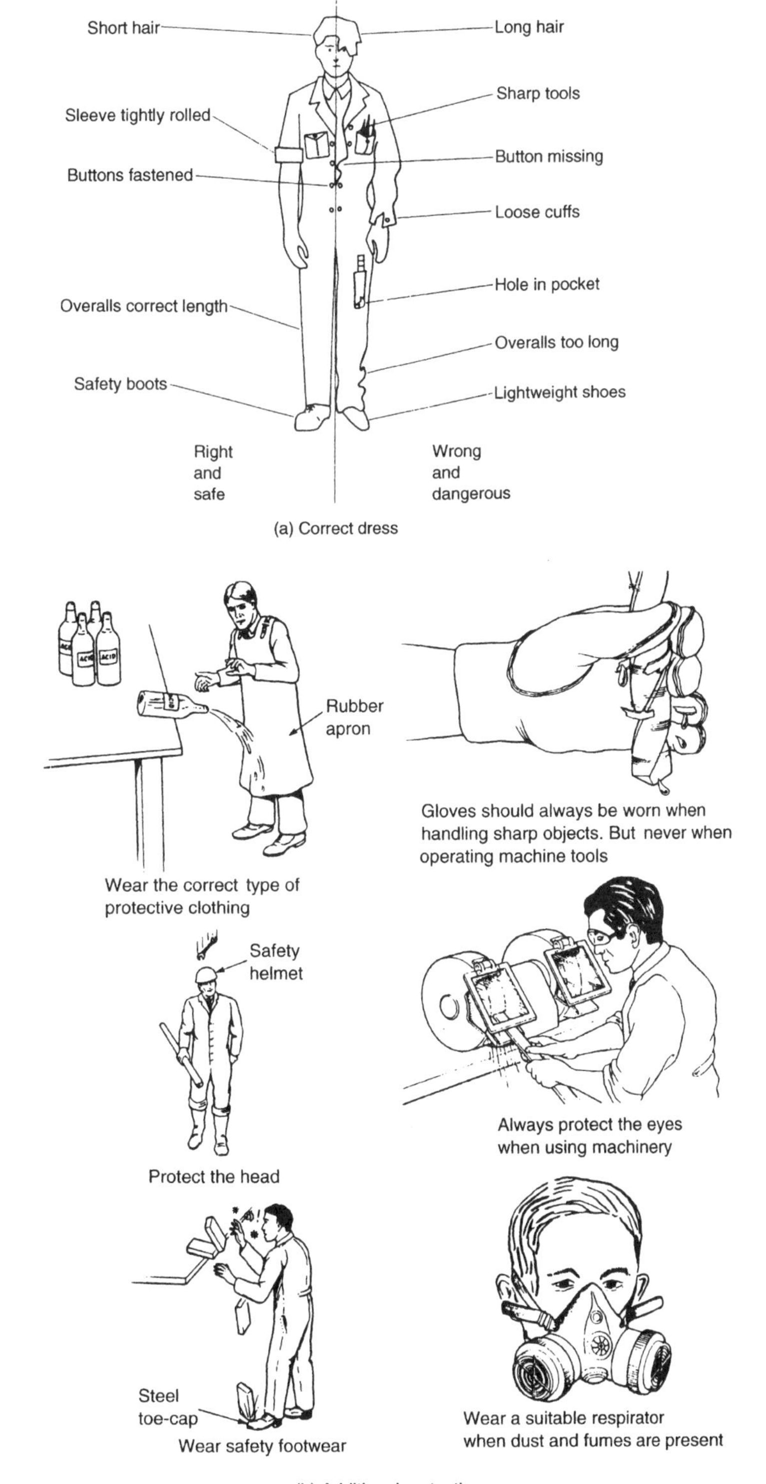

Figure 3.52 Safety clothing and equipment. Courtesy of Training Publications Ltd.

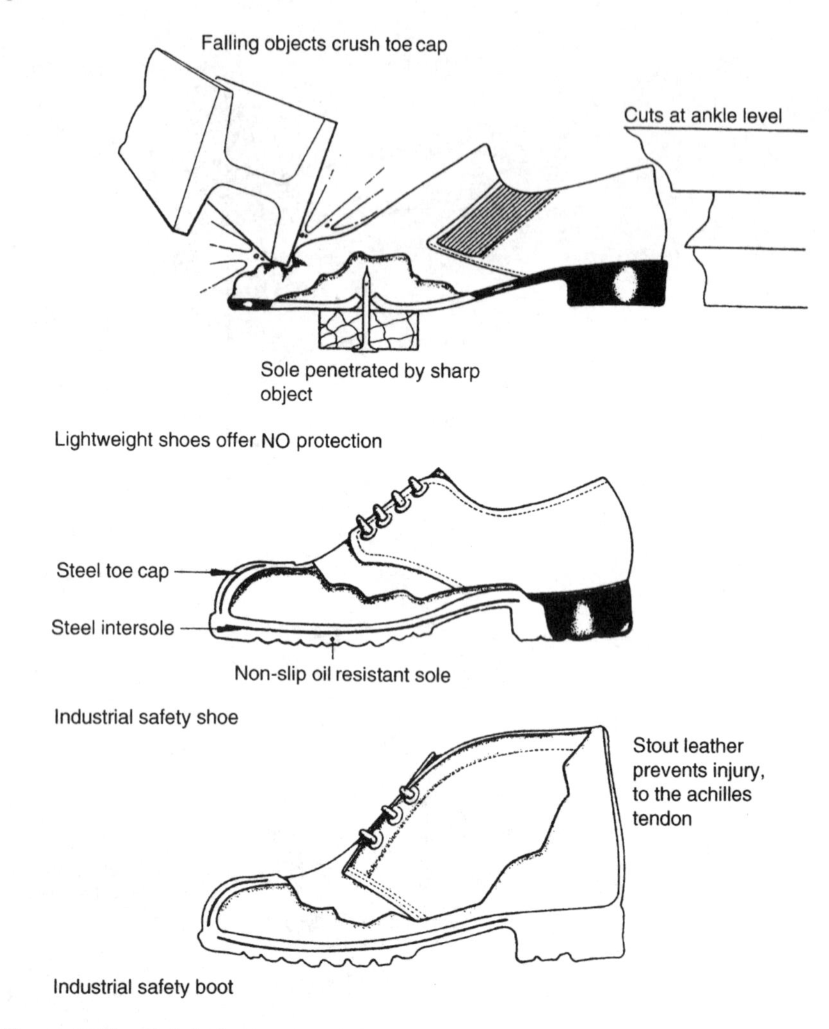

Figure 3.53 Safety footwear.

Safety footwear

Safety footwear includes boots, shoes, overshoes and slipovers and should include one or more of the following features which are also shown in Fig. 3.53:

- Steel toe caps
- Oil and solvent resistant soles
- Heat-resistant soles
- Antistatic soles
- Steel sole plates
- Anti-slip soles.

Regular checks should be carried out to ensure that the footwear is in good condition and gives adequate protection, for example the soles should not be worn smooth and the steel toe-caps must be free from dents or other damage.

Eye protection

Due to the many different types of eye protection available it is vital to wear the correct type of protection for work to be carried out. Safety spectacles with side guards provide adequate protection for general industrial use and, depending on the impact rating of the lens, may also be used when machining or handling small quantities of chemicals. In more hazardous environments, for example where there is a possibility of molten metal or chemical splashes, spark showers or projectiles, safety goggles, full face visors or face-shields should be worn. These are made of special safety glass or plastic.

Hand-held or helmet-type screens fitted with the appropriate dark glass filter should be worn when arc welding in order to protect your eyes from UV light, as well as the heat and excessively bright visible light radiated from the weld pool. Goggles fitted with filter screens should be worn when gas welding. Gas welding goggles only give protection from heat and excessively bright visible light, as no UV rays are present. Therefore gas-welding goggles are unsafe and unsuitable for arc welding. Regular inspection of eye protection should be carried out to ensure that its serviceability is maintained.

Safety helmets

Safety helmets (hard hats) should always be worn when there is a risk from falling objects, or when entering low structures, as when working around scaffolding. The wearing of such head-gear is compulsory in such areas and access will be refused if this rule is not complied with. Regular checks should be made to ensure the serviceability of helmets and that they have no dents or cracks and that the lining is intact. The harness of safety helmets must be adjusted to the individual wearer in order to give the correct clearance. For reasons of hygiene, hats should not be passed from one person to another, and the harness should be regularly washed and cleaned.

Respirators

Several different types of respirator are available, ranging from the simple dust mask that traps fibres and particles of material, to the canister, cartridge, valve and vapour type which absorb organic, inorganic vapours gas and trap particles. Air-fed coverall suits give full protection from vapours, toxic gases and fine particulate material. As with all safety equipment, respirators must be checked and tested regularly for serviceability and canisters or elements replaced after the required period.

Gloves

Several types of protective glove are available for different working environments and examples are shown in Fig. 3.54. Worn and torn gloves should be discarded. They offer little protection and can become entangled with machinery.

Ear defenders

Ear defenders may be of the disposable type made from soft or mouldable material so that they can fit snugly in the ear. Alternatively they may be of the type designed to cover the whole ear and are worn rather like headphones. Ear defenders are worn as protection against noises that affect concentration

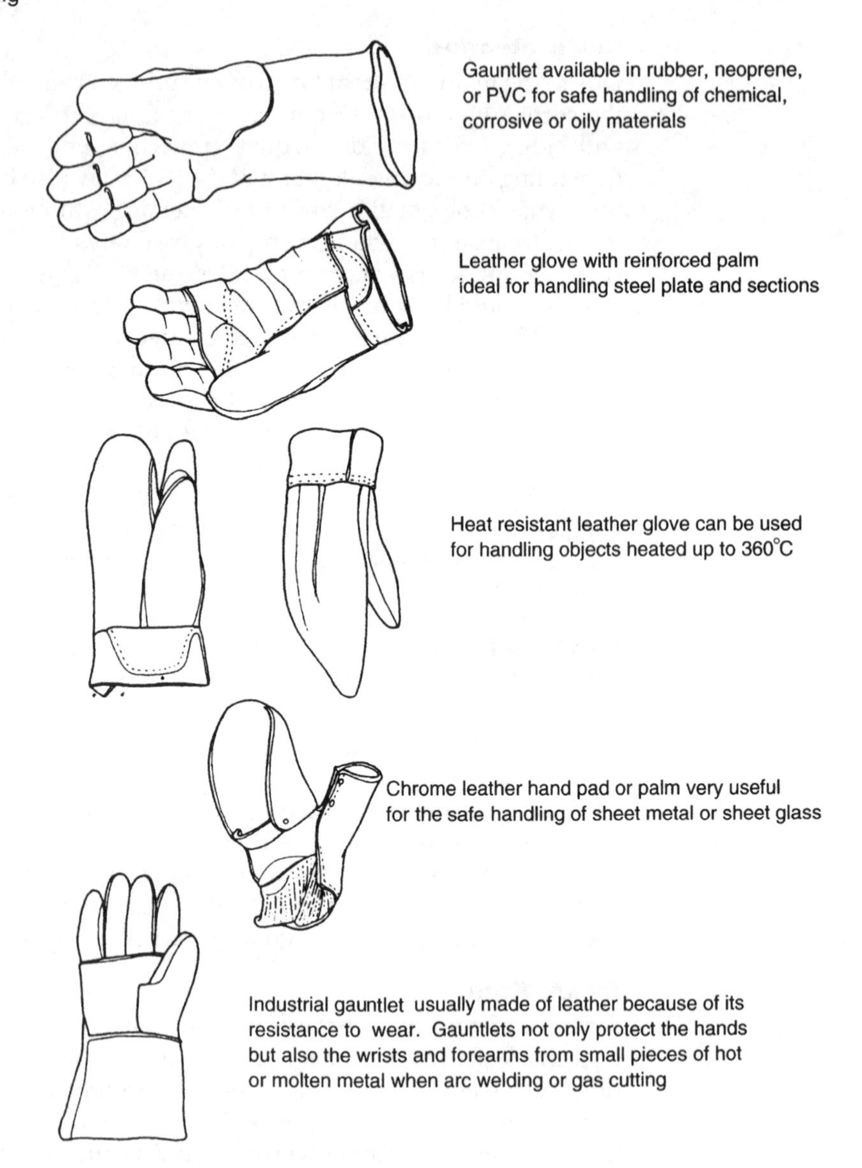

Figure 3.54 Gloves. From *General Engineering* by R. L. Timings, Longman. Reprinted by permission of Pearson Education Ltd.

and cause impairment to hearing. For reasons of hygiene, they should never be swapped between people and must be checked, cleaned, disinfected, and tested regularly.

Hats and caps

Hats and caps may be worn for two purposes. They may be worn for personal protection to prevent you from becoming entangled in machinery. This type of accident can result in serious, scalping type injuries. Hats and caps are worn in the catering and food processing industries to protect the product being manufactured, for example processed foodstuffs, from becoming contaminated with loose hairs and/or dandruff.

3.11.5 Hazardous environments

Appropriate safety clothing must be worn in all hazardous environments, so let's consider what constitutes a hazardous environment.

Hot environments

Hot environments include foundries, heat treatment, moulding, and metal dip coating shops, bakeries and confectionery processing plants. Any form of clothing will give some protection from heat. However in the hot environments just described, thick clothing made of insulating materials will delay the penetration of heat to the body. Such clothing must be made from flameproof or flame retardant materials. Most synthetic fibres are unsuitable since they melt when heated and may stick to the skin causing burns that are difficult to treat.

Metallized suits and aprons give greater protection against radiant heat. Gloves will protect the hands against heat and hot materials, but the gloves must be made from a material to suit the material being handled or process being carried out. Prechilled garments, such as ice vests, are worn as protection against excessive heat. Air or water-cooled suits can also be used but they require external air and water services in the form of trailing hoses that limit mobility and radius of operation in intensely hot environments. Under extreme conditions a proprietary brand of protective clothing known as a 'Vortex' suit can be worn to protect the body.

Cold environments

Cold environments include refrigeration plants and cold stores for meat and food products. Cold environment clothing, necessary to prevent hypothermia and frostbite, must not only provide thermal insulation but also allow the evaporation of perspiration thus avoiding it freezing on the body, these requirements cannot be provided by a single material. Several layers of protection are preferable so that individual layers of clothing can be removed to suit the environment and the bodily heat generated while working.

Contaminated environments

Working in contaminated environments such as those associated with paint spraying, and where toxic, asbestos and radioactive materials are present, require the use of high levels of protection. Full protective (safety) clothing for such hostile environments include respirators, gloves and masks. Respirators are used to filter out harmful particles, gases and fumes but only in oxygen bearing atmospheres. Breathing apparatus with its own air supply such as that worn by firemen are also used on a short term basis in dangerous manufacturing environments, for example, the cleaning and repair of degreasing and chemical plants. For long-term exposure, such as when shot, grit and vapour blasting, an uncontaminated source of air via an airline is provided. Exposed skin should be covered as much as possible to prevent absorption of the contaminant. Radiation intensity can be detected and measured with a Geiger counter. Preferably industrial robots in place of human operatives should be used wherever possible in such high-risk areas.

Physically dangerous environments

Physically dangerous environments include areas containing machinery, fabrication equipment, lifting gear or catwalks. The various types of safety clothing and or equipment worn to protect various parts of the body have already been described and must be selected and worn according to the hazard or hazards likely to be encountered. Again, it is preferable to use industrial robots in place of human operatives wherever possible in such high-risk areas.

Electrically dangerous environments

Most workplaces in manufacturing can be electrically dangerous environments with electric shock being the most common danger. Electric shock is the effect on the nervous and muscular systems of the body caused by an electrical current passing through it. For normal, healthy persons equipment working from a 110-volt supply is considered safe. Most portable industrial equipment is now designed to run off this voltage via a transformer. An electric shock causes the muscles to contract which in itself may not be serious, but it could cause injury or death if a person is working high up on a ladder, gantry, catwalk or platform and the muscular convulsion causes them to fall. More dangerous is the effect of the shock on the functioning of the heart as this can lead to unconsciousness and death.

The following safety equipment should be used when working with electricity to prevent electric shock:

- Rubber-soled shoes or boots
- Rubber gloves
- Insulated mats and sheets
- Insulated tools
- Lightweight aluminium ladders and steps must not be used. Only wooden ladders and steps should be used when working with electrical circuits that cannot be reached from the ground.

Hygienically dangerous environments

Hygienically dangerous environments are those in which workers can contaminate or be contaminated by the products on which they are working. Such environments include food processing areas and pharmaceutical plants. When working with food, the following should be used to ensure that food is not subject to the risk of contamination.

- Head covering, completely enclosing the hair such as hats, caps and hairnets
- Clean, washable, light-coloured protective overalls, preferably without pockets into which particles of contaminants may collect. Protective clothing should be changed and cleaned/laundered regularly
- Ordinary clothing should be completely enclosed by protective clothing
- Thin, transparent plastic gloves used when handling food should be of the disposable type and worn only once.

3.11.6 Safety equipment and systems

Safety equipment and systems include the following:

- Guards
- Simultaneous (two-handed) control devices
- Trip devices

- Visual and audible warning devices
- Mechanical restraint devices
- Overrun devices
- Safety valves
- Fuses
- Circuit breakers
- Residual current circuit breakers
- Emergency stop buttons
- Limit switches.

Guards

Guards are used to protect operators from the dangerous moving parts of machinery. They may be fixed, in which case they must be tamper-proof; or removable in order to adjust or change items, in which case the machine must be isolated from the electrical supply. This is usually achieved by means of a cut-out switch mounted on the guard and connected to the electrical mains supply to the motor so that the machine cannot run when the guard is open. Some guards are adjusted prior to machine operation, such as those mounted on drilling machines, power-saws and presses. Other types of guard keep the dangerous part out of reach, such as those fitted to guillotines and cropping machines. Guards must be maintained in good condition by regular inspection and adjustment as necessary. Some examples are shown in Fig. 3.55.

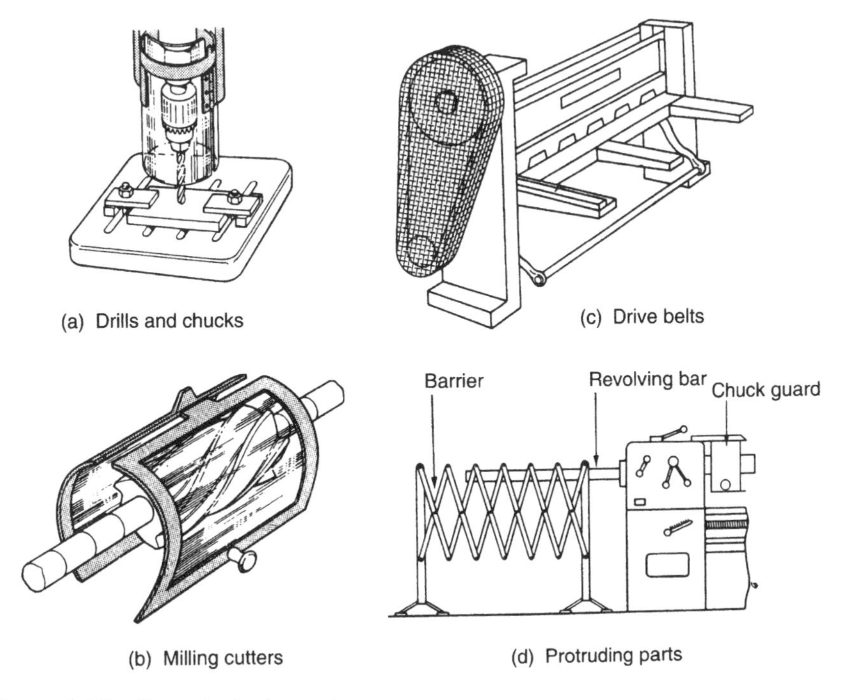

Figure 3.55 Some typical guards.

Simultaneous control devices

A simultaneous or two-handed control device is a device that requires the operator to use both hands at the same time to operate the controls of a machine. In this way both hands of the operator have to be removed from the hazard zone and, therefore, cannot possibly be at risk. This type of device is used on presses and shoe-making machinery.

Trip devices

Trip devices stop a machine when any part of the operator's body goes beyond the safe working limit. They may be pressure, mechanically or photo-electrically activated. Pressure sensitive mats are positioned adjacent to dangerous machinery, such as industrial robots, and anyone stepping on to such a mat causes a switch to be activated resulting in the machine being stopped. A light curtain may also be used. Photo-electric cells are positioned around the hazard zone at a safe working distance. Beams of light shine continually on to the photo-electric cells. If any of the light beams are broken, for example by someone passing through one, the machine is automatically stopped. This is a *fail-safe* device since the machine will be stopped if the lights fail for any reason and remove their protection.

Visual and audible warnings

Visual and audible warning devices such as lights, two-tone sirens and buzzers are used to attract the operators' attention and warn them that something needs their immediate attention.

Mechanical restraint devices

Mechanical restraint devices are provided to stop a machine when it is not working correctly. This type of device is used on the locking system on pressure die-casting and moulding machines. Interlocked guards are fitted to power presses. Here, the machine cannot operate while the tools are being loaded or unloaded. It can only be operated following the operator closing the guard.

Overrun devices

Overrun devices are fitted on machines with interlocking guards. Although the power to the machine is switched off while the guard is open, the momentum of the moving parts prevents the machine from coming to rest immediately. Under these circumstances, the operator may have access to the moving parts. An overrun device delays the opening of the guard and prevents access until the machine has stopped. Overrun devices of a different type are fitted to lifting gear such as pulley blocks and hoists. This prevents the lifting device going into reverse, allowing the load to descend, when the operator releases the rope or chain.

Safety valves

Safety valves are used to prevent pressure exceeding a preset maximum on boilers and air receivers. They should be tested regularly by a suitably qualified person and the result of the test and its date should be logged. Safety valves should have sufficient capacity so that they can release the steam faster than it can be generated, and release air or hydraulic fluid faster than it can be pumped.

Fuses

Fuses are used to protect electrical circuits from current overload. The fuse fitted should be of the correct type and rating for the appliance or circuit it is to protect. Fuses do not blow at their stated rating. The rated current is the maximum current that the fuse can carry continuously. The fusing current may be as high as 1.5 times or twice the rated current. This is called the fusing factor. Fuses only provide *coarse protection.*

Circuit breakers

The circuit breaker is a device that trips a switch to the off position when a current overload occurs. Although it can be easily reset to the on position, it is best to determine the reason for the current overload before doing so. The circuit breaker fitted should be of the correct type and rating for the appliance or circuit it is to protect. Unlike fuses, circuit breakers provide *close protection.* The tripping current is only very slightly greater than the rated current that can be carried continuously. Further the tripping time in the event of an overload is only a matter of a few milliseconds.

Circuit breakers with residual current detection (RCD)

Circuit breakers with residual current detection (RCD) protect against earth leakage faults. They should be tested regularly, to ensure that they are operating correctly. RCD protection is used to protect circuits where a fault causing a current leakage to earth is insufficient to trip the circuit breaker but more than enough to cause an electric shock. The device monitors the current flow in the live and neutral conductors and if there is any difference, however slight, due to some of the current leaking to earth the contact breaker is tripped immediately. Such devices are used in the supply system for portable power tools especially where they are used out of doors and in damp environments.

Emergency stop buttons

All manufacturing plants with electrically powered machines and equipment will have conveniently positioned emergency stop buttons. When pushed all electrical machinery and equipment in the plant will stop. The button is bright red and may have a bright yellow surrounding panel, and be sited in prominent positions. This device isolates the plant at the main switchboard, and individual machines or pieces of equipment cannot be restarted by pressing their start buttons. All employees in the plant should be familiar with the position of these buttons. Only an authorized person can switch the electricity back on again.

Test your knowledge 3.17

1. Explain briefly what equipment you would use to deal with:

   (a) A fire in a chip pan
   (b) A fire in an office stationery cupboard
   (c) A fire in a photocopier.

2. Explain what emergency action you would take if a workmate's clothing caught fire.
3. Explain the action you would take, and the order in which you would take it, if you found a major outbreak of fire in a hazardous area of your works such as a paint store.
4. State what action you would take if you suffered a minor cut to your hand. State the reasons for your actions.
5. List the protective clothing and equipment you would wear in:
   (a) An engineering machine shop
   (b) A food processing plant
   (c) A drop forging shop
   (d) A woodworking (joinery) shop.
6. Explain the precautions you would take when using portable electric power tools out of doors.
7. State the action you would take if you found that the guard on your machine was not properly adjusted.

3.12 Safe working practices

The *Health and Safety at Work Act* provides a comprehensive, integrated system of law dealing with the health, safety and welfare of workpeople and the general public as affected by work activity. Not only are employers and employees *equally responsible* under the Act for ensuring safety at work, so also are manufacturers of equipment and suppliers of goods and materials. Employers are responsible for the health and safety of their employees by the provision and maintenance of safe working conditions. Equally, employees must obey the safe working practices as laid down by the employer, with guidance from the HSE. These safe working practices are based on the following:

- Established and proven practices
- Makers' instructions and guidance notes
- Data sheets
- Health and safety regulations
- Training courses.

Employees must receive supervision and training, especially when new processes of methods are introduced. You must never operate any equipment until you have received training in its correct and safe use. Even then you must not use any equipment without the permission of your supervisor.

3.12.1 Maintenance procedures

The Health and Safety at Work Act makes employers responsible for the safe health risk-free provision and maintenance of plant and systems at work. To meet this obligation, managers must ensure that safety equipment and systems are regularly inspected and maintained in the correct manner at all times by instituting appropriate maintenance procedures, that is, planned maintenance. All manufacturing organizations will have a maintenance section or department, either 'in house' or 'contracted out', whose function is to carry out these tasks, and will include skilled workers in the fields of mechanical, plant, electrical engineering, building and building services.

3.12.2 Fire alarms

Fire alarms are used to warn all staff of the danger of a fire or other emergency. All staff should understand the action to be taken when a fire alarm sounds, and evacuate the workplace safely and quickly. Fire instruction notices should be placed at key points around the workplace, and in each room. Fire drills should be carried out at least once each year and the fire alarms sound tested regularly. The result of the fire drill (e.g. the time taken to evacuate the premises) should be logged.

3.12.3 Training

All operators need to be fully trained in the safe and correct use of the machines and equipment used in their place of work. They will also need to be retrained when new equipment is installed in their place of work. Sometimes this training will be carried out 'in-house' but sometimes the operators will need to attend training courses organized by the manufacturers of the equipment or systems. In some instances the operator will need to be independently certificated. For example only a trained and certificated person may change a grinding wheel.

3.12.4 Supervision

To do their job properly and command the respect of the workers in their charge, supervisors not only require wide experience and knowledge of the process skills involved, they also need to be properly trained in management and interpersonal skills. They should also be competent to train new recruits to their area of responsibility.

3.12.5 Safety equipment and systems

Safety equipment and systems specific to various processes have already been discussed in this chapter. Safety procedures must be strictly adhered to and safety equipment must not be abused or tampered with.

3.12.6 Protective clothing and equipment

The use of protective clothing and equipment has already been introduced. It is essential to co-operate with the management in using the equipment provided not only to protect yourself but also, in some instances, to render the product free from contamination. Sometimes the protective clothing may be uncomfortable and inconvenient to wear. To neglect its use for such reasons is short-sighted since you not only put yourself at risk of physical injury, you also put yourself and your employer at risk of legal prosecution.

3.12.7 Cutting tools

Even simple hand tools such as scissors, hand shears (tin-snips) and knives can cause a nasty cut if not used correctly. Always carry them point downwards when walking about. Always cut *away* from yourself and *away* from your free hand. Files should be used with both hands to give proper control and must be fitted with a handle in good condition. Scrapers should also be used so that the scraping action is away from your body. Saws should be carefully guided at the beginning of a cut with the minimum of force to prevent it slipping. There is more danger associated with a wood saw than a metal cutting hacksaw since only one hand is used and the teeth are much coarser. Once the cut has been started keep your free hand away from the line of the cut and the teeth of the saw.

The cutters on machine tools should be properly guarded not only to protect your hands but, in the case of drilling machines, to prevent you being scalped. Milling machines and guillotines are particularly dangerous and must not only be fully guarded, they must only be used by fully trained operators.

3.12.8 Assembly tools

Sewing needles should be used with a thimble and when using a sewing machine you should keep your fingers well away from the line of sewing. Soldering irons need to be treated with care as they operate at temperatures sufficient to inflict a nasty burn and also start a fire if laid down on cloth, paper or wood. The blades of screwdrivers should be matched to the head of the screw to reduce the risk of the screwdriver slipping and marking the work and/or damaging the head of the screw. Never hold the workpiece so that the screwdriver is pointing towards your hand. A slip could result in the screwdriver stabbing into your hand.

Spanners and socket screw keys should fit the hexagon head accurately so that they cannot slip. An oversize spanner with packing should never be used. If the spanner slips, you can receive a nasty injury and adjacent components may be damaged. Never extend a spanner as the increased leverage can over-stress or break the fastener. Where it is important to achieve the exact tightness specified by the designer, you should use a torque spanner set to the required value.

Test your knowledge 3.18

1. Under the Health and Safety at Work Act, name the groups of persons responsible for safety.
2. Name the organization that implements the Health and Safety at Work Act.
3. Under what conditions is it legally acceptable for you to operate a piece of equipment or a machine at your place of training or your place of work.
4. In addition to the time taken to evacuate the premises, suggest other important information that you think should be logged on completion of a fire drill.
5. Find out what is meant by:
   (a) A prohibition order
   (b) An improvement notice.
6. Explain briefly why operators should be properly trained in the use of manufacturing equipment.
7. Explain how supervisors can contribute to the safety of the workplace for which they are responsible.

Key notes 3.5

- Risk assessment must be carried out whenever a new process, material or item of plant is introduced to determine the level of risk present and suggest ways to lessen or reduce the level of risk.
- Fire extinguishers, hose reels, axes, fire blankets and sprinkler systems must be regularly inspected and tested – preferably by a representative from the local fire service.

- Whenever a fire appliance is used it must be immediately recharged.
- Extinguishers should only be used in the event of a small incident. Always evacuate the area and call the professional service if the incident cannot be contained.
- First aid equipment should be maintained, regularly checked and replenished by a qualified first aid practitioner.
- Only a trained first aid practitioner should administer first aid.
- Appropriate personal safety clothing and equipment should always be worn when working.
- Hazardous environments may be very hot, very cold, or contaminated.
- Physically dangerous environments are found where machinery and lifting equipment is in use.
- Electrical equipment should be regularly checked and maintained by a qualified electrician.
- Electrical equipment used in a factory or on site should be supplied at 110 V AC via a step-down transformer to render accidental shocks less liable to be fatal.
- Machines and their cutters must be guarded and the guards and other safety devices must not be removed or tampered with by the operator.
- The Health and Safety at Work Act applies equally to employers, employees, makers of machine tools and suppliers or hirers of plant and materials.
- Inspectors from the HSE must be notified whenever there is a major incident. They have the right to administer an improvement notice, a prohibition notice or, in extreme cases, to remove and destroy a potentially hazardous piece of equipment or material.
- Operator skill training must include safety in the use of the tools, equipment and machines associated with his or her trade.

Assessment activities

1. Write a report for a product of your own choosing in *detail*. The report should include:

   - A list of the key characteristics of the materials used
   - A summary of your use of ICT to source and research the materials chosen
   - A description of the plant and equipment needed for the manufacture of your product and how that equipment was prepared
   - A description of the processing methods used
   - A description of the finishing processes used
   - A description of the methods of handling and storing the materials used and packaging the finished product
   - A description of the quality control techniques you used
   - A list of the safety and hygiene precautions you took while handling the materials, processing the materials and handling and packaging the finished product.

2. Write a second report for another product of your own choosing in *outline only*. This product should use a contrasting group of materials and processes. It should outline the same key elements listed in activity 1 above.

Chapter 4 Applications of technology

Summary

When you have read this chapter you should be able to:

- Appreciate the importance of information and communications technology (ICT) in manufacturing
- Understand how systems and control technology can be used to organize, monitor and control production
- Apply modern and 'smart' materials to design and manufacture
- Appreciate the impact of ICT and control technology on all sectors of manufacturing
- Appreciate the advantages and disadvantages that the use of modern technology has brought to society
- Understand the significance of the key stages of production
- Understand how to investigate products and research information.

4.1 The use of information and communication technology

Modern information and communications technology (ICT) made possible by today's widespread use of computers has revolutionized the world of commerce and industry. It has made possible the globalization of manufacturing. Figure 4.1 shows how broad is the spread of computer technology. Applications range from word processing packages for letter writing to complete management packages capable of controlling a large, multinational manufacturing organization. ICT can be used to design products, to program computer controlled machines, to operate the machines and integrate them

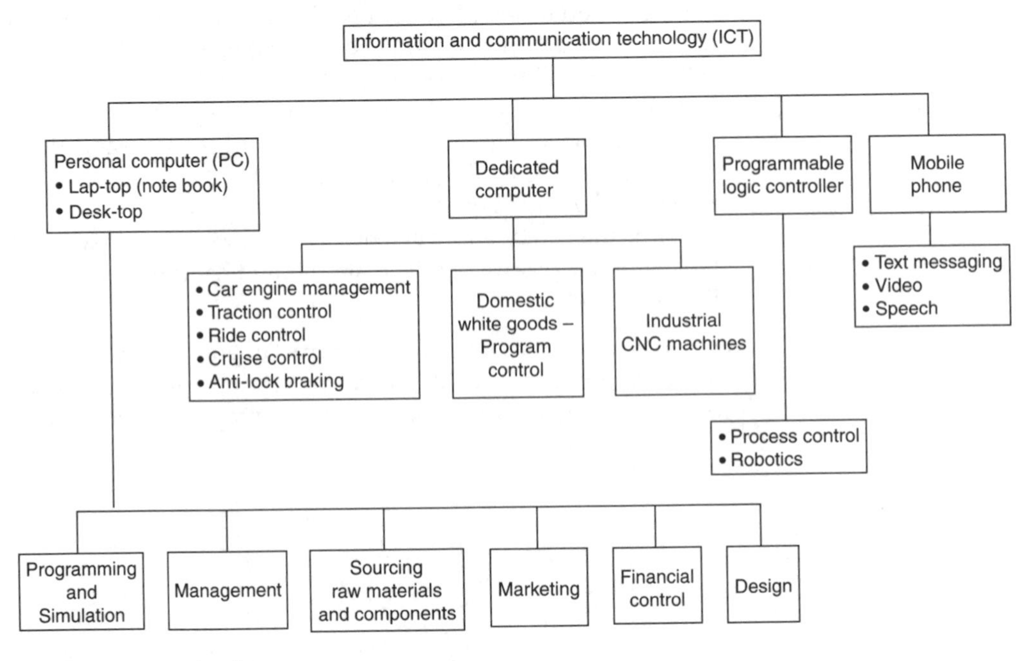

Figure 4.1 Computer technology.

with industrial robots. ICT can be used to source materials, to control the cost of manufacture, to control the finances of multinational companies and market a company's products via Web sites and television advertising using multimedia techniques.

4.1.1 Sourcing and handling information and data

Traditionally, all business communications involved letter writing. Such letters relied on a postal system for the delivery of the information. Before the advent of ITC, commercial transactions with an overseas customer or supplier could take days and even weeks. Nowadays communication by electronic mail (e-mail) or fax has largely replaced letter writing. E-mails can be sent anywhere in the world in a matter of seconds and a reply received equally quickly. When lengthy documents need to be sent in this manner, or when diagrams are involved, this information can be sent as an *attachment* to the e-mail simply by calling up the appropriate files. If the information is not already on file then the documents required can be *scanned into the computer* using a flatbed scanner. Digital camera files can also be included. E-mails rely upon a network of interlinked computer systems called the *Internet.*

The Internet and the World Wide Web

The *Internet* is a term used to describe an interconnection of worldwide computer networks and, as such, it is not one entity but a co-operative system of networks linking many institutions and organizations. The *World Wide Web* (WWW) is a term used to describe the hypertext interface to the information available on the Internet. It is a subset of the Internet – a collection of interlinked documents that work together using specific Internet protocol (IP) called *hypertext transfer protocol* (HTTP). In other words the Internet exists independently of the WWW, but the WWW cannot exist without the Net. Hence any computer with a link to a host server will have the ability to link to any document held on any host server in any location around the world. For any manufacturing organization, the Internet provides a way to communicate with external customers and suppliers.

With the sheer volume of information stored by national and global companies and corporations – including reams of printed information such as computer documentation, procedures, specifications and reference documents – the argument for making *information on-line* available becomes increasingly valid. Users no longer have the time to find some obscure item of information from a shelf full of manuals. Companies can no longer justify the cost of printing all this information without any guarantee that users are actually going to read it. There is also the problem, in terms of time and cost, in keeping large amounts of printed information up to date and withdrawing out-of-date information, for out-dated information to remain in circulation can have serious consequences.

Intranets

Many companies now use *Intranets* to deliver information to internal users. The term *Intranet* refers to the fact that this internal Web is run inside a private network, often without direct connection to the Internet (the external Web). Like the WWW, an Intranet also uses browsers, Web servers and Web sites but on a smaller scale and under the direct control of the individual company. Every company has large amounts of confidential business information that it does not wish to make public – regrettably industrial espionage is 'big business' – and by its very nature the Internet is not a secure means of communication for transacting business. An Intranet allows companies to put all documentation into one source (database). This source can be constantly updated to ensure that users always get up-to-date information. Information stored on an Intranet is much more secure and easier to control.

Extranets

Extranets are several Intranets linked together so that businesses can share information with their customers and suppliers. For example, consider the production of a European aircraft by a number of major aerospace companies located in different European countries. They might connect their individual company Intranets (or part of their Intranets to each other so that they can quickly and easily exchange vital information for the successful design, manufacture and marketing of their products. The information may be exchanged using private leased lines (for security) or the public Internet for less sensitive material. The companies may also decide to set up private newsgroups and bulletin boards with discussion threads so that employees from the different companies involved can exchange ideas and share information. The ability to research, convey and share information is essential to the successful survival of manufacturing companies in today's highly competitive environment.

Case study 4.1: The Internet and the World Wide Web

Web sites are made up of a collection of Web pages. Web pages are written in *Hypertext Markup Language* (HTML) which tells a *Web browser* (such as Netscape's Communicator or Microsoft's Internet Explorer) how to display the various elements of a Web page. Just by clicking on a *hyperlink* you can be transported to a site on the other side of the world.

A set of unique addresses is used to distinguish the individual sites on the WWW. An IP address is a 4- to12-digit number that identifies a specific computer connected to the Internet. The digits are organized in four groups of numbers (which can range from 0 to 255) separated by full stops. Depending on how an Internet service provider (ISP) assigns IP addresses, you may have one address all the time or a different address each time you connect.

Every Web page on the Internet and even each object that you see displayed on a Web page, has its own unique address, known as a *uniform resource locator* (URL). The URL tells a browser exactly where to go to find the page or object that it has to display. It is important to realize that, when a Web page is sent over the Internet, what is actually sent is not a complete page (as it might appear in a word processor) but a set of instructions that allow the page to be reconstructed by a browser. Figure 4.2 shows a typical engineering company's Web page displayed in a Web browser and Fig. 4.3 shows the HTML code responsible for generating the page shown in Fig. 4.2.

Being able to locate the information that you need from a vast number of sites scattered across the globe can be a daunting prospect. However, since

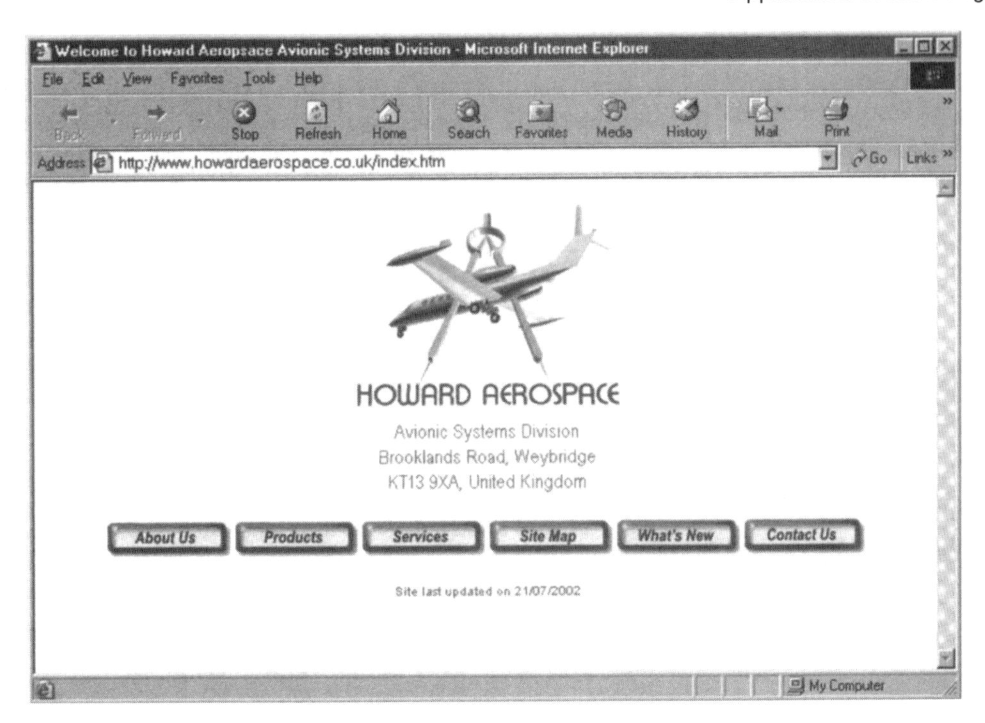

Figure 4.2 A typical engineering company's Web page displayed in Web browser.

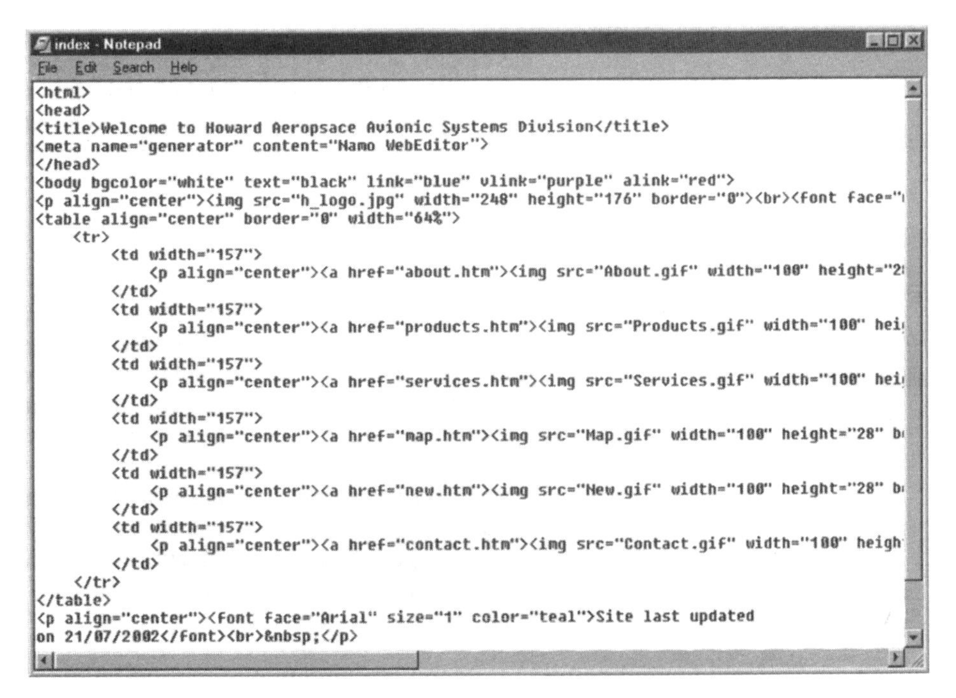


```
<html>
<head>
<title>Welcome to Howard Aeropsace Avionic Systems Division</title>
<meta name="generator" content="Namo WebEditor">
</head>
<body bgcolor="white" text="black" link="blue" vlink="purple" alink="red">
<p align="center"><img src="h_logo.jpg" width="248" height="176" border="0"><br><font face="
<table align="center" border="0" width="64%">
    <tr>
        <td width="157">
            <p align="center"><a href="about.htm"><img src="About.gif" width="100" height="2
        </td>
        <td width="157">
            <p align="center"><a href="products.htm"><img src="Products.gif" width="100" hei
        </td>
        <td width="157">
            <p align="center"><a href="services.htm"><img src="Services.gif" width="100" hei
        </td>
        <td width="157">
            <p align="center"><a href="map.htm"><img src="Map.gif" width="100" height="28" b
        </td>
        <td width="157">
            <p align="center"><a href="new.htm"><img src="New.gif" width="100" height="28" b
        </td>
        <td width="157">
            <p align="center"><a href="contact.htm"><img src="Contact.gif" width="100" heigh
        </td>
    </tr>
</table>
<p align="center"><font face="Arial" size="1" color="teal">Site last updated
on 21/07/2002</font><br> </p>
```

Figure 4.3 The HTML code responsible for generating the page shown in Figure 4.2.

this is a fairly common requirement, a special type of site, known as a *search site*, is available to help you with the task. There are several different types of search site on the Web: *search engines*, *Web directories* and *metasearch sites*.

Search engines such as Excite, HotBot and Google use automated software called Web crawlers or spiders. These programs move from Web site to Web site, logging each site title, URL, and at least some of its text content. The object is to hit millions of Web sites and stay as current with them as possible. The result is a long list of Web sites placed in a database that users search by typing in a keyword or phrase.

Web directories such as Yahoo and Magellan offer an editorially selected, topically organized list of Web sites. To accomplish that goal, these sites employ editors to find new Web sites and work with programmers to categorize them and build their links into the site's index. To make things even easier, all the major search engine sites now have built-in topical search indexes, and most Web directories have added a keyword search.

4.1.2 Databases

Companies build up a vast amount of useful data over the years. Data about customers, about suppliers, financial information, personnel information, process information, information about their competitors, and much else. However they do not require all this information all the time. They need to be able to access the data selectively. For example, they may require tenders from their woven fabric suppliers but not from the suppliers of knitted fabrics or from suppliers of buttons and zip fasteners. Let's see how this can be achieved.

Case study 4.2: Databases

A database is simply an organized collection of data. These data are mainly organized into a number of *records*, each of which contains a number of *fields*. Because of their size and complexity and the need to be able to search for information quickly and easily, a database is usually stored within a computer. A special program – a *database manager* or *database management system* (DBMS) – provides an interface between users and the database itself. The DBMS keeps track of where the information is stored and provides an index so that users can quickly and easily locate the information they require (see Fig. 4.4).

The database management software will also allow users to search for related items. For example, a particular component may be used in a number of different products. The database will allow you to quickly identify each product that uses the component as well as the materials and processes that are used to produce it.

The structure of a simple database is shown in Fig. 4.5. The database consists of a number of records arranged in the form of a table. Each record is divided into a number of fields. The fields contain different information but they all relate to a particular component. The fields are arranged as follows:

Field 1 Key (or index number)
Field 2 Part number
Field 3 Type of part
Field 4 Description or finish of part.

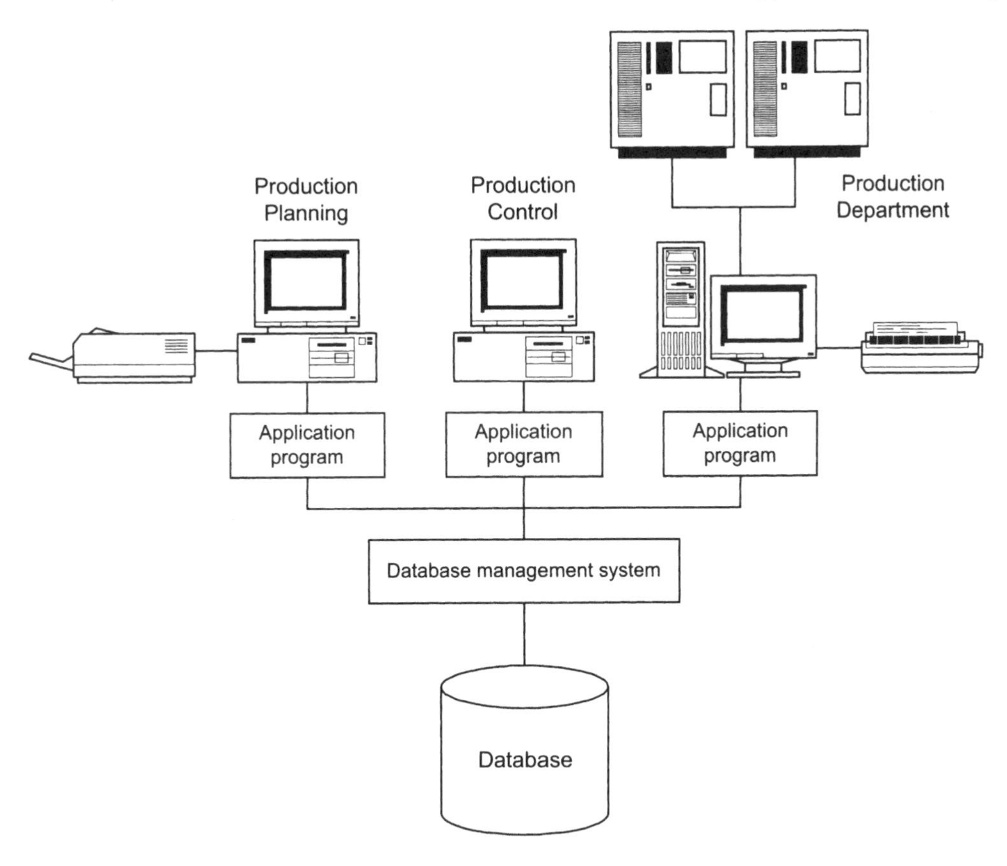

Figure 4.4 A database management system (DBMS).

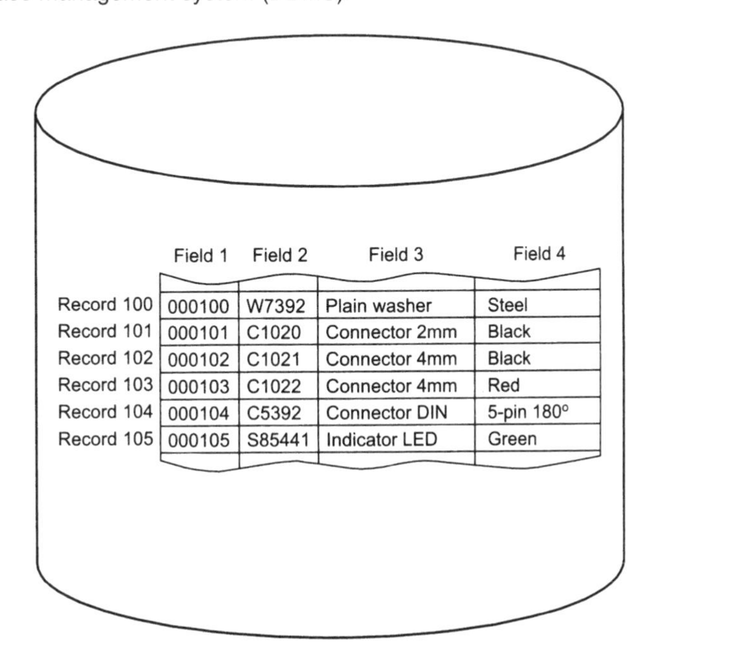

	Field 1	Field 2	Field 3	Field 4
Record 100	000100	W7392	Plain washer	Steel
Record 101	000101	C1020	Connector 2mm	Black
Record 102	000102	C1021	Connector 4mm	Black
Record 103	000103	C1022	Connector 4mm	Red
Record 104	000104	C5392	Connector DIN	5-pin 180°
Record 105	000105	S85441	Indicator LED	Green

Figure 4.5 The structure of a simple parts database.

Consider Record 100 in the database. This shows the data held for a steel washer with a part number W7392. Record 101, on the other hand, is for a black 2 mm connector with a part number C1020.

In most engineering companies several databases are used including a product database, customer database, and a spare parts database. Recently, however, there has been a trend towards integrating many of the databases within a company into one large database. This database becomes central to all of the functions within the company. In effect, it becomes the 'glue' that holds all the company's departments together.

The concept of a centralized manufacturing database is a very sound one because it ensures that every function within the company has access to the same data. By using a single database, all departments become aware of changes and modifications at the same time and there is less danger of data becoming out of date.

4.1.3 Spreadsheets

Traditionally the accounts department maintained handwritten records of a company's transactions in cash-books, ledgers, and analysis books. Rows of clerks busily entered columns of figures in these books, carefully transferring the information (data) from one book to another. Inevitably mistakes occurred and qualified accountants had to carry out checks (*an audit*) to identify and correct any errors that may have occurred. To assist the clerks, the pages in the account books were ruled so that the entries could be kept in neat and easily read columns. Computer *spreadsheets* have now taken the place of manual bookkeeping and the computer does the computations automatically and more accurately provided the correct information is keyed in. Remember the expression GIGO – garbage in, garbage out – a computer is only as accurate as the information keyed into it. Let's consider spreadsheets in more depth.

Case study 4.3: Spreadsheets

A *spreadsheet* is a document compiled with the help of a software package such as Excel or Lotus. The document consists of the following items as shown in Fig. 4.6:

- A *worksheet* divided into rows and columns
- Cells
- Numbers
- Labels
- Formulas.

- A worksheet is like a blank page into which you can keyboard numbers and labels. Each worksheet can contain up to 256 vertical columns and up to 16 384 horizontal rows, a great deal of information. However they are rarely this big. The vertical columns are labelled A, B, C, D, etc. The horizontal rows are numbered 1, 2, 3, 4, etc.
- A *cell* is the 'box' created by the intersection of a row and a column. The data you type into a worksheet are typed into a cell. Cells are identified

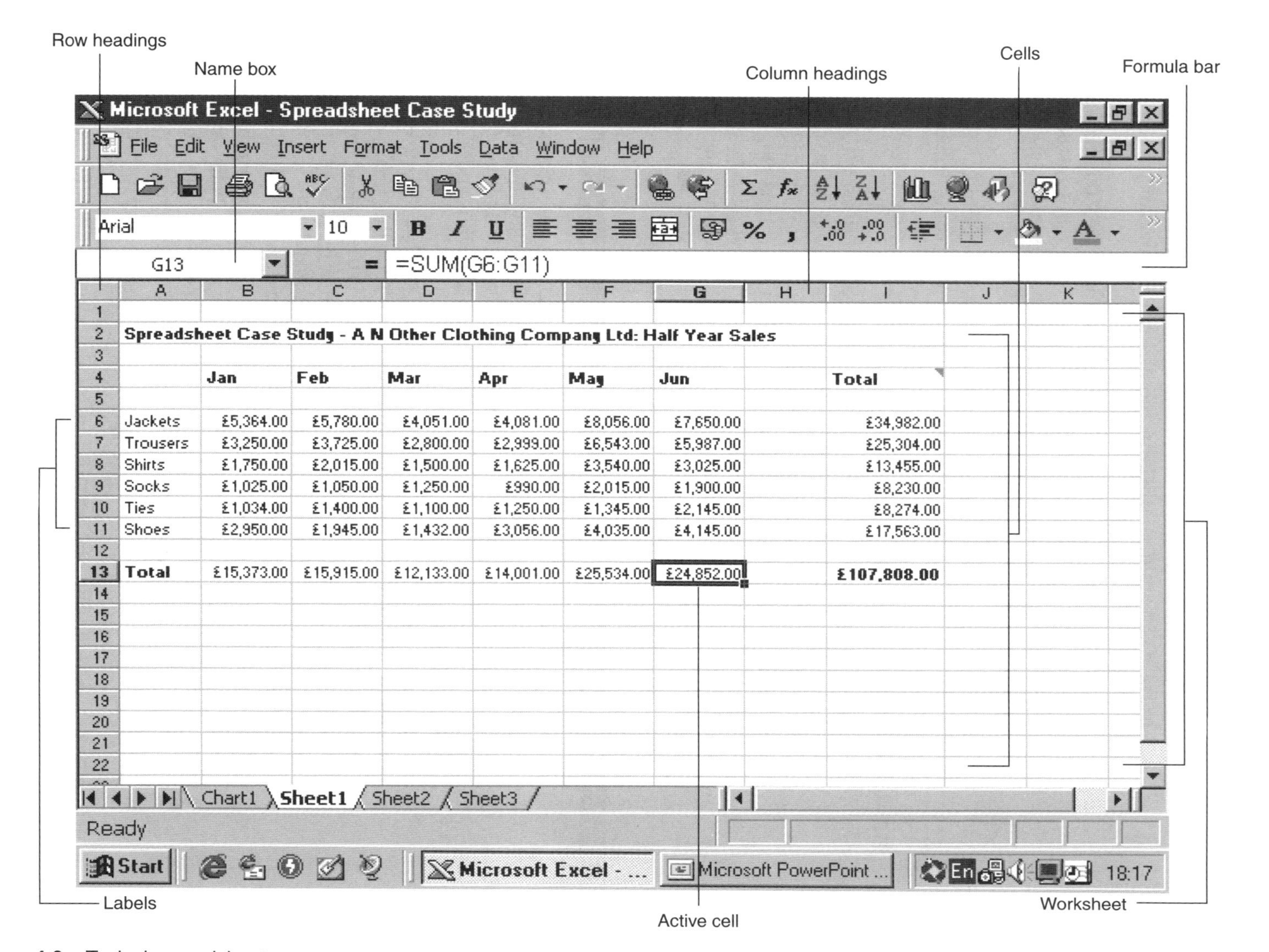

Figure 4.6 Typical spreadsheet.

by their column letters followed by their row numbers, for example the cell at the intersection of column D and row 14 is called cell D14.

- *Numbers* can represent sizes or quantities such as 56.4 mm or 12 shirts, etc., or amounts such as £1200.00.
- *Labels* identify what your spreadsheet entries refer to in case you forget. In Fig. 4.6 the column labels are months and the row labels are the products sold.
- Formulas allow you to calculate new results based on the numbers entered into the cells. For example as simple as monthly totals or, on the other hand, they can be extremely complicated such as differential equations or complex statistics forecasting.

Therefore spreadsheets can simply replace traditional manual accounting books or they can offer additional information such as forecasting and budgeting.

4.1.4 Computer-aided engineering (CAE)

In Chapter 2 you were introduced to computer-aided design (CAD), however this is but one aspect of computer-aided engineering (CAE). CAE is about applying computer technology to *all* the stages that go into providing an engineered product or service. Although developed for the engineering manufacturing sector, the same or similar techniques can be applied to the joinery industry for the automated design and volume production of such products as window frames. It can also be applied to the automated design and large-scale manufacture of woven and knitted fabrics and garments.

Case study 4.4: Computer-aided engineering

When applied effectively, CAE ties all of the functions within a manufacturing company together. Within a true CAE environment, information (i.e. data) is passed from one computer-aided process to another. This often involves computer simulation, computer-aided drawing and design, and computer-aided manufacture (CAM).

The single term CAD/CAM was originally used to describe the integration of computer-aided drawing and design with CAM. Another term – computer-integrated manufacture (CIM) is used nowadays to describe an environment in which computers are used as a common link that binds together the various stages of manufacturing a product, from the initial design and drawing to final product testing. Whilst all these abbreviations can be confusing (particularly as some are frequently interchanged) it is worth remembering that 'computer' appears in all of them. What we are really talking about is the application of computers within manufacturing where the boundaries between the strict disciplines of CAD and CAM are becoming increasingly blurred.

CAD is often used to produce the drawings and designs used in manufacturing. Several different versions of CAD are used in manufacturing and some examples were shown in Figs. 2.41, 2.42 and 2.43. Some further examples are now shown in Figs. 4.7 to 4.10 inclusive. Although engineering examples are shown they could equally well have been fabric designs or joinery products made from wood.

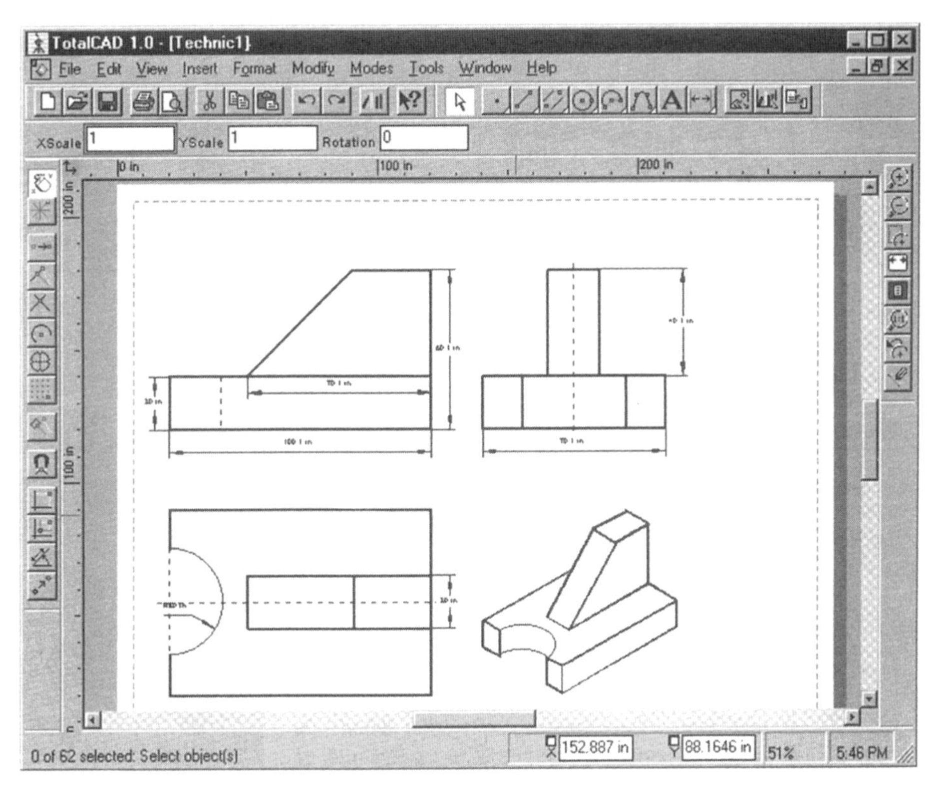

Figure 4.7 A conventional engineering drawing produced by a CAD package.

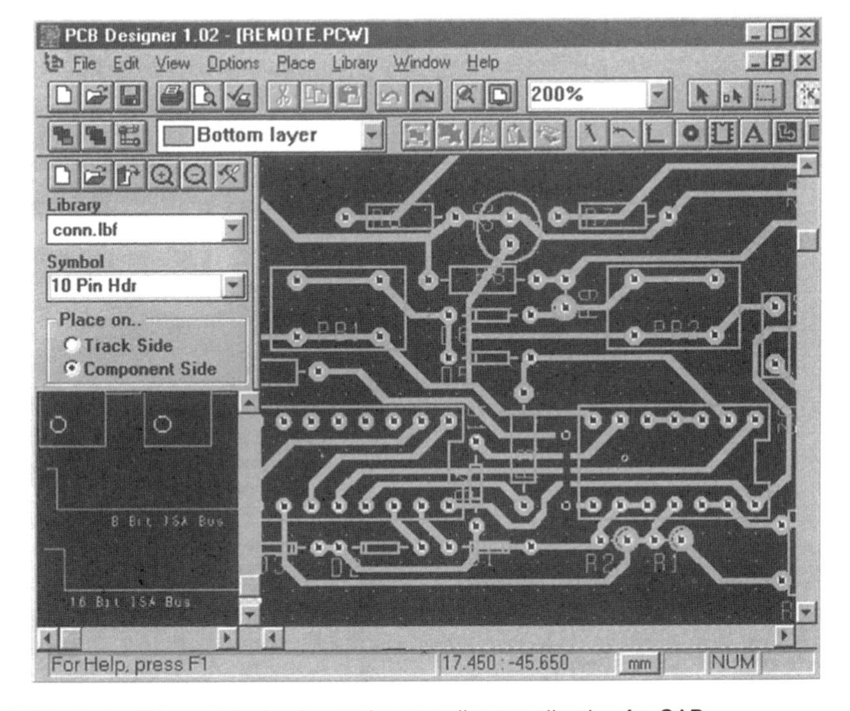

Figure 4.8 Printed circuit board design is another excellent application for CAD.

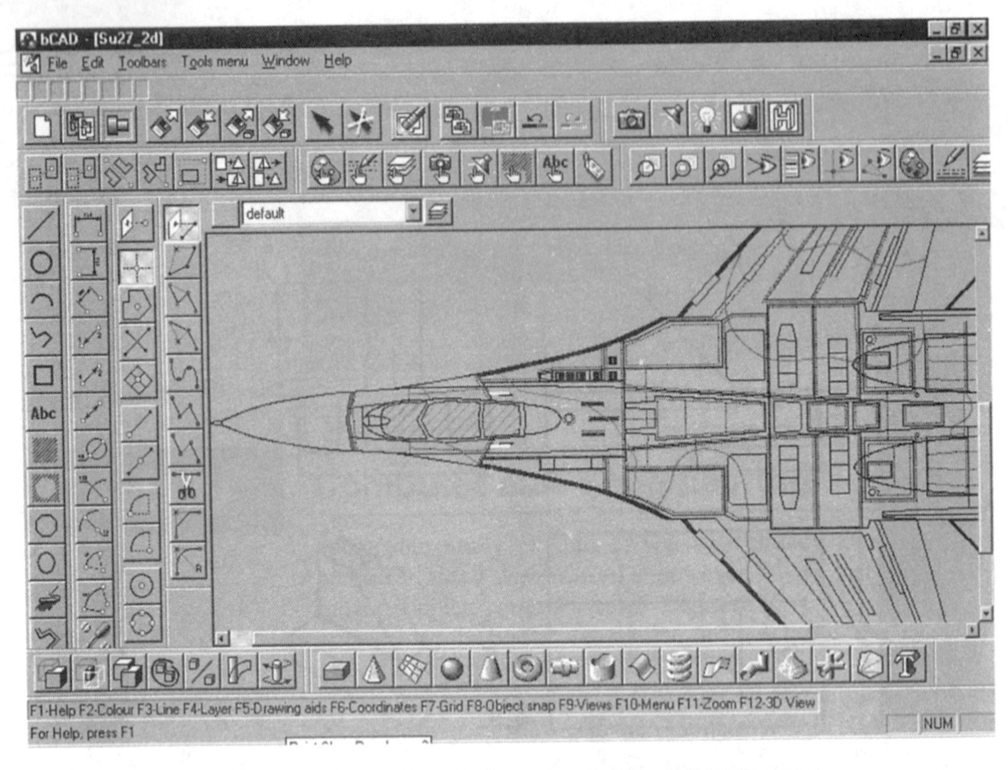

Figure 4.9 Complex drawings can be produced prior to generating solid 3-D views.

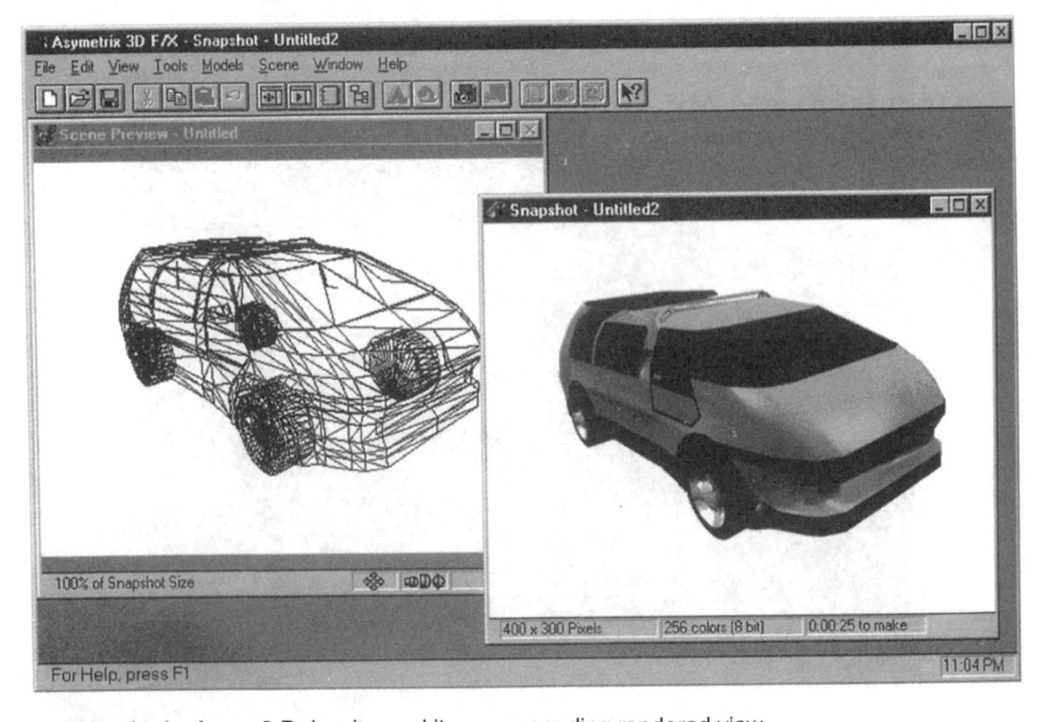

Figure 4.10 A wire frame 3-D drawing and its corresponding rendered view.

CAM covers a number of more specialized applications of computers in manufacturing including CIM, manufacturing system modelling and integration, systems integration, artificial intelligence (AI), applications to manufacturing control, CAD/CAM, robotics and metrology (the science of fine measurement). CAE analysis can be conducted to investigate and predict mechanical, thermal and fatigue stress, fluid flow and heat transfer, and the vibration and noise characteristics of design concepts to optimize final product performance. In addition metal and plastic flow, solid modelling and variation simulation analysis is performed to examine the feasibility of manufacturing a particular part.

In addition, all the computer numerically controlled (CNC) machine tools within a particular manufacturing company may be directly linked to the CAE network. Indeed, most modern manufacturing plants rely heavily on CIM systems.

Some of the most advanced automated systems are employed by those industries that process petrochemicals, pharmaceuticals and bulk food and drink processing such as baking and brewing. Automated systems involving robot devices are used in the manufacture of cars and trucks. Robots are also used in a huge range of applications that involve assembly or manipulation of components in the electronics industry assembling components on printed circuit boards.

Another development that has greatly affected the manufacturing industries is the use of computers to integrate design and manufacturing into one continuous automated activity. The introduction of CIM has significantly increased productivity and reduced the time required to develop new products (see also Section 4.1.5 Rapid prototyping (RP), and Section 4.1.6 Reverse engineering). When using a CIM system an engineer develops the design of a component directly on the computer screen. Information about the component and how it is to be manufactured is then passed from computer to computer automatically within the CAD/CAM system.

After the design has been tested and approved, the CIM system prepares instructions for CNC machine tools and places orders for the required materials and 'bought-in' parts, such as nuts and bolts or adhesives. Computer simulation of the CNC program prior to manufacture ensures that there are no snags and that expensive cutting tools are not going to be smashed by accidentally running into work clamps. Quality control is maintained by built-in measuring devices (sensors) in the CNC machine. These allow the machine to be automatically adjusted for wear and will also call up and automatically install back-up tooling when adjustment is no longer possible. A supervisor would be automatically alerted that the original tooling is in need of refurbishment. The machine can work under 'lights-out' conditions, unattended, overnight. In the event of a major fault arising the machine will close itself down.

Test your knowledge 4.1

1. Describe the differences between the Internet, an intranet and an extranet and give an example where each could be used to advantage.
2. Explain what is meant by the WWW and explain how you would 'surf' it in order to obtain information.

3. Describe what is meant by a database and where it could be used.
4. Describe a spreadsheet and where it could be used.
5. Explain what is meant by the terms CAE, CAD, CAM, CIM.

4.1.5 Rapid prototyping

RP makes it possible to respond to the demands of niche markets and reduce development times, essential when agile manufacturing is required to keep up with the constantly and rapidly changing demands of today's highly competitive markets. RP can be used for such applications as:

- Marketing models
- Casting patterns
- Electrodes for the electrodischarge machining (EDM) of plastic moulding dies and forging dies.

As we saw in the previous case study, data can be transferred from one system to another. This is achieved by exporting data from one application and importing these data into another application. The ability to export and import files from one system to another has led to large-scale system integration. Modern industry and working practices have adopted such system integration to increase efficiency by, for example, using common data.

RP is a modern technique that uses this approach. For example, geometry imported from virtual modelling or directly from a 3D computer-aided design (CAD) geometry can be used to produce solid objects by the process of RP. This has the following advantages:

- It allows designers to view their work very quickly.
- There is no need for fully dimensioned detail drawings.
- The digital data can be transmitted by e-mail from the designer directly to a company specializing in RP or to a customer if that customer has RP facilities.

There are, obviously, many advantages in being able to produce a physical model quickly and relatively cheaply that can be handled by both the designer and the client. For example RP:

- Produces models for market research, publicity, packaging, etc.
- Reduces 'time to market' for a new product
- Generates customer goodwill through improved quality
- Expands the product range
- Reduces the cost and fear of failure
- Improves design communication and helps to eliminate design mistakes
- Converts 3D CAD images into accurate physical models at a fraction of the cost of traditional methods, since there are no tooling costs
- Can be used as a powerful marketing tool, because the actual prototype rather than an illustration can be seen.

Let's now see how RP can be achieved.

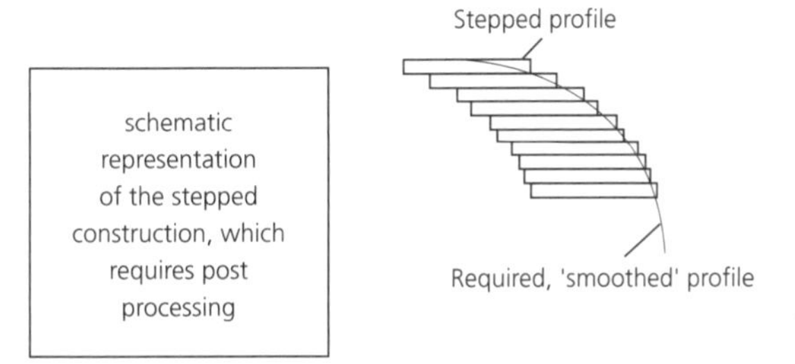

Figure 4.11 Stepped construction.

Case study 4.5: Rapid prototyping

RP software deals with geometry by importing *stereolithography* (STL) files. These files are a list of triangular surfaces that describe a computer generated solid model in ASCII format. It is the standard input for most RP machines and is one of the basic file types that such packages use either to rapid prototype an object using desktop milling or for some other prototyping technique such as *stereolithography apparatus* (SLA).

The CAD data are processed by slicing the computer model into layers, each layer being typically 0.1 to 0.25 mm thick. The RP machine then uses these data to construct the model layer by layer, each layer being bonded to the previous one until a solid object is formed. Due to the mechanics of this laminated method a stepped surface is developed on curved surfaces as shown in Fig. 4.11. The smoothing of this stepped surface is essential if the maximum advantage of the process is to be realized.

RP apparatus may use material additive techniques or material subtractive techniques. One of several material additive techniques is STL. The STL file of the component to be modelled is 'sliced' by the device's software. Each slice is then etched on to the surface of a photosensitive ultraviolet-curable resin using a 'swinging laser'. Where the laser beam strikes the surface, the resin is cured. Each layer is typically 0.13 mm thick. At the end of each pass, which covers the whole surface of that layer, the platform descends to allow liquid resin to flow over the previously cured layer. A recoating bar passes over the surfaces to ensure that a consistent layer thickness is achieved. This bar ensures that no air is trapped between the layers. This technique is shown in Fig. 4.12. On completion, the model is carefully removed and washed in a solvent to remove any uncured resin. The model is then placed in an ultraviolet oven to ensure all the resin is cured. The overhanging sections are supported on slender 'stilts' that can be easily removed from the finished model. The software package anticipates the need for these 'stilts' and provides them automatically. The designer does not have to provide them.

As an alternative to building up the model, a material subtractive process such as *desktop milling* using DeskProto can be used. Desktop milling is usually achieved by using a small but extremely fast traverse CNC type machine, for example the Roland 3100. It plugs into the parallel port of a PC, like a printer. The tool paths are generated directly from 3D geometry, such

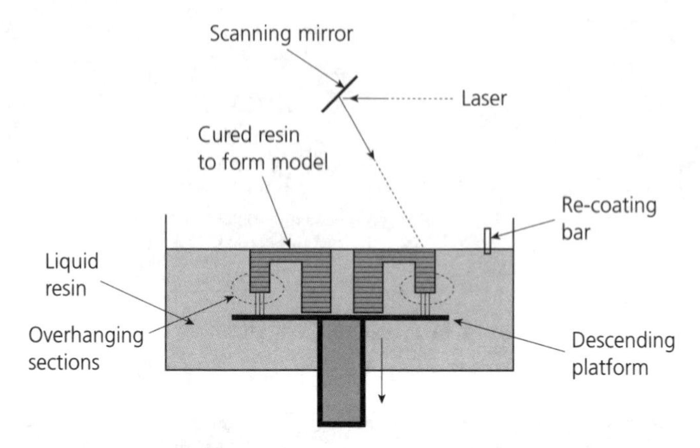

Figure 4.12 Schematic diagram of the stereolithography (SLA) apparatus.

as 3DS, DXF or STL files. The machining process is fully automated using simple tooling and feed rate parameters. The cutter paths are generated automatically and follow a given path direction.

The main advantages of desktop milling over other RP techniques are as follows:

- The prototype can be made from more robust material, i.e. aluminium.
- Small series production is possible by casting in silicone-rubber moulds, created from a milled master pattern.
- Milled moulds in 'tooling board' for vacuum forming and hand lay-up in polyester. Two typical uses in this field are chocolate moulds and blister packs.
- Models in foam or wax for the lost foam/wax casting processes, for example in aluminium using lost foam or in gold and silver jewellery using lost wax.
- Milled aluminium tools (moulds) for both cavity and cores of small injection moulded plastic parts. The moulds only have a limited life but are adequate for a short run to test the market response to a new product before entering upon the manufacture of high cost alloy steel moulds.
- Electrodes for *spark erosion* to be used for creating production tolling in alloy steel.

4.1.6 Reverse engineering

As its name implies, reverse engineering works back from the finished product. The 3D geometry is created by accurately measuring an existing physical product. This is achieved by probing or scanning to produce a cloud of point data. Packages such as DeskProto, previously mentioned, can be used to convert the cloud point data into an STL file ready to produce a replica object by a typical RP technique or sinking a die by CNC machining in which replicas can be cast or moulded. Let's see how reverse engineering is achieved.

Case study 4.6: Reverse engineering

There are two methods of achieving a cloud of point data, probing and scanning.

Probing

Coordinate measuring machines have evolved from simple layout machines using manually operated techniques into highly sophisticated and extremely accurate automated inspection systems capable of checking the profile of components with complex shapes such as the turbine blades used in jet engines. A major factor in this evolution has been the touch-trigger and other forms of inspection and subsequent innovations such as the *motorized probe head* and the *automatic probe exchange system* for unmanned flexible inspection. The touch-trigger probe was originally developed by Renishaw Ltd, in collaboration with Rolls-Royce plc, when a unique solution was required for the accurate measurement of key components in the Anglo-French Concorde aircraft's engines. The touch-trigger probe is a widely used 3D sensor capable of rapid and accurate inspection with a very low trigger contact force. Touch-trigger probes are used for tool-setting and in-cycle inspection in CNC machines, as well as being used to determine 3D geometrical points in space.

Scanning

Scanning is the term used to describe the process of gathering information about an undefined three-dimensional surface. It is used in such diverse fields as tool and die making, press-tool making, aerospace, jewellery, medical appliances and confectionery moulds. Scanning is used whenever there is a need to reproduce a complex free-form shape. During the scanning process an *analogue scanning probe* (not a touch-trigger probe) is commanded to contact and move back and forth across the unknown surface. During this process the system records information about the surface in the form of numerical data. These data are then used to create a CNC program that will machine a replica or a geometric variation of the scanned shape. Alternatively the collected data can be exported in various formats to a CAD/CAM system for further processing. A typical scanning process is shown in Fig. 4.13. Optical scanning using laser light can be used for scanning non-rigid surfaces.

4.1.7. Computer numerical control (CNC)

The essential differences between the manual control of machines and automated machines were considered briefly in Chapter 3 (Figs. 3.9 and 3.10 together with an abstract from a CNC part program in Fig. 3.22). CNC was originally developed for the engineering industry and the first machines were CNC lathes (turning centres) and CNC milling machines (machining centres). However CNC spread to other machines such as surface and cylindrical grinding machines, centreless grinding machines, laser cutting machines and turret punching presses. It spread beyond engineering to control woodworking machines, fabric weaving machines, knitting machines and sewing machines (particularly for embroidery, badges, etc.). It is a widely used and important method of control throughout all sectors of manufacturing, so let's consider it further.

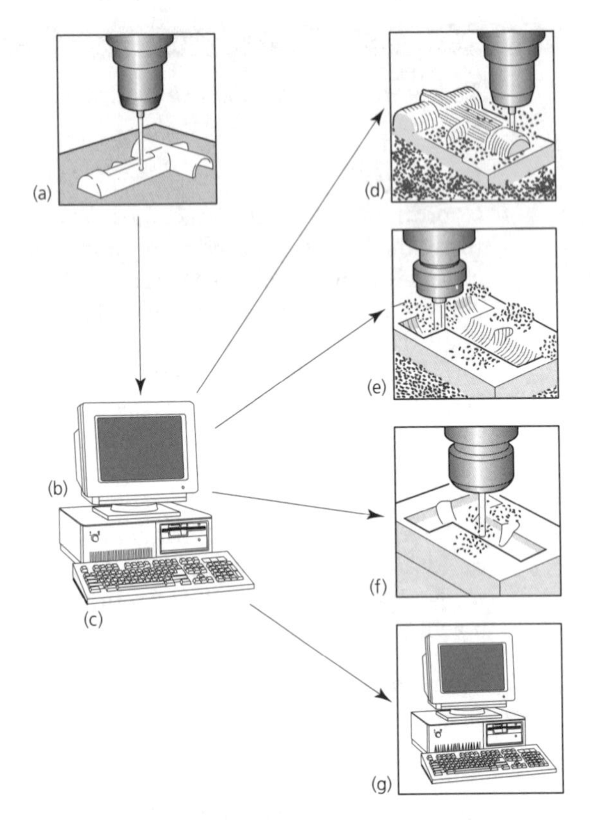

Figure 4.13 The scanning process: (a) scan, a range of devices capture data from an initial part/model; (b) manipulate data, create variant of original; (c) create output, CNC program or export data to CAD system; (d) replica; (e) rough machine mould (CNC mill); (f) finish mould cavity by CNC-type electro-discharge machining (EDM); (g) CAD display if required.

Axis nomenclature

An important feature that is supplied to the control system is slide displacement. Most machines have two or more slides (usually perpendicular to each other and, in addition, these slides can be moved in one of two directions, i.e. left or right, up or down). It is important therefore that the control system knows which slide is to move, how far it is to move, and in which direction it is to move. British Standard BS3635 provides axis and motion nomenclature and Fig. 4.14 shows how the axis notation is applied to a turning centre (lathe) and a machining centre (milling machine).

- The *Z-axis* is always the main spindle axis and is positive away from the work. This is a safety feature, so that should the programmer forget the directional sign (the negative sign) in front of the numerical positional data the tool will move away from the work and not crash into it.
- The *X-axis* is always horizontal and parallel to the working surface.
- The *Y-axis* is always perpendicular to both the X- and Z-axes.

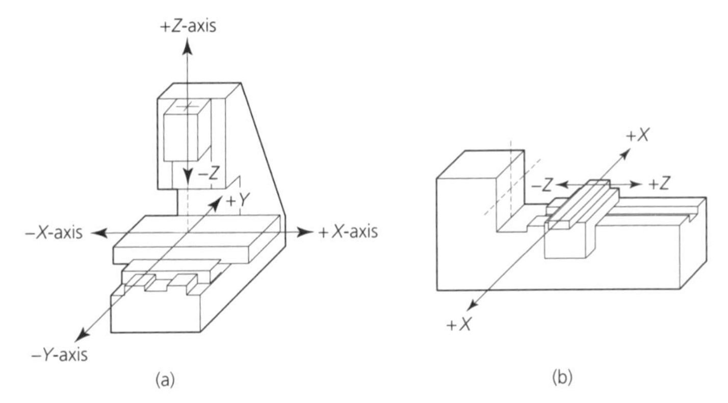

Figure 4.14 Examples of axis notation: (a) axes for vertical milling machines and drilling machines; (b) axes for lathes, and horizontal boring machines.

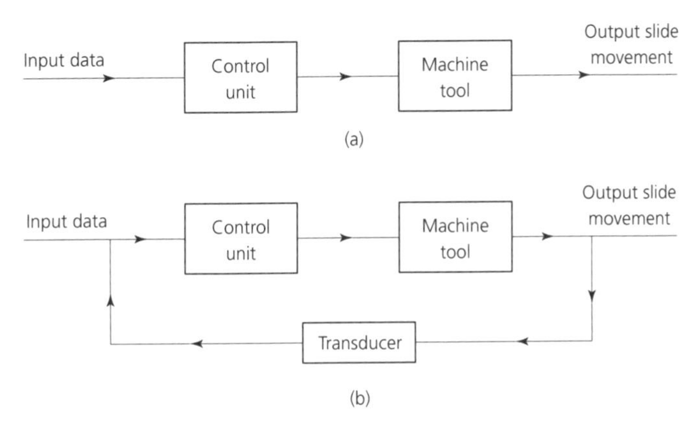

Figure 4.15 Control systems: (a) open-loop control, no feedback loop; (b) closed-loop control, the transducer is part of a feedback loop.

Control systems

Control systems can be *open-loop* or *closed-loop* as shown in Fig. 4.15. In the open-loop system, Fig. 4.15(a) the machine slides are moved according to information loaded from the part program into the control system, without any feedback signal to monitor the slide positions. The most common method of moving the slides is by a lead screw driven by a *stepper motor* either directly or via a toothed belt drive. A *stepper motor* is an electric motor energized by a train of electrical pulses rather than a continuous electrical signal. Each pulse causes the motor to rotate through a small discrete angle. Thus the motor rotates in a series of steps according to the number of pulses supplied to it from the controller. Stepper motors have only limited torque and are usually found on small machines where the loads are limited, for example, small machine tools used in teaching, knitting machines, computer printers and scanners, etc. Although open-loop systems are lower in cost than closed-loop systems, they are intrinsically less accurate since, if the load on the motor causes it to stall, the controller continues to supply pulses to the motor unaware that no movement is taking place.

In closed-loop systems a signal is sent back to the control unit from a measuring device (called a transducer) attached to the slideways, indicating the actual movement and position of the slides. Until the slide arrives at the required position, the control unit continues to adjust its position until it is correct. Such a system is said to have feedback and is shown in Fig. 4.15(b). Although more complex and costly than the open-loop system, close-loop systems give more accurate positioning especially on medium and large machines where the forces involved in moving the machine elements can be high. For this type of system powerful *servomotors* having a high starting torque are used to drive the lead screws instead of the stepper motors used on smaller and cheaper machines.

Programming

A CNC machine is controlled by a part-program loaded into the machine's control unit that contains a dedicated computer. The program can be written manually or program writing can be computer assisted. Unfortunately the program language varies between makes of machine and even between various machines of the same make but of different generations. One advantage of computer assisted programming is that only one language has to be learnt and the program can then be converted into the machine's language automatically by a suitable post-processor software package. This is a considerable advantage where programs have to be written for a large number of different machines of different makes bought over a period of time.

Dimensional words

A *character* is a number, letter or symbol that is recognized by the controller. A *word* is an associated group of characters that define one complete element of information, for example, F120. There are two types of word: *dimensional words* and *management words*. We are going to consider dimensional words first.

These are any words related to linear dimensions. That is, any word commencing with the letters X, Y and Z (note that I, J and K are also used to refer to circles and arcs of circles but are beyond the scope of this book). For example, X35.4 specifies a movement of 35.4 mm along the X-axis of a machine. Similarly Y15.7 specifies a movement of 15.7 mm along the Y-axis of a machine.

In addition to stating the axis along which the machine element must move, and the distance it must move, the direction must also be specified. To indicate the direction of movement the dimension is made positive or negative. If there is *no* sign in front of the digits, positive movement is assumed. Negative movement is indicated by placing a negative (−) sign in front of the dimension. This is shown in Fig. 4.16. As previously mentioned, *omitting* the negative (−) sign from a Z axis dimension will result in the tool moving away from the work as a safety precaution.

Management words

These are any words not related to a dimension. That is, any word commencing with the character N, G, F, S, T and M or any word in which the above characters are implied. Here are some examples:

- *N is a block or sequence number*. A block or sequence is a line of data and commands recognized by the controller. The character N is followed by

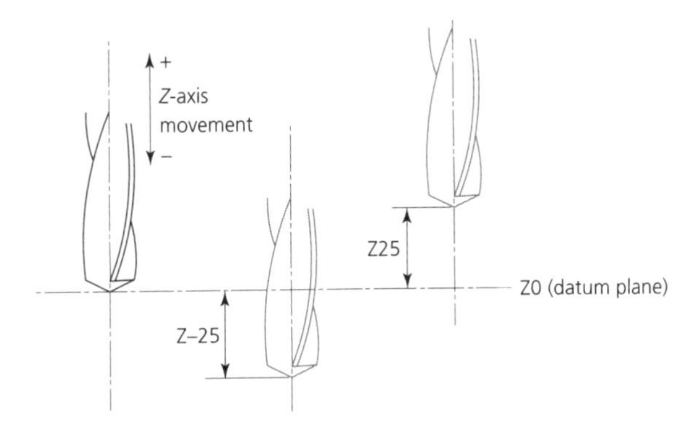

Figure 4.16 Z-axis movement.

up to four digits. A block word is usually the first word that appears in any block and identifies that block. Blocks are usually numbered in steps of 5 or 10 so that additional blocks can be inserted later if found to be needed.

- *G codes are preparatory codes or functions*. The character G is followed by up to two digits. These are used to prepare or inform the machine controller of the functions required for the next operation. In the examples in this chapter, the G codes referred to are taken from BS 3635. Different machines and controllers use different G codes. Always consult the machine manual before writing a program. Some preparatory codes are *modal* that is, once they have been selected they will remain in operation until changed or cancelled. This is the same as the 'caps lock' on your computer keyboard. Some typical G-codes are: G00 indicates rapid positioning, G01 indicates positioning at a normal feed rate, G81 to G89 inclusive indicates fixed (canned) cycles, and G80 cancels any fixed cycle currently in operation.
- *F is a feed rate command.* The character F followed by up to four digits indicates to the controller the desired feedrate for machining and may be defined in terms of: millimetres per minute, or millimetres per revolution.
- *S is a spindle speed command.* The character S followed by up to four digits defines the spindle speed and may be defined in terms of revolutions per minute or metres per minute.
- *T is a tool number.* The character T is followed by up to two digits and identifies which tool is to be used. Each tool used will have its own tool number and the computer, as well as memorizing the tool number, also memorizes such additional data as the tool length offset and/or the tool diameter/radius compensation for each tool. In a machine fitted with automatic tool changing, the position of the tool in the tool magazine is also memorized by the computer.
- *M is a miscellaneous command.* The character M followed by up to two digits provides for a number of miscellaneous or machine management commands. Apart from the pareparatory functions (G-codes), there are a number of other commands that are required throughout the program. For example, starting and stopping the spindle; turning the

coolant off and on; changing speed and changing tools. The M-codes referred to in this chapter are taken from BS 3635. As for the G-codes these vary from machine to machine and the maker's programming manual should always be consulted. Some typical M-codes are M03 which commands the spindle to rotate clockwise, M08 which commands coolant on and M09 which commands coolant off.

Canned cycles

To assist program writing, numerical control computers have certain pre-programmed standardized cycles called *canned cycles* embedded in their memories. Let's consider the sequel of events of a canned drilling cycle G81 is called up in the program. This is shown in Fig. 4.17.

1. Rapid traverse to centre of first hole
2. Rapid traverse to clearance plane height
3. Feed to depth of hole
4. Rapid up to clearance plane height
5. Rapid traverse to centre of next hole – repeat for as many holes as required.

The only data the programmer has to provide are as follows:

- The position of the hole centres
- The spindle speed
- The feedrate
- The tool (drill) number.

There are many more canned cycles such as cutting screw threads, scaling, mirror imaging, rotation and cutting internal and external screw threads. This is only a simple introduction to CNC programming and there is a great deal more to learn. For example *cutter diameter compensation* when milling. This automatically adjusts the program to allow for the diameter of the cutter. *Tool nose radius compensation*, when turning on a lathe, automatically adjusts the programme to allow for the radius on the tip of the tool. When you write the program you assume the tool has a sharp point which it never

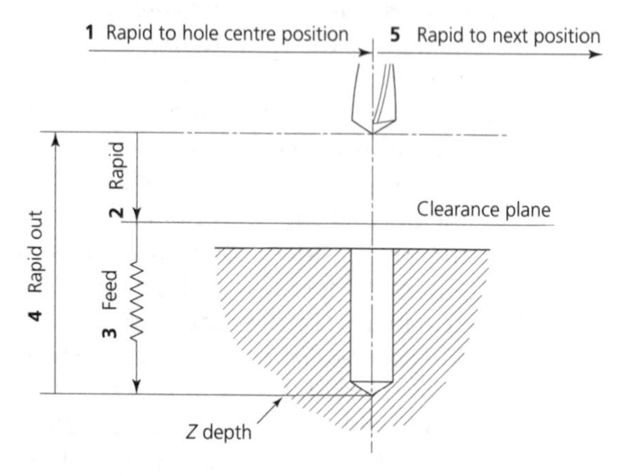

Figure 4.17 Canned drilling cycle (G81).

does in practice. Tool length offset allows tools of various lengths to be used with a common datum. It also allows tools to be changed when they become worn for tools of a different length without having to rewrite the program. If you are interested, you should consult the programming manual associated with one of the machines in your school or college.

Case study 4.7: A simple milling program

Figure 4.18 shows a simple component to be made on a CNC milling machine. The material is low carbon steel and the blank to be machined is 10 mm thick. A typical part program based on BS 3635 is shown in Fig. 4.19.

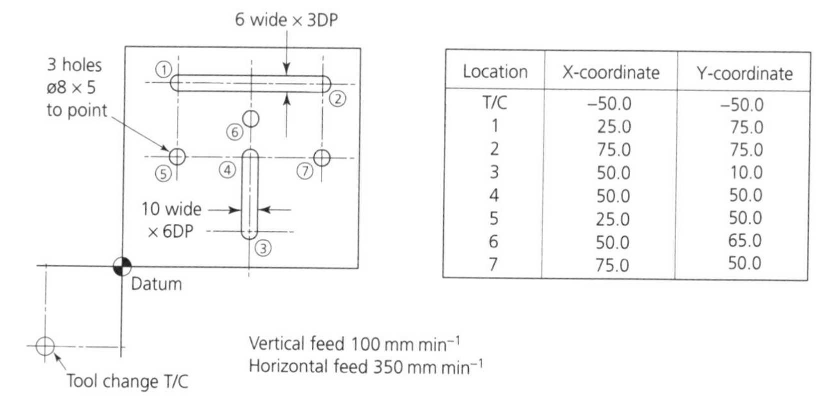

Location	X-coordinate	Y-coordinate
T/C	−50.0	−50.0
1	25.0	75.0
2	75.0	75.0
3	50.0	10.0
4	50.0	50.0
5	25.0	50.0
6	50.0	65.0
7	75.0	50.0

Figure 4.18 Component to be milled and drilled on a CNC machine.

Sequence number	Code program	Explanation
%	G00 G71 G75 G90	Default line
N10	X–50.0 Y–50.0 S1000 T1	
N20	M06	TC posn spindle/tool offset (ø6 mm)
N30	X25.0 Y75.0 Z1.0	Rapid posn (1)
N40	G01 Z–3.0 F100	Feed to depth
N50	G01 X75.0 F350	Feed to posn (2)
N60	G00 Z1.0	Rapid tool up
N70	X–50.0 Y–50.0 S800 T2	
	M06	TC posn spindle/tool offset (ø10 mm)
N80	X–50.0 Y10.0 Z1.0	Rapid posn (3)
N90	G01 Z–6.0 F100	Feed to depth
N100	G01 Y50.0 F350	Feed to posn (4)
N110	G00 Z10	Rapid tool up
N120	X–50.0 Y–50.0 S1100 T3	
	M06	TC posn spindle/tool offset (ø8 mm drill)
N130	X25.0 Y50.0 Z1.0	Rapid posn (5)
N140	G81 X25.0 Y50.0 Z–7.0 F100	Drill on restate depth (inc) feed
N150	X50.0 Y65.0	Posn (6)
N160	X75.0 Y50.0	Posn (7) drill
N170	G80	Switch off drill cycle
N180	G00 X–50.0 Y–50.0 M02	TC posn end of prog
E		End of tape

Figure 4.19 Typical part program for machining the component in Fig. 4.18.

Case study 4.8: A simple turning program

Figure 4.20 shows a simple turned component to be made on a Hardinge turning centre fitted with a GE1050 controller. The material is free cutting low-carbon steel. A typical part program for the component in Fig. 4.20 is shown in Fig. 4.21.

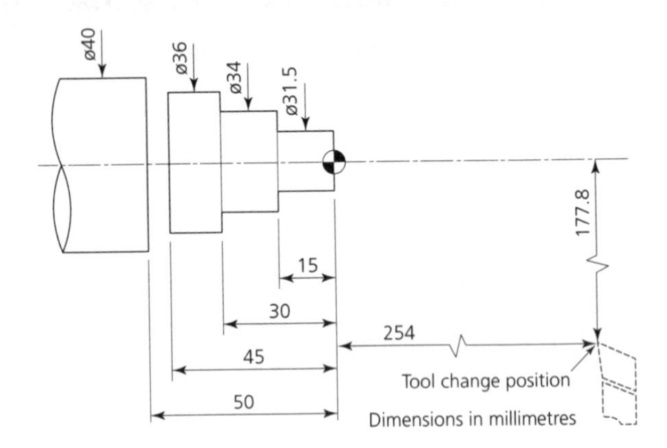

Figure 4.20 Component to be turned on a CNC lathe (dimensions in millimetres).

Sequence number	Code program	Explanation
N10	G71	Metric
N20	G95	Feed in mm rev
N30	G97 S1000 M03	Direct spindle. 1000 rev min^{-1} Spindle on CW
N40	G00 M08	Rapid mode. Coolant on
N50	G53 X177.8 Z254 TO	To tool change position
N60	M01	Optional stop
N70	T100	Rotate turret Pos 1
N80	G54 X0 Z2 T101	Move to start with tool 1
N90	G01 Z–0.5 F0.2	Move to depth prior to face end
N100	X31.5	Face end
N110	Z–15	Turn 031.5
N120	X34	Face edge
N130	Z–30	Turn 034
N140	X36	Face edge
N150	Z–50	Turn 036
N160	G53 X177.8 Z254 TO	To tool change
N170	T400	Rotate turret Pos 4
N180	G54 X40 Z–45 T404	To part off position. Tool 4 offset 04
N190	G01 X–1.0 F0.1	Part off
N200	G00 X40	Retract
N210	G53 X177.8 Z254 TO	To tool change
N220	M02	End program

Figure 4.21 CNC part program for turning the component in Fig. 4.20. Machine: Hardinge lathe; controller: GE1050.

Test your knowledge 4.2

1. Describe briefly what is meant by 'RP'.
2. Describe briefly what is meant by 'reverse engineering'.
3. Describe briefly what is meant by 'CNC'.
4. Explain what is meant by a canned cycle and how it can help the programmer.
5. Compare the advantages and limitations of *open loop* and *closed loop* systems when applied to automated machines.

Key notes 4.1

- ICT stands for information and communications technology.
- The internet is a term that describes a worldwide computer network. It is not a single entity but a co-operative system of networks linking many institutions and organizations.
- An Intranet is a private network within an organization or institution. It is similar to the Internet but is used for confidential communications with limited access.
- Extranets are several intranets linked together so that any number of users can collaborate for a project but retain the security afforded by a private network.
- The WWW is a term used to describe the hypertext interface to the information available on the Internet. It is a subset of the Internet.
- HTTP stands for hypertext transfer protocol.
- HTML stands for hypertext markup language for writing Web pages for a Web site.
- HTML tells a Web browser (such as Netscape's Communicator or Microsoft's Internet Explorer) how to display the various elements of a Web page).
- IP stands for Internet protocol address: a 4- to 12-digit number that identifies a specific computer connected to the Internet.
- ISP stands for Internet service provider. The ISP assigns IP addresses.
- URL stands for uniform resource locator.
- Search engines, such as Google, use automated software called Web crawlers or spiders to collect and log the information requested.
- Web directories, such as Yahoo, offer an editorially selected, topically organized list of Web sites.
- Database, the name given to an organized collection of useful data stored in the memory of a computer.
- A spreadsheet is a document compiled with the aid of a software package such as Excel or Lotus to replace manual accounting and management documents.
- CAE stands for computer-aided engineering.
- CAD stands for computer-aided design or computer-aided drawing.
- CAM stands for computer-aided manufacture.
- CAD/CAM was originally used to describe the integration of design and manufacture. This has now largely given way to CIM.
- CIM stands for computer integrated manufacture.
- RP is the manufacture of models, prototypes and even dies and moulds by the transfer of numerical data from the CAD image. The data from the CAD image are transferred directly to SLA that builds up the design in 3D layer by layer or to a desktop CNC type milling machine that cuts the design out of a block of soft material.
- Reverse engineering is a process that obtains digital data from an existing product by probing or scanning and then processing the data on a computer so that they can be scaled or modified and then used to produce either a positive or a negative replica.
- CNC is the automatic control of machine tools and other equipment, such as woodworking routers, knitting machines and weaving looms, by means of a computer program.
- An automated control system that feeds control data to the stepper motors moving the machine elements but which *provides no feedback* to ensure its commands have been performed is said to be an *open loop* system.

- An automated control system that feeds control data to the servomotors moving the machine elements and which *provides feedback* to ensure its commands have been performed is said to be a *closed loop* system.
- A program is an alphanumeric set of instructions that can control a CNC machine so that it manufactures a required product.
- A canned cycle is a pre-programmed standardized cycle embedded in the controller's memory so that the programmer can call it up by a simple code as and when required. For example, G81 is a canned drilling cycle.

4.2 Control technology and robotics

Our daily lives are affected directly by electronic automation, from the bread we eat to the cars we drive. The global economy is very competitive with contracts for the purchase of goods going to countries where labour and material costs are lowest. That is why, in order to remain competitive, so many companies have had to adopt high levels of automation. For example:

- Bakeries where the ingredients are mixed automatically in powerful mixers fed from giant hoppers. The dough is automatically weighed, cut to size to within critical limits and then dropped into baking tins. These tins, containing the dough, are carried through an infra-red oven on a slowly moving conveyor to emerge baked to a consistent standard ready for the customer.
- Pharmaceutical products are made in class 100 clean room conditions where the equipment and the environment have to be biologically clean to avoid contamination. High levels of automation limit the risk of contamination from human beings. Many critical phases such as the filling and heat-sealing of glass injection phials take place automatically on equipment housed behind sealed glass screens, where even higher levels of clean conditions exist.
- The production of volatile and hazardous chemicals is a highly controlled and automated process. This not only reduces the risk of humans coming into contact with potentially dangerous products that may be corrosive, toxic, flammable or even explosive but reduces the risk to the environment by removing the chance of human error.
- The manufacture of cars and the assembly of other complex products, such as circuit boards, are also highly automated nowadays. Again this reduces the risk of human error and the tedium of the operator having to repeat the same process over and over again. This removes the opportunity for fatigue errors and improves product quality.

Let's now see how such high levels of system automation can be achieved.

4.2.1 Automation and control

Many automated systems are used to control manufacturing processes, for example the production and packaging of chemicals, pharmaceuticals, food and drink products, toiletries and paint products. No matter what the application, they all have the same building blocks of control.

- *Inputs* – signals from sensors, both digital and analogue (digital signals can only be *on* or *off* but analogue signals can *vary* in magnitude).

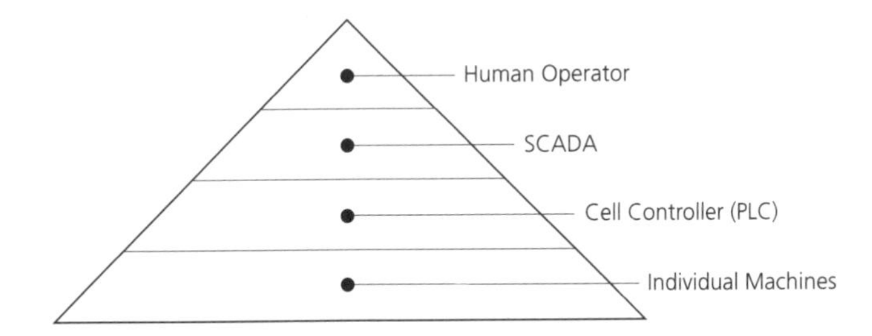

Figure 4.22 Computer-integrated manufacture (CIM) pyramid of command.

- *Outputs* – signals to switch solenoid valves or start and stop motors.
- *Feedback* – monitoring of the process in order to compare the desired result with the actual result, for example, a safety check.

The complexity of modern automated systems means that individual items of equipment all have their own, on-board, dedicated computer which communicates with other items of equipment, for example robots communicating with machine tools to see if they are ready to be loaded or unloaded. The individual systems can then be regarded as *cells of automation*. If they cannot communicate outside the cell to the larger system then they are known as *islands of automation*.

The main area of impact that ICT has had on many sectors of industry is its ability to network together diverse systems and machines. For example, distributed control can be used within a factory where intelligent machines can communicate with one another. It is the same as the CIM pyramid of command that is shown in Fig. 4.22. This has a human operator at the top level with ultimate control over the computerized supervisory and control and data acquisition (SCADA), which in turn is in command over the cell controller (PLC), right down to individual machines and sensors. Remember a PLC is a *programmable logic controller*. These devices are widely used in the control of systems and robots so let's now learn something about them.

Case study 4.9: Programmable logic controllers

PLCs are microprocessor devices that are used for controlling a wide variety of automatic processes, from operating an airport baggage handling system to brewing a pint of lager or controlling a robot to assemble a computer mother board. PLCs are physically and electrically rugged and they are designed specifically for operating in an industrial process control environment.

The control program for a PLC is usually stored in one or more semiconductor memory devices. The program can be entered (or modified) using a simple hand-held programmer, a laptop computer, or downloaded from a local area network (LAN). PLC manufacturers include Allen Bradley, Siemens, and Mitsubishi.

A block schematic diagram of a typical PLC is shown in Fig. 4.23. The heart of the PLC is a central processor unit (CPU) and its immediate support devices, ROM, RAM and timers/counters. In addition, two dedicated areas of RAM (known as the process input image and the process output image)

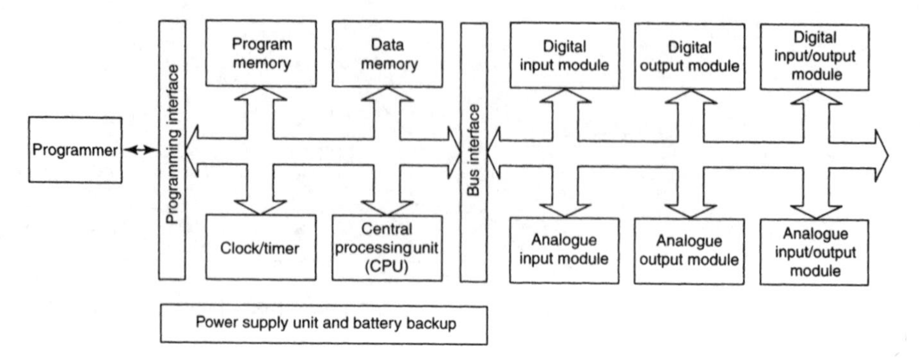

Figure 4.23 Block schematic of a PLC.

help to keep track of the state of all the system's inputs and outputs (more about this later). Also, in addition, most PLCs incorporate some on-board input/output (I/O) modules.

A series interface provides a means of programming the PLC (either by means of a hand-held programmer as shown in Fig. 4.24(a) or by downloading programs from a PC). In addition to the internal I/O module, as number of external I/O modules may be connected to a common data path (or 'bus') linked to the controller. These modules contain the necessary interface circuitry (including level shifting, optical isolation, analogue to digital conversion and digital to analogue conversion as appropriate). A typical PLC arrangement based on a Siemens S5-95U PLC is shown in Fig. 4.24(b). This shows a number of features in common with all PLCs – notably the modular construction base on expandable base units.

Although most PLCs have some form of on-board I/O, this is often somewhat limited (typically to a few digital I/O lines plus one or two analogue inputs). Most 'real-world' applications require a PLC to be concerned with a much larger number of inputs and outputs and thus additional I/O modules must be provided. The following types of I/O modules are available:

- Digital input and output modules
- Analogue input and analogue output modules
- Timer devices
- Counter modules
- Comparator modules
- Simulator modules
- Diagnostic modules
- Other modules, including stepper motor controllers, and communication modules, etc.

In operation, the PLC's control program repeatedly scans the state of each of its inputs in order to determine whether they are *on* or *off*, before using this information to decide on what should happen to the status of each of its output lines. The PLC usually derives its inputs from *make* and *break* devices such as switches, keypads, buttons, or limit switches. It can also derive its input from sensors (for example liquid level sensors, temperature sensors, position sensors, motion sensors, etc.).

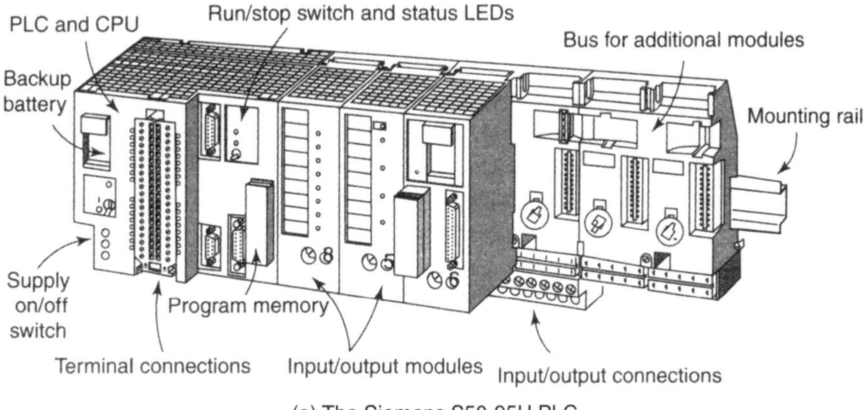

(a) The Siemens S50-95U PLC

(b) A typical hand-held PLC programmer

Figure 4.24 A typical PLC.

4.2.2 Industrial robots

There are two main types of industrial robot.

- The *general-purpose robot* possessing certain anthropomorphic (human-like) characteristics. The most anthropomorphic characteristic of an industrial robot is its arm and wrist. This combination of arm and wrist together with the capability of the robot to be programmed, makes it ideally suited to a variety of production tasks, including machine loading and unloading, welding, spray painting and assembly. The robot can be programmed to perform a sequence of mechanical motions, and it can repeat that motion sequence indefinitely until reprogrammed to perform some other task.
- The *pick and place* dedicated robot is less complex and is made up of standard components to perform a specific task. It costs a great deal less than the general-purpose robot and is also much more accurate. It is usually driven by a PLC. An example of a pick and place robot is considered in case study 4.10.

Robots, used in conjunction with other robots or with other machines, have three major advantages. First, they allow almost total automation of production processes, leading to increased rates of production, better quality control and an increased response to varying demand. Second, they permit the adaptability, at speed, of the production unit. The production line can be switched rapidly from one product to another, for example from one model of a car to another. Third, manufacturers use robots to reduce manning levels and reduce their dependence on the availability of a large pool of skilled labour. This allows manufacturing to become global so as to take advantage of the lower costs associated with developing countries. This also assists the economies of such countries to develop.

Robot programming methods

As already stated dedicated robots are driven by PLCs that can be programmed from a hand-held programmer or from a computer. General-purpose robots, however, can be programmed in a number of ways. They share a number of software and hardware similarities with computer-controlled machines. They use the same linear and rotary encoders for positional control. They use the same stepper and servo drives and they can have open or closed loop (feed-back) control systems. General-purpose robots are controlled by dedicated computers that can be linked with other computer-controlled devices in order to build up an automated manufacturing cell. However the method of programming can be substantially different to that used with CNC machine tools.

'Lead-through' programming

This is done manually by a skilled operator who leads the robot end effector through the required pattern of movements. For example, the operator may hold the spray paint gun on the end of the robot arm and guide it through the sequence of movements necessary to paint a car body panel. The robot arm joint movements needed to complete this operation are automatically recorded in the computer memory of the robot and can be repeated when and as often as required.

'Drive-through' programming

The robot is programmed by driving it through the required sequence of movements, under power. The operator controls the speed and direction, etc., by means of a teaching pendant. This is a small hand-held keypad connected to the robot controller by a trailing lead. The motion pattern is recorded in the computer memory and can be repeated when and as often as required.

'Off-line' programming

'Off-line' programming has the same benefits that CNC simulation enjoys, i.e.:

- Checking that the robot simulation does not collide with any object within its cell.
- Expensive robots do not need to be taken out of production whilst the operator develops the robot program.

- Only one generic language is required to develop a program for a number of different robots; the program is then post-processed into the required specific robot language for each robot.

Robot arm geometry

A robot must be able to reach workpieces and tools. This requires a combination of an arm and a wrist subassembly, plus a 'hand' usually called the *end effector*. The robot's sphere of influence is based upon the volume into which the robot's arm can deliver the wrist subassembly and end effector. The end effector can be a welding head, a spray paint gun, a gripper for loading and unloading work or any other device. Let's consider the geometric configurations normally available.

Cartesian co-ordinate robots

Cartesian co-ordinate robots consist of three orthogonal linear sliding axes (orthogonal, at right angles to each other; linear, slide in a straight line). The control systems are similar to CNC machines, therefore the arm resolution and accuracy is also of the same order of magnitude. An important feature of a Cartesian robot is its equal and constant spatial resolution; that is, the resolution is fixed in all axes of motion and throughout its work volume of the robot arm. This is not the case with other co-ordinate systems. Figure 4.25 shows the basic geometry of a Cartesian robot.

Cylindrical co-ordinate robots

Cylindrical co-ordinate robots consist of a horizontal arm mounted on a vertical column, which, in turn, is mounted on a rotary base. The horizontal arm

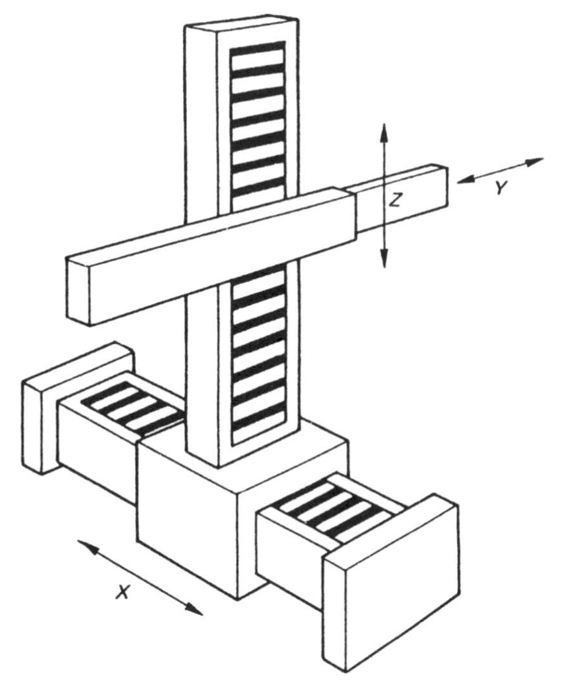

Figure 4.25 Cartesian co-ordinate robot.

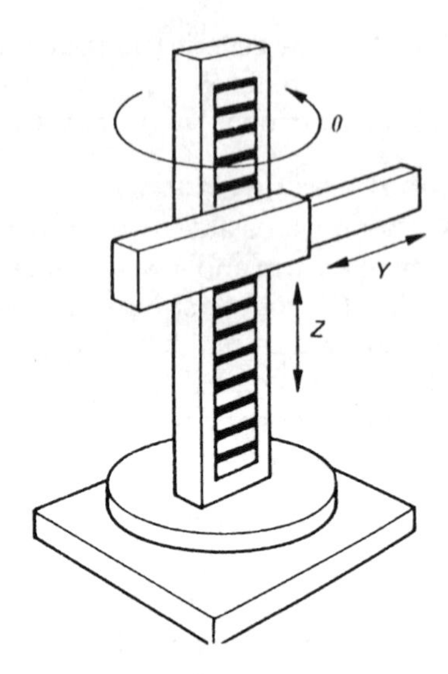

Figure 4.26 Cylindrical co-ordinate robot.

moves in and out, the carriage moves vertically up and down the column, and these units rotate as a single assembly on the base. The working volume is therefore the annular space of a cylinder. The resolution of cylindrical co-ordinate robots is not constant and depends on the distance between the column and the wrist along the horizontal arm. Cylindrical geometry robots as shown in Fig. 4.26 do offer the advantage of higher positional speed at the end of the arm compared with Cartesian co-ordinate machines.

Spherical co-ordinate robots

These consist of a rotary base supporting an elevated pivot that, in turn, carries a retractable arm as shown in Fig. 4.27. The working envelope at the wrist end of the arm is a thick spherical shell. The main advantage of spherical robots over the Cartesian and cylindrical ones is better mechanical flexibility and speed of positioning. Motions with rotary axes are much faster than motions along linear axes. The main disadvantage of spherical robots compared with their Cartesian counterparts is that there are two axes having a low resolution that varies with the arm length. Therefore spherical robots, although faster, are less accurate.

Revolute co-ordinate robots

The revolute, or articulated, arm robot consists of three rigid members connected by two rotary joints mounted on a rotary base. It closely resembles the human arm. The end effector is analogous to the hand which is attached to the forearm by the wrist. The elbow joint connects the forearm to the upper arm and the shoulder joint of the robot connects the upper arm to the base as shown in Fig. 4.28. Since the revolute robot has three rotary axes it has a relatively low resolution compared with the robots discussed previously. Further, errors occurring at the arm joints are cumulative (they add up). The main advantages of revolute co-ordinate robots are their mechanical

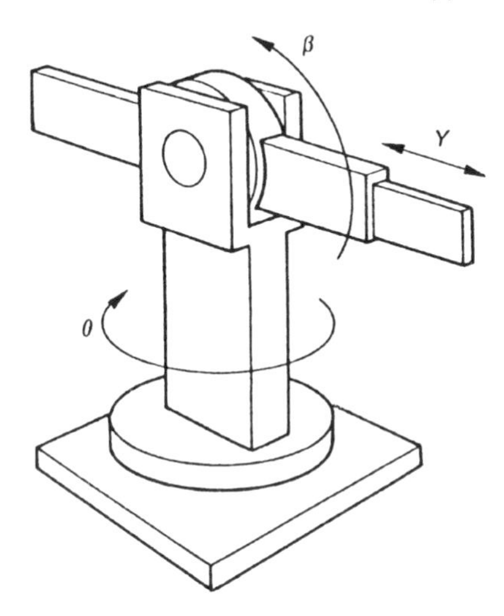

Figure 4.27 Spherical co-ordinate robot.

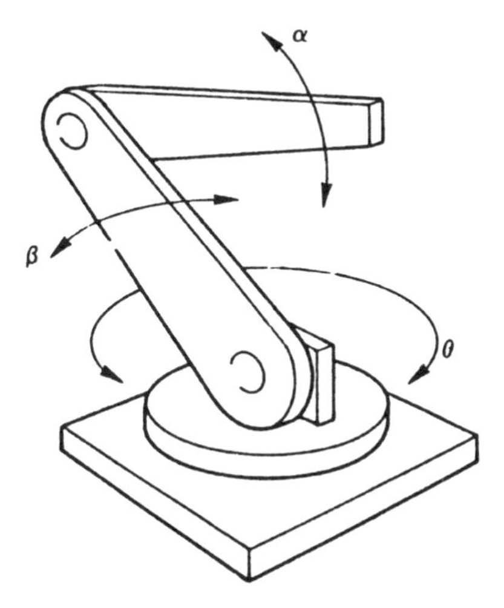

Figure 4.28 Revolute co-ordinate robot.

flexibility and their positional speed. This has made them the most popular of the medium-sized robots.

The wrist

The *end effector* (spray-paint gun, welding head, nut-runner, gripper, etc.) is connected to the main frame of the robot through the wrist. The wrist includes three axes of movement referred to as roll, pitch and yaw as shown in Fig. 4.29. In order to reduce the mass at the wrist, the wrist drives are sometimes located in the base of the robot, and the motion is transferred to the wrist by rigid links or chains. Reduction of mass at the wrist increases the maximum allowable load and reduces the moment of inertia that, in turn, improves the dynamic performance of the robot. Figure 4.29 shows the configuration of a typical industrial robot.

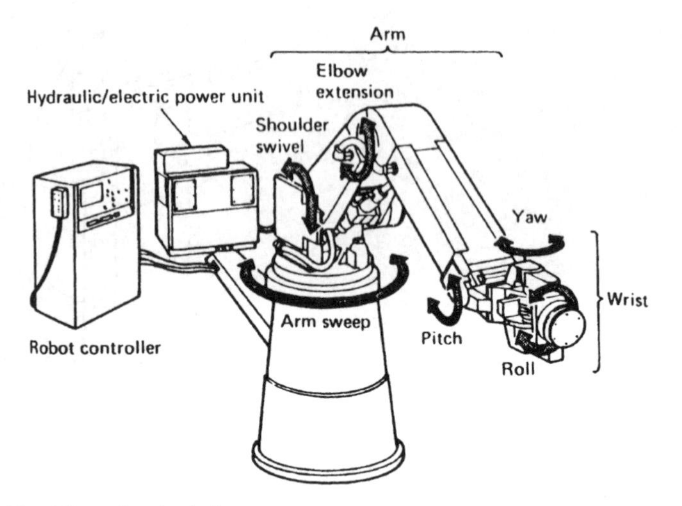

Figure 4.29 The robot 'wrist' movements.

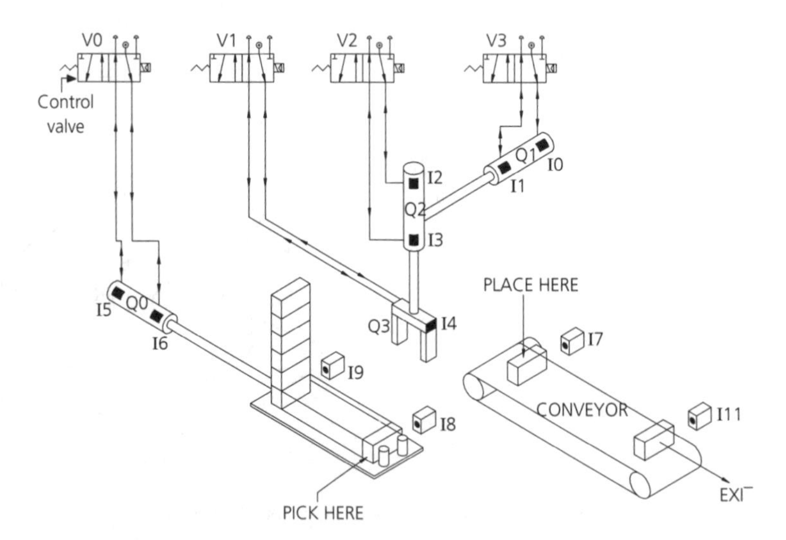

Figure 4.30 Pick and place robot schematic diagram.

Case study 4.10: Pick and place robot

This describes a simple pick and place robot, the function of which is most easily understood by comparing the schematic diagram Fig. 4.30 with the flow chart shown in Fig. 4.31. The robot is constructed from standard pneumatic piston and cylinder units (Q0 to Q8 inclusive) and driven by a PLC responding to proximity sensors (I0 to I19 inclusive).

The proximity sensors inform the PLC of the status of the pistons and whether it is safe for the next move in the sequence to be executed. If all is safe and the pistons are in the correct positions, then the PLC instructs the solenoid valves V0 to V3 inclusive to initiate the next move.

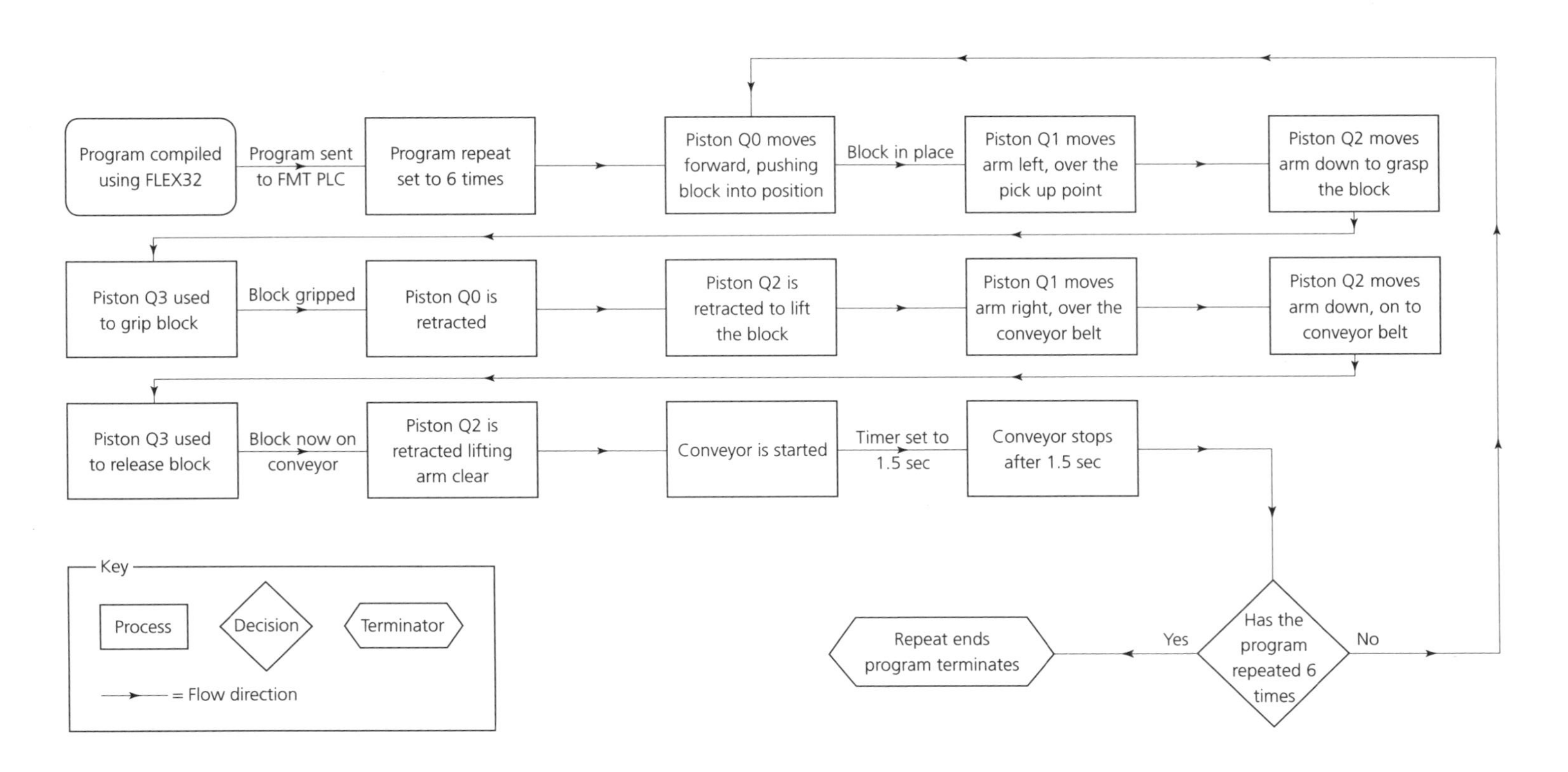

Figure 4.31 Flowchart for pick and place operation.

Case study 4.11: Wood machining

It has been stated previously that automation has spread from engineering to many other manufacturing sectors. An interesting example is SMC 5S flexible production line for wooden window production manufactured in Italy. This is shown in Fig. 4.32.

The System 5S window processing line takes the raw timber and fully integrates the following wood machining operations: planing, moulding, tenoning, boring, profiling, machining slots for hinges, automatic weather strip fitting and glazing bead mitre cut. Finally the machine automated processing line delivers complete sets of wooden window components ready for assembly.

The processing line is fully automated and computer controlled using ASTROCAD software. Once the window design has been drawn using ASTROCAD the program is generated immediately and automatically since the software 'knows' CNC codes for all the machines that have been integrated into the processing line. There are three levels of control.

Level 1

Level 1 is the machine level and uses integrated and modular CNC/PLCs that are generally available in industry in order to optimize the performance of the individual machines, to increase reliability and to facilitate the procurement of spare parts in the event of a breakdown.

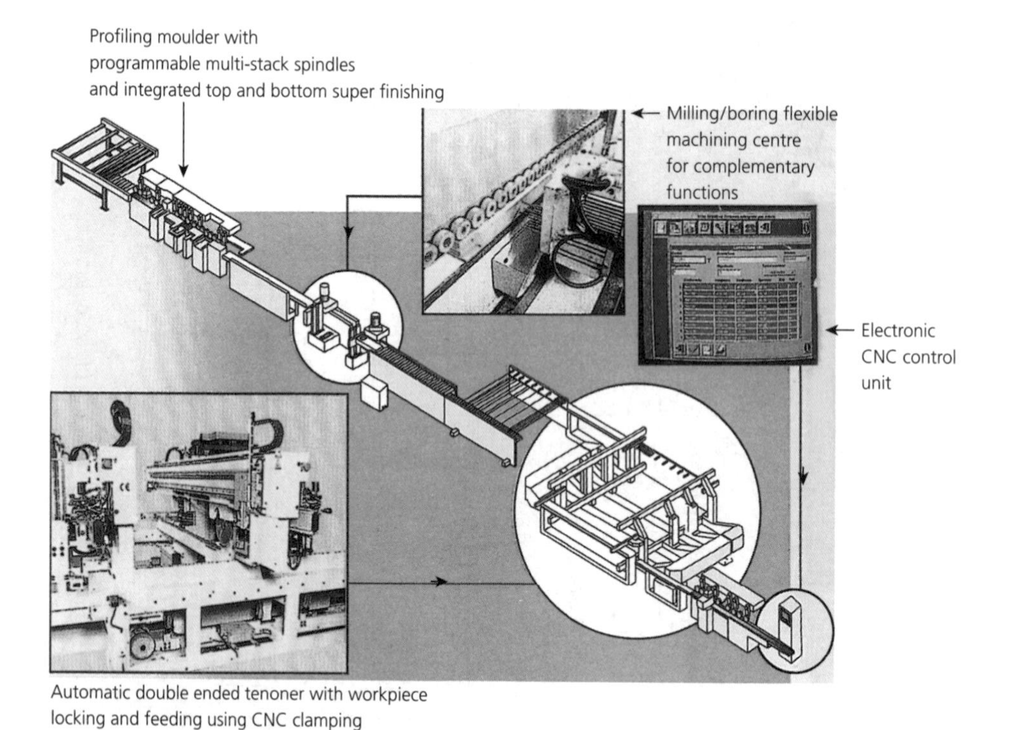

Figure 4.32 SMC system 5S flexible window processing line.

Level 2

The line supervisor makes use of an industrial PC with user specific software that has been designed to offer XWINDOW/MOTIF with graphics to assist the operator to handle all functions, i.e.:

- Centralized diagnostics to reduce intervention time when machine problems occur
- The reporting of executed functions on the production line assisted by graphic monitoring
- Pre-arrangement for remote service
- Pre-arrangement for network link with network processing systems.

Level 3

Level 3 provides supervision of all operating activities in the company. This can be provided with appropriate software packages that handle production orders and automatically elaborates on the jobbing list of the pieces to be produced, according to a customer purchase order. From the raw material fed into the first machine in the line, complete sets of machined components are automatically delivered in quantities and shapes required ready for assembly by the customer. The product mix can be pre-planned on a daily basis.

Test your knowledge 4.3

1. Describe the main differences between general-purpose industrial robots and special purpose, pick and place robots.
2. Describe the essential differences between a CNC controller and a PLC.
3. Suggest two typical applications for a PLC.
4. Explain what is meant by 'lead through' programming and 'drive through' programming and state where each would be used.
5. With the aid of sketches show what is meant by TWO of the following: cartesian co-ordinate robots, cylindrical co-ordinate robots, spherical co-ordinate robots and revolute co-ordinate robots.

4.3 Smart materials, modern materials and components

In addition to the application of modern technology to the manufacture of goods for consumers worldwide, we must not forget that the materials being used are also constantly being improved and new materials and combinations of materials are being developed. So, also, are the components that are used in the assembly of modern products.

4.3.1 Shape memory materials

Strange as it may seem, quite a number of materials appear to have a *memory*. For example, if a spring made from a hardened and tempered medium- or high-carbon steel, or from an alloy spring steel, is compressed or stretched it will return to its original size and shape when the deforming force is removed. The spring 'remembers' its original size and shape providing it is not overloaded.

You will no doubt have seen adverts on television for spectacle frames that can be twisted out of shape but which return to their correct shape when the deforming forces are removed. One group of alloys called *Flexon* contain 50–80 per cent copper, 2–8 per cent aluminium and the remainder titanium. This alloy is springier than steel and it is also stronger and more fatigue resistant. Another *shape memory effect* (SME) alloy contains 40 per cent nickel and 50 per cent titanium. It is eight times springier than spring steel, lighter and more fatigue resistant.

A group of commercial alloys are the SME brasses. Here the 'memory effect' is based upon temperature. The alloy is able to exist in two distinct shapes or crystal structures, one above and one below a critical *transformation temperature* (T_m). This reversible change in structure is linked to a dimensional change and the alloy thus exhibits a 'memory' of its high and low temperature shapes. These alloys contain 55–80 per cent copper, 2–8 per cent aluminium and the remainder is zinc. By choosing a suitable composition transition temperatures between -70°C and $+130$°C can be achieved. The force associated with the changes in shape can be used to operate temperature-sensitive devices, such as thermostats, in place of the more conventional bimetallic strip.

Plastics can also have a memory, for example the coiled-up cable that connects the handset to the body of your telephone. You stretch the cable out when you use the phone, only for the cable to return to its original coiled up state when you replace the handset on its cradle. The plastic insulation has a memory.

Use is made of the memory effect of certain plastics in the packaging industry by heat shrinking protective foil around pre-packed foods and other commodities. The commodity to be packed is wrapped loosely in the foil that is then heated to above its glass *transition temperature* (T_g), the temperature at which the material ceases to be brittle like glass and becomes flexible, but well below its *melting temperature* (T_m). This results in the plastic material returning to its original disorientated polymer structure and shrinkage occurs resulting in the commodity becoming tightly packed.

Fabrics can also have a memory, for example 'drip-dry' materials do not need to be ironed after being washed. They are simply hung up to dry and, as soon as the moisture in them has evaporated, they become smooth and smart again, free from creases. On the other hand some fabrics are treated to retain their creases, for example the creases down the legs of trousers. You only have to look at photographs of men wearing wool fabric trousers from the nineteenth and early twentieth century to see how baggy and shapeless were those trouser legs.

4.3.2 Plastic coatings and adhesives

Thermosetting and thermoplastic materials for mouldings, extrusions, sheets and foils have already been considered earlier in this book. Some of these plastic materials can also be modified for use as coatings and adhesives.

We have already considered the coating of metal objects by fluidized bed dipping and by the use of plastisols. Glossy magazines are printed on paper that has been coated with a glossy plastic. Fabrics can also be coated. Let's now consider some coated and 'smart' fabrics.

Case study 4.12: Coated and smart fabrics

One of the earliest was Mackintosh rubber used for rainwear. It consisted of a cotton or satin fabric coated with latex rubber. Unfortunately it tended to perish and had only a limited life. As well as being impervious to rain, the rubber lining was also impervious to perspiration so, on a hot day, they could become uncomfortable to wear. Nowadays rainwear fabrics make use of synthetic rubbers such as neoprene and coatings in the form of silicone waxes. An interesting fabric that is impervious to rain, yet comfortable to wear and provides heat insulation in cold environments is Gore-tex WindStopper®. This is shown in Fig. 4.33. There is an inner fleece that can be made from polyester fibres, a WindStopper® membrane that is microporous, and an outer layer of fleece or knitwear. The microporous membrane is impervious to wind and rain yet allows moisture (perspiration) from the body of the wearer to pass through and evaporate.

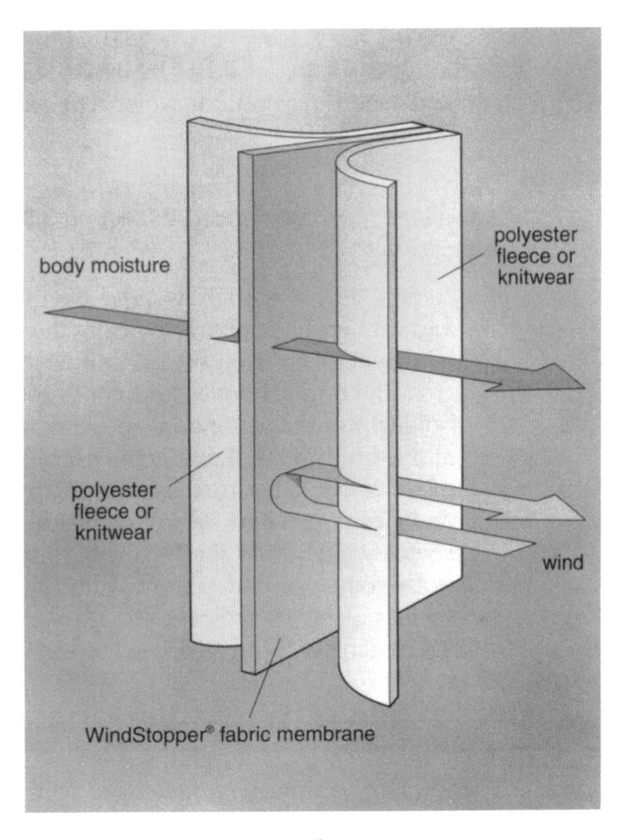

Figure 4.33 How Gore-tex WindStopper® fabric works.

Truly smart fabrics can be:

- Passive smart – that can only sense environmental conditions
- Active smart – that can sense, react and adapt to environmental conditions.

Applications of smart materials include:

- *Smart fabrics* (*skins*) are used for sound absorption and vibration control in transport seats, for example cars, trains and aeroplanes.
- *Heat generating fibres* that respond to the environment by changing colour to fit in with the surroundings are used for combat uniforms in the army.
- *Thermo-chromic dyes* react and respond to temperature and beachwear made from fabrics treated with such dyes can warn the wearer of over-exposure to sunlight.
- Intelligent polymer systems are available that measure and respond to the movement of the wearer.
- *Kevlar* is a fibre that is five times stronger than steel and is resistant to cutting. It is used for body armour for the police and military, and as a reinforcing fabric in some car tyres and drive belts.
- *Nomex* is a fire resistant fabric used for protective clothing for fire-fighters, racing car drivers and plant operatives.
- *Synchilla* is an artificial fleece made from recycled polypropolene (discarded fizzy drinks bottles). The fibres are spun into yarns and knitted into a fabric that is then brushed to raise the surface. It is used for sports and leisure-wear and is water repellent.

4.3.3 Adhesives

Adhesives were introduced in Section 3.8.4. They can be divided into two main groups:

- *Natural adhesives* such as *gums* obtained from vegetable matter, resins and rubbers being extracted from the sap of trees and starch derivatives being extracted from the byproducts of flour milling. *Glues* are derived from the horns, hooves and bones of animals. Such glues soften at the boiling point of water and set again on cooling. They were widely used, at one time, in the manufacture of furniture and wooden toys.
- *Synthetic adhesives* are based on thermosetting, thermoplastic polymers and elastomers and some examples are given in Table 4.1. These adhesive are much stronger that the natural adhesives and are more versatile in the range of materials that they will join. Some high strength synthetic adhesives are referred to as 'superglues'. These are now considered in case study 4.13.

Case study 4.13: Specialized adhesives

The specialized adhesives developed by the Loctite Corporation are reactive polymers that change from liquids to solids through various chemical polymerization (curing) reactions. The company has developed numerous

Table 4.1 Main categories of adhesives

Origin	Basic type	Adhesive material
Natural	Animal	Albumen, animal glue (inc. fish), casein, shellac, beeswax
	Vegetable	Natural resins (gum arabic, tragacanth, colophony, Canada balsam, etc.); oils and waxes (carnauba wax, linseed oils); proteins (soyabean); carbohydrates (starch, dextrines)
	Mineral	Inorganic materials (silicates, magnesia, phosphates, litharge, sulphur, etc.); mineral waxes (paraffin), mineral resins (copal, amber); bitumen (inc. asphalt)
	Elastomers	Natural rubber (and derivatives, chlorinated rubber, cyclized rubber, rubber hydrochloride)
Synthetic	Elastomers	Synthetic rubbers and derivatives (butyl, polyisobutylene, polybutadiene blends (inc. styrene and acrylonitrile), polyisoprenes, polychloroprene, polyurethane, silicone, polysulphide, polyolefins (ethylene vinyl chloride, ethylene polypropylene) Reclaimed rubbers
	Thermoplastic	Cellulose derivatives (acetate, acetate-butyrate, caprate, nitrate, methyl cellulose, hydroxy ethyl cellulose, ethyl cellulose, carboxy methyl cellulose) Vinyl polymers and copolymers (polyvinyl-acetate, alcohol, acetal, chloride, polyvinylidene chloride, polyvinyl alkyl ethers) Polyesters (saturated) (polystyrene, polyamides, nylons and modifications) Polyacrylates (methacrylate and acrylate polymers, cyano-acrylates, acrylamide) Polyethers (polyhydroxy ether, polyphenolic ethers) Polysulphones
	Thermosetting	Amino plastics (urea and melamine formaldehydes and modifications) Epoxides and modifications (epoxy polyamide, epoxy bitumen, epoxy polysulphide, epoxy nylon) Phenolic resins and modifications (phenol and resorcinol formaldehydes, phenolic-nitrile, phenolic-neoprene, phenolic-epoxy) Polyesters (unsaturated) Polyaromatics (polyimide, polybenzimidazole, polybenzothiazole, polyphenylene) Furanes (phenol furfural)

adhesives with special curing properties for unique situations. They are popularly known as 'superglues'.

- *Adhesives cured by anaerobic reactions.* These are single component adhesives that remain inactive as long as they are in the presence of atmospheric oxygen. When the adhesive is deprived of oxygen by bringing the adhesive coated surfaces together, curing occurs rapidly.
- *Adhesives cured by ultraviolet (UV) light.* The cure times of these adhesives depends upon the intensity and wavelength of the UV light. Curing is prevented during storage by the presence of chemicals called photo-inhibitors. The photo-inhibtors are 'split' by the UV light releasing free radicals that, in turn, start the curing process.
- *Adhesives cured by anionic reaction (cyanoacrylates).* Single component cyanoacrylates cure on coming into contact with slightly alkaline surfaces. In general, the ambient humidity of the air in the workplace and on the bonding surface is sufficient to initiate curing to handling strength within a few seconds. The optimum conditions are achieved

when the relative humidity value is 40–60 per cent at room temperature. Parts must be quickly joined after application of the adhesive since curing commences within a few seconds. The adhesive should be applied to only one of the surfaces being joined.

- *Adhesives cured with activator systems* (*modified acrylics*). These adhesives cure at room temperature with activators such as acetone and heptane-isopropanol. Adhesives and activators are applied separately to the bonding surfaces. Adhesive to the one joint and activator to the other joint surface. Curing takes place as soon as the joint surfaces are brought together.
- *Adhesives cured by ambient moisture.* These adhesives/sealants polymerize (cure) in most cases through a condensation effect that involves a reaction with ambient moisture.

1. *Silicone rubbers* vulcanize at room temperature by reacting with ambient moisture. The solid rubber silicone has excellent thermal resistance, is flexible and tough, and is an effective sealant for a variety of fluid types.
2. *Urethanes.* Polyurethanes are formed through a mechanism in which water reacts (in most cases) with a formulative additive containing isocyanate groups. These adhesives/sealants have excellent toughness and flexibility and excellent gap filling (up to 5 mm) properties.

4.3.4 Biodegradable plastics

One third of all plastic production is used for disposable products such as bottles, bags and packaging. We have already seen how thermoplastics can be ground up and recycled since they soften ready for moulding every time they are heated. We have also seen how discarded polypropylene fizzy drink bottles can be chopped up and converted into warm, shower-proof fleece fabrics. The rest are buried in landfill sites where they cause a major environmental problem since most plastic materials do not degrade.

Biodegradable plastics are being developed that will break down safely over a relatively short time when exposed to the micro-organisms in the ground or in water with no toxic byproducts being produced. There are three main types of bioplastics available at present:

- Starch-based bioplastics
- Lactic-based bioplastics
- Bioplastics made from the fermentation of sugars.

The starch-based system is the most widely used commercially. Here the starch granules are processed into a powder, which is heated until it becomes a sticky liquid. The liquid is then cooled, formed into pellets and processed in a conventional plant like any other thermoplastic. The lactic-based system uses fermenting corn or other similar feedstocks to produce lactic acid. This is then polymerized to form a polyester resin that is biodegradable. In the third type, organic acids are added to a sugar feedstock and the resulting reaction produces a highly crystalline and very stiff polymer which, after further processing behaves in a manner similar to more conventional polymers, yet remains biodegradable.

PVC used for building applications (doors, door frames, window frames, gutters, and facia boards is often formulated to make it resistant to ultraviolet light). However for many other applications the reverse characteristics are required so that a 'photo-activator' is added to the plastic. This absorbs ultraviolet light and destroys the polymer by making it biodegradable.

4.3.5 Technology applied to food and drinks

Once mankind had discovered how to make fire, it was a short step to discover that cooking made many foods more digestible and palatable. The next step up the culinary ladder was the discovery of *additives*, for example salt (sodium chloride). Then came yeast to leaven bread. As has been stated earlier in this book, the primitive ales brewed by the monks in the Middle Ages were much improved by the addition of hops, both as a preservative and as a flavouring. Nowadays we have a whole range of additives, both natural and synthetic, to enhance our cooking skills, for example:

- Saccharin, aspartame and acesulfame-k as artificial sweeteners in 'slimline' foods and drinks
- Sodium benzoate, sodium sorbate, sodium metabisulphite and sulphur dioxide as preservatives
- Ascorbic acid as an antioxidant
- Sodium alginate as a thickener in salad creams and low-fat spreads
- Mono- and di-glycerides as emulsifiers in salad creams and low-fat spreads
- Sodium pyrophosphate and sodium bicarbonate as raising agents when baking
- Artificial colourings and flavourings (E-numbers)
- Added vitamins and trace elements to reinforce foods low in these essential dietary aids
- Monosodium glutamate as a flavour enhancer.

The ever-increasing range of commercially available convenience foods and processed foods available on the shelves of our supermarkets is testimony to the continuing development of food and drink technology. In addition to the development of convenience foods, modern technology has entered the catering industry. Developed by the electronics industry for radar installations, microwave technology is now used for reheating and cooking a range of foods both fresh and frozen.

4.3.6 Computer technology

The development of computers has advanced with the miniaturization of components. The first leap forward was the replacement of relatively unreliable thermionic valves by reliable, solid-state devices such as transistors and diodes. These reduced the size of the computer from something that occupied a large room and dissipated prodigious amounts of heat energy to a much smaller and more reliable device. The next advance was the introduction of the integrated circuit in which very many components were built on a silicon chip. Such integrated circuits could be designed as amplifiers, oscillators, processors, and computer memory devices.

Case study 4.14: The microprocessor

The information revolution has largely been made possible by developments in electronics and in the manufacture of integrated circuits in particular. Ongoing improvements in manufacturing technology have given us increasingly powerful integrated circuit chips. Of these, the microprocessor (a chip that performs all the essential functions of a computer) has been the most notable development.

A microprocessor is a single chip of silicon on to which are built all the components that perform all the essential functions of a computer *CPU*. Microprocessor are found in a huge variety of applications including engine management systems, computer controlled machine tools and woodworking machines, looms (weaving) machines, knitting machines, sewing machines, printing machines, embroidery machines, office machines, PLCs and many others.

The CPU performs three functions: it controls the system's operation; it performs algebraic and logical operations; and it stores information (*data*) while it is processing that information. The CPU works in conjunction with other chips, notably those that provide random access memory (RAM) and read-only memory (ROM) and input/output (I/O).

The key process in the development of increasingly powerful microprocessor chips is known as microlithography. In this process the circuits are designed and laid out using a computer before being photographically reduced to a size where individual circuit lines are 1/100 the size of a human hair. Early miniaturization techniques, which were referred to as large-scale integration (LSI), resulted in the production of the first generation of 256 K-bit memory chips. Such a chip actually has a storage capacity of 262 144 bits (where each bit is 0 or 1). Today, as a result of very large scale integration (VLSI), chips can be made to contain more than a million transistors. A typical VLSI device is shown in Fig. 4.34.

One of the most popular microprocessors to appear in the last 20 years was originally conceived in 1977 by a project team at Motorola. This moved the technology forward from the 8-bit microprocessors available at that time

Figure 4.34 A VLSI chip.

(circa 1977) to a 16-bit microprocessor that was extendible to a full 32-bits. Furthermore this microprocessor would require a 64-pin package (only 40-pin packages had been used at that time) and the semiconductor chip would have to be very much larger than anything that had been used before. The 68 000 chip employs what has become commonly known as a complex instruction set. It got its name from the fact that it contained 68 000 transistors on the same chip. This may seem very small compared, say, with the latest generation of Intel Pentium chips but, at the time, it was a major leap forward and a huge act of faith on the part of the design and development team who laid the foundations for today's powerful microprocessors. Computers that employ complex instruction sets are known as CISC (complex instruction set computer) machines.

4.3.7 Microprocessor systems

In the previous case study we discussed the development of microprocessors. Even the latest VLSI would be useless, however, if left unsupported by a number of other electronic devices connected together in to what is called a *microprocessor system*. The next case study will consider such a system.

Case study 4.15: Microprocessor systems

The first microprocessor systems, developed in the 1970s, were simple and crude by today's standards, but they found immediate application in the automotive industry for engine management and antilock braking systems. Today, microprocessor systems are found in a huge variety of applications from personal computers to washing machines.

Figure 4.35 shows a block schematic diagram of a complete microprocessor system. All such systems, regardless of their complexity, conform to this basic arrangement.

The CPU is generally the microprocessor chip itself and contains the following units:

- Storage locations (called *registers*) that can be used to hold instructions, data, and addresses during processing.

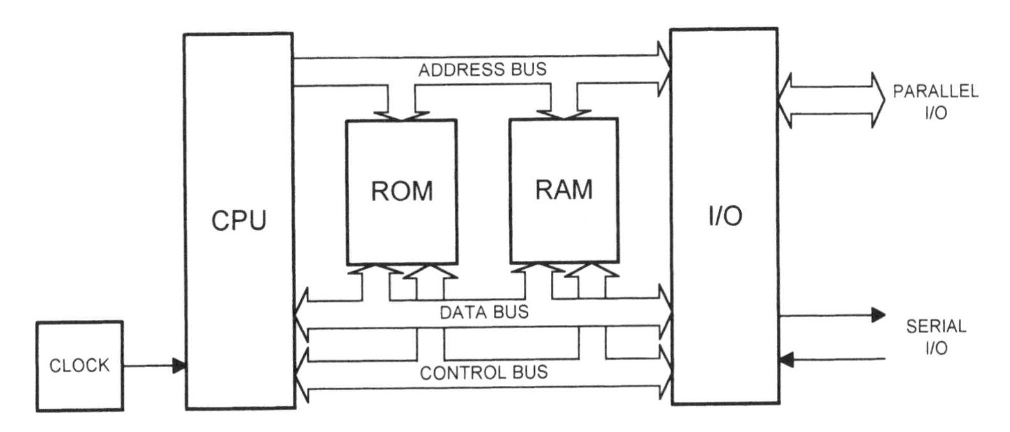

Figure 4.35 A block schematic diagram of a microprocessor system.

- An *arithmetic logic unit* (ALU) that is able to perform a variety of arithmetic and logical functions (such as comparing two numbers).
- A *control unit* that accepts and generates external control signals (such as *read* and *write*) and provides timing signals for the entire system.

In order to ensure that all data flow within the system is orderly, it is necessary to synchronize all the data transfers using a clock signal. This signal is often generated by a clock circuit (similar to the clock circuit in a digital watch but much faster). To ensure accuracy and stability the clock circuit is usually based on a miniature quartz crystal.

All microprocessors require access to read/write memory in which data (for example, the results of calculations) can be temporarily stored during processing. Whilst some microprocessors (often referred to as *microcontrollers*) contain their own small read/write memory, this is usually provided by means of a semiconductor *RAM* devices.

Microprocessors generally also require more permanent storage for their control programs and, where appropriate, operating systems and high-level language interpreters. This is usually provided by means of semiconductor *ROM* devices.

To fulfil any useful function, a microprocessor system needs to have links with the outside world. These links are usually provided by means of one, or more, VSLI devices that may be configured under software control and are therefore said to be programmable. The I/O devices fall into two general categories: *parallel* (where a byte of 8-bits is transferred at a time along 8 wires), or *serial* (where one bit after another is transferred along a single wire).

The basic components of a microprocessor system (CPU, RAM, ROM, Clock and I/O) are linked together using a multiple connecting arrangement known as a *bus*. The *address bus* is used to specify memory locations (i.e. addresses). The data bus is used to transfer data between devices and the control bus is used to provide timing and control signals (such as *read* and *write*, *reset* and *interrupt*) throughout the system. A typical microprocessor system, based on an 8-bit processor, is shown in Fig. 4.36. This particular example is used to control a computer printer.

Figure 4.36 This microprocessor system (based on an 8-bit microprocessor) is used to control a computer printer.

Note how it is built up on a printed circuit board where, in addition to all the integrated circuits, there are a large number of discrete devices, such as wire-ended resistors and capacitors. The wire ends are bent at right-angles to the device body and passed through holes in the appropriate track ends of the circuit board and soldered into position. If you refer back to Fig. 4.34 you will see, in the top left-hand corner, that surface mounted components such as resistors have been used. These are soldered directly to pads on the track ends. This reduces the size of the circuit board required and speeds up production as no holes have to be cut in the circuit board and the wire ends of the components don't have to be bent and inserted in the holes prior to soldering. It is such attention to detail, the continuing miniaturization of discrete components, and the adoption of back-lighted liquid crystal displays (LCDs) that enables today's lap-top computers to be as small as they are, yet have the computing power of a desk-top computer.

Test your knowledge 4.4

1. Explain what is meant by 'shape memory effect' (SME) as applied to materials.
2. Describe two examples of coated fabrics.
3. Describe two examples of 'smart' fabrics.
4. Explain the essential differences between natural glues, natural gums and synthetic adhesives and give examples where each could be used.
5. Explain why biodegradable plastics need to be developed.
6. Name six typical food additives and the type of product where each is likely to be used.
7. State what is meant by a VLSI chip.
8. Explain what is meant by a microprocessor system.

Key notes 4.2

- PLCs are free-standing units capable of being used to control a range of devices and processes, unlike CNC control units that are integrated into and dedicated to one particular machine.
- General-purpose industrial robots have anthropomorphic (human-like) characteristics in their movements. They can be programmed for loading and unloading machines and carrying out such tasks as spray painting and welding and have a CNC type integrated control unit.
- Pick and place robots are designed for a single purpose and are controlled by a standard PLC. They operate more rapidly and are cheaper in first cost. When they are no longer required, the sensors and piston and cylinder units together with the PLC can be adapted for use elsewhere.
- Lead through programming. A general-purpose industrial robot can be programmed by a skilled operator holding the device (for example, a spray paint gun) on the wrist of the robot and guiding the robot through the sequence of movements necessary to perform a particular job. The robot stores the movements in the memory of its controller and repeats them as often as required.
- Drive through programming. The robot is programmed by driving it through the required sequence of movements by use of a hand-held keypad connected to the robot controller through a trailing cable.
- Off-line programming. The program is written and tested by simulation on a computer. The files are saved to disk and transferred to the robot controller.
- An end effector is any device that can be mounted on the 'wrist' of a robot to perform a particular job.

- The term *shape memory effect* (SME) applies to the ability of a material to *remember* its shape and return to it when its shape has temporarily been disturbed, for example a spring that returns to its original length after being compressed or stretched.
- Adhesives may be natural or synthetic. Natural gums are obtained from vegetable matter and have the advantage of being non-toxic. Natural glues are made from the hooves and horns of animals. Such glues melt at the boiling point of water and become solid again when cooled.
- Synthetic adhesives are much stronger and more versatile than natural adhesives. They are mainly byproducts of the plastics industry.
- *Superglues* for specialist applications have been developed that set through a number of different mechanisms such as anaerobic reactions, exposure to ultraviolet light, anionic reactions and ambient moisture.
- Biodegradable plastics are plastic materials that are being developed to ease the burden of waste disposal. Conventional plastics lie in the ground indefinitely and never decompose. Biodegradable plastics decompose through the action of sunlight and also through the action of the natural bacteria in the soil.
- Microprocessor*s* are single integrated circuit chips carrying all the components needed to carry out all the essential functions of a computer CPU. Modern microprocessors are very powerful and have millions of components on a single chip. They are the result of VLSI technology.
- Microprocessor systems are the circuits built up around the microprocessor itself and consist of the CPU, the RAM, the ROM, clock and I/O device.

4.4 Impact of modern technology

So far in this chapter we have considered the impact of modern technology on the range, types and availability of some products. Without the development of solid state devices such as transistors and diodes, and the further development of these devices into integrated circuits on single chips, computers would never have reached today's level of computing power, reliability and small size in relatively low cost devices.

4.4.1 Range, types and availability of products

Without powerful, compact, reliable and low-cost computers we would not have been able to develop CAE and automation. It is hard to imagine manufacturing without CNC machine tools and robots, designing without CAD and the ability to carry out complex stress analysis at the touch of a button and simulation to test our designs without the high cost of manufacturing models to test to destruction.

It's equally hard to imagine life in the home without all the computer-controlled devices we now take for granted. Devices such as digital radio and television, CD and DVD players, computer-controlled clothes washing and washing-up machines, lap-top computers to send and receive e-mails to our friends throughout the world and the ability to surf the internet for information, and mobile telephones.

We must remember that the development of the transistor and related solid-state devices has led to the miniaturization and micro-miniaturization of the associated components. For example, in the days before transistors when radio sets had thermionic valves, voltage and current values were very high compared with modern radio and hi-fi equipment resulting in bulky components to provide the necessary insulation and current carrying capacity. Also room had to be provided in the cabinet for the ventilation of radios and amplifiers that dissipated upwards of a hundred watts of heat energy.

4.4.2 Safety and efficiency in manufacturing

The adoption of automated manufacture through the use of CNC machine tools and robots has led to increased safety and efficiency by:

- Lowering the risk of accidents related to operator fatigue. Robots can work day and night without becoming tired and do not require the levels of light and heat in the working environment that human operators require.
- Lowering the risk of accidents by the use of robots to remove human operators from hazardous environments and the need to perform potentially hazardous processes manually.
- Reduced energy consumption not only in the working environment by being able to use robots in environments with reduced light and heat but even working under 'lights out' conditions. Energy consumption is reduced by miniaturization – fewer raw materials are required and, in turn, less energy is required to extract and manufacture the reduced amounts of raw materials used. Less energy is required to transport miniaturized components and the devices that use them. Less energy is usually required to power the products using modern designs, materials and components.
- Robots can work faster. If you have ever seen a pick and place robot building up a circuit board for a computer you would realize that the most skilled manual worker would be unable to approach such rates of production – very definitely 'no contest'. Since 'time is money' then the faster rates of production achievable with automated machines and processes results in cost reduction and increased cost-efficiency.
- Automated manufacture results in products that are of higher quality and that are more reliable since machines do not become tired and make mistakes, so reject components are largely eliminated. Less scrap means less raw material used and less wasted energy. It also results in lower costs and higher profits.

4.4.3 Improved characteristics of manufactured products

The use of computer-aided design and 3-D modelling together with computer-aided stress analysis has enabled manufacturers to improve the performance of their products by increasing the reliability and improving the strength to weight/density ratio. In the case of road and rail vehicles, and aircraft the use of computer-aided design has also enabled manufacturers to improve the power to weight/density ratio. You only have to compare modern machines with their earlier counterparts to be seen in museums, to realize how grossly over-engineered were the earlier examples. For example, the use of light alloys and plastic materials in cars coupled with computerized engine management systems has greatly increased their performance but, at the same time, reduced the amount of fuel they burn. Note that catalytic converters may remove some noxious substances from the exhaust fumes but, unfortunately, they reduce engine performance and increase the amount of fuel used.

Research into side impact bars and 'air bags' to protect the occupants of a car in the event of an accident, together with front and rear crumple zones that reduce the impact in the event of an accident has greatly increased the safety of modern vehicles. This research has greatly increased the chances of survival of the passengers in the event of an accident. Again, this research would have been very difficult to perform without the availability of computer-aided design and stress analysis, and computer simulation and 3D-modelling.

4.4.4 Marketing

No matter how sophisticated a new design or an improved design may be, or what good value for money it offers the buyer, it would all be a waste of time and money without a good marketing strategy. Marketing has also become increasingly dependent on computer techniques. Computers are used for:

- *Market analysis* to determine what the market demand is likely to be and what a company's competitors are offering. Also it can investigate and determine market trends so that new products can be developed in advance of a company's competitors.
- *Advertising* as can be seen by the increasing use of multi-media techniques used to create eye-catching adverts on television.
- *Web sites*. Firms both large and small cannot afford to miss out on the potential sales that can be generated by a good Web site. Web sites may be simple or they may be animated. They may contain digital photographs of a company's products or they may contain virtual imagery.

4.5 The advantages and disadvantages of the use of modern technology

The effects of modern technology have been as far reaching on society as the earlier industrial revolution and the advent of steam power. The age of the computer and modern technology has brought with it many changes for good, but it has also brought hardship as well.

4.5.1 Changes in the type and size of the workforce

Traditionally manufacturing in all its guises developed wherever there was a supply of skilled labour, suitable material resources, and good transport facilities. For example, in the days of steam and heavy engineering, coal was required in large quantities together with iron ore. Until the coming of the railways, industry had to be established alongside sources of these raw materials. However, with the advent of electricity and a modern road system industry could be sited anywhere.

People of all levels of ability and education were needed in large numbers as most of the work was manual. It was often hard and tiring, but there were jobs for large numbers of people. Some employers were hard taskmasters but others were enlightened and provided schools, hospitals and housing for their employees. With the coming of modern technology, automation has cost many jobs and resulted in much unemployment. The loss of coal mining and iron smelting to developing countries has brought ruin and deprivation to many old industrial communities.

The new industries have not provided the mass employment and relatively high wages of the ones they have replaced. Many replacement industries pay lower wages. This, at a time, when people's social expectations are constantly being raised by advertising, has resulted in both parents having to work. The work itself often involves unsocial hours resulting in an increasing breakdown in home life.

Many of the new jobs are highly skilled – such as computer and CNC programming. This has resulted in the need for smaller workforce of very skilled and highly qualified people being able to satisfy all the needs of manufacturing. Many displaced workers have had to retrain and find employment in the service industries, often for lower wages.

4.5.2 Changes in the working environment

Many of the old industries were carried out in dirty and hazardous environments that would not be tolerated either by the workers or the Health and Safety Executive today. Modern manufacturing may provide fewer jobs, but the working conditions are much improved. Obviously the preparation of frozen processed foods and pharmaceuticals must be carried on in scrupulously clean and hygienic conditions. Many of the industrial diseases associated with the old industries have now been swept away. However some new ones have emerged. Computer operators can develop spinal problems due to poor posture whilst working in one position for long periods of time. Repetitive strain injury (RSI) can be developed by constant use of a keyboard and mouse.

Modern workplaces tend to be light and airy compared with their older counterparts. The exception being some smaller 'start-up' industries taking advantage of cheap and old premises. Hard physical work has been much reduced by the widespread use of mechanical handling devices and the more general use of smaller and lighter commodities be they raw materials or finished products. Heavy lorries delivering the manufactured goods will have power-assisted steering and automatic gearboxes with sound insulated, comfortable cabs to reduce driver fatigue. They will often be fitted with cranes or tailgate lifts to aid loading and unloading.

Case study 4.16: The impact of modern technology on the global environment and sustainability

Much of the manufacturing industry rests in the hands of the big, international (global) companies. Three aspects of this 'globalization' should be considered:

1. The effect on employment
2. The effect on the global environment
3. Sustainability.

Employment

We have already discussed how the use of automated machines and industrial robots has reduced the demand for skilled labour. It is no longer necessary to locate factories where there is an abundant supply of skilled workers. At the beginning of this book we saw how footwear could be produced on a global basis to exploit the cheapest materials and lowest wages. This means that design, development and marketing will most likely be carried out in Europe or the USA and the actual manufacturing will take place in the Far East or in Central or South America. This enables the big, multinational companies to continue to make profits in an increasingly competitive world economy. Unfortunately it also means that many basic manufacturing industries have been moved out of the UK as being too expensive, thus creating unemployment and the loss of many highly skilled and highly paid jobs.

There are of course still many jobs that are labour intensive, for example in the clothing industry. Fashions vary from country to country. Clothes for use in Europe tend to be designed in Europe and marketed in Europe. Clothes for America tend to be designed in America and marketed in America. However, the sourcing of materials for these clothes and their actual manufacture will take place wherever the operating costs are lowest.

The effect on the global environment

The global environment is affected by the release of 'greenhouse gases' mainly carbon dioxide, carbon monoxide and methane (a natural hydrocarbon gas). Apart from natural sources such as volcanoes and other geophysical activity, these gases are the result of burning fossil fuels.

We have already seen that much heavy industry has moved away from the developed countries to the developing countries of the world where operating costs are lower and environmental controls are more relaxed. Although this has reduced many of the problems associated with pollution in the developed countries, it has not improved the situation globally. In fact it has, to some extent, worsened the situation since there is less control over environmental pollution in many of the developing countries. In addition to the burning of fossil fuels (coal, oil and natural gas) for the extraction of metals from their ores, fossil fuels continue to be burnt in large amounts for the generation of electricity for heating and lighting and for driving the machinery used in manufacturing. Further, fossil fuels are used for all forms of transport. Even the electrification of our railways has not helped the situation since fossil fuels have to be burnt in the generation of electricity. It has merely moved the problem to new sites but, globally, there has been no improvement in the total volume of greenhouse gases released into the atmosphere.

Another problem is the destruction of the ozone layer by the release of nitrogen oxides (NOXs) and refrigerants (CFCs) into the upper atmosphere. The use of CFCs has now been banned in many countries, but the generation of nitrogen compounds has increased as fossil fuels are burnt more efficiently at higher temperatures. The only long-term solution lies in developing technologies that reduce our dependency on fossil fuels globally.

Pollution

Although the generation of greenhouse gases and the destruction of the ozone layer can, in the long term, have catastrophic effects on the environment

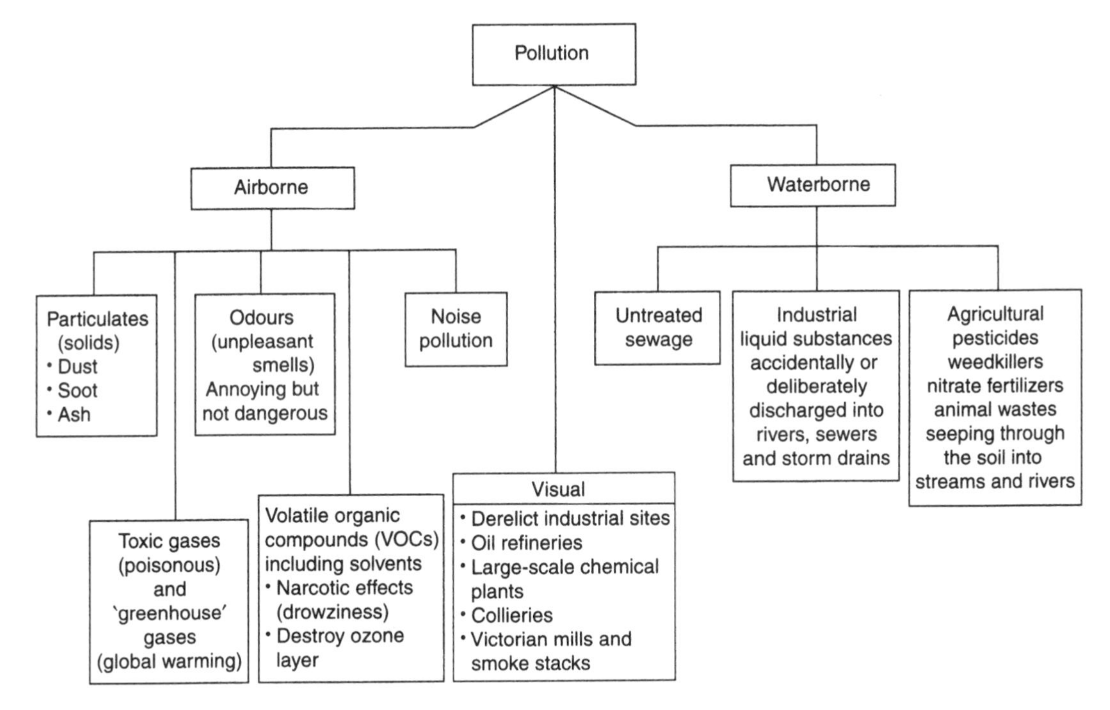

Figure 4.37 Sources of pollution.

and the weather systems upon which we depend for our survival, pollution has much more immediate effect on our *quality of life*. Figure 4.37 shows some of the more common sources of pollution.

Sustainability

Metallic ores and fossil fuels are *finite resources*. That is, once used they have gone forever and can never be replaced. Since plastic materials are byproducts of oil refining they are also a finite resource. We can delay using up these finite resources by recycling waste, and by reducing waste. Although recycling waste is expensive, so is waste disposal. The development of recycling technology and the use of biodegradable materials helps but only by postponing the inevitable when the finite resources start to run out. The categories of waste are outlined in Fig. 4.38.

Sustainable resources are resources that can be bred or grown. For example wool and cotton will be available as long as there are sheep to shear and cotton plants to grow. Timber obtained from quick growing conifer trees is also a sustainable resource, but we must reduce our dependency on the slow growing deciduous hardwoods from the tropical rainforests. These rainforests must be preserved as they are the source of renewable plant species that are found nowhere else. Sustainable sources of energy must be found and developed, not only because the fossil fuels are a finite resource but also to reduce pollution levels in the global environment. Table 4.2 sets out the advantages and disadvantages of sustainable (renewable) energy sources. However, not only must we develop sustainable energy resources that are *environmentally friendly*, we must endeavour to use less energy as well, for

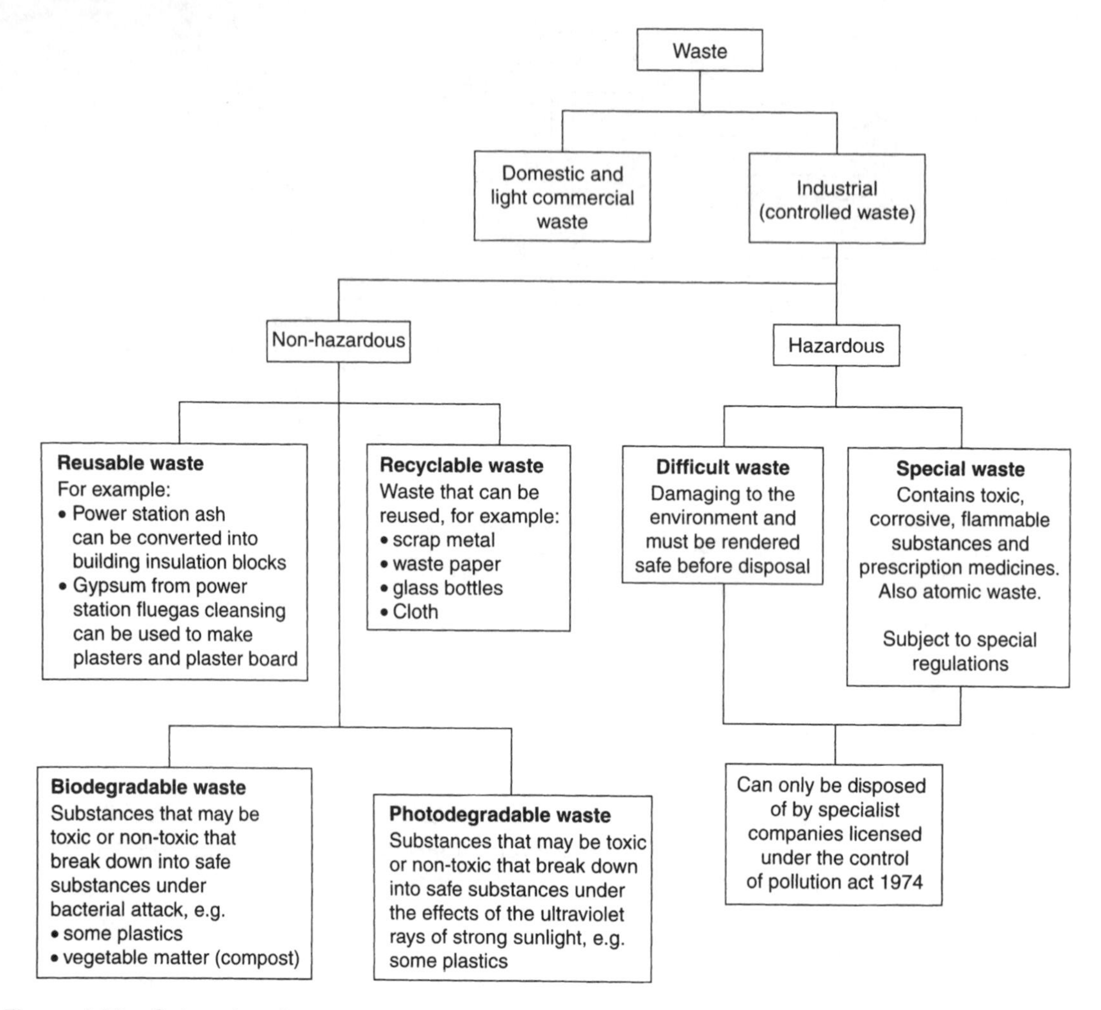

Figure 4.38 Categories of waste.

example the micro-miniaturization of the goods we make, the use of lightweight materials to reduce the energy used by vehicles and the goods they carry. We must seek to increase efficiency in our use of energy, for example we have already moved on from the incandescent electric light to fluorescent lights and the next development will be even more efficient high-intensity light emitting diodes. The relatively power-hungry computer screens of desktop computers based on the thermionic cathode ray tube is giving way to the much more energy efficient and compact liquid crystal diode (LCD) displays that have been developed for lap-top computers.

4.6 The stages in manufacturing a product

Throughout this book we have examined many aspects of manufacture. We have considered the structure of many diverse industrial sectors, the design process, the materials required and their properties, the processes, marketing,

Table 4.2 Advantages and disadvantages of infinite (renewable) energy sources

Advantages	Disadvantages
Hydroelectricity • Clean, no atmospheric pollution • Cheap and constant supply	• High capital cost constructing dams • Ecological damage to flooded areas • Often remote from the users • Only 2% hydroelectricity produced in UK for geographical reasons
Tidal energy • Could be a constant source of electricity • Clean, no atmospheric pollution	• High capital cost makes such generation uneconomical at present • Ecological damage to wetlands surrounding estuaries across which barrages would have to be built
Wave energy • Clean, no atmospheric pollution • Could be a constant source of electricity	• Experimental work in progress but many technical problems to be overcome
Solar energy • Clean, no atmospheric pollution • Silent in operation (see wind power) • Safe in operation • Mainly used for water heating in buildings at present	• Costs make it uncompetitive with conventional generation at present • Large areas of land required • Generation zero at night when demand is at a maximum • Sunlight insufficiently constant in UK
Wind energy • Clean, no atmospheric pollution • Wind is strongest in winter when demand is greatest • Technology relatively simple	• Large wind turbines are noisy and have a limited output • Large areas of land are required for the many turbines required for wind 'farms' • Visual pollution – many large turbines clustered on hill tops are unsightly • Rotating blades can cause local disruption to TV reception
Geothermal energy • Cleaner than conventional energy sources • Some hot springs and geysers are used for district heating schemes but not in the UK	• No suitable sources in UK expect possibly in Cornwall where hotrock formations exist deep underground • Long-term effects of removing heat energy from the core of the earth are at present unknown • Generation of electricity by geothermal energy only in the experimental stage at present
Biomass energy (i) Biogas (landfill gas) LFG • Contains up to 60% methane and has a similar calorific value to propane, butane and natural gas • Produced when vegetable matter rots. The residual compost is a useful fertilizer (ii) Burning waste vegetable matter • Waste vegetable matter and waste forestry products are cheap and plentiful • Burning reduces volume of matter to be disposed of	• Atmospheric pollution occurs whenever burning takes place • LFGs are dangerous if allowed to accumulate – should be flamed off or used for electricity generation • Atmosphere pollution occurs whenever burning takes place • Soil erosion and climatic changes occur if proper woodland management and a policy of reforestation is not carried out

the impact of computer technology for communications and process control, and the impact on the global environment and the sustainability of energy and material resources. The key stages in manufacturing any product can be summarized as:

- Marketing
- Design
- Production planning
- Purchasing, material supply and control
- Production processing
- Assembly and finishing
- Packaging and despatch.

When you examine and investigate any product, you need to be able to identify all the key stages in its manufacture as set out in the above list. Let's consider a bicycle.

The marketing department will have already done its research and identified a requirement for a new model. It will have drawn up a design brief and discussed a number of design solutions with the engineering design staff. Once a design has been agreed, the marketing department will turn its attention to promotional advertising.

Since the bicycle under consideration is to be propelled by the rider, the design must be kept as light as possible. However, the designer must ensure that the frame is adequately strong to bear the weight of the rider and to resist the impact loading when travelling over rough ground if the design is for a mountain bike. A bicycle designed only for use on smooth roads could be built more lightly. The bearings and transmission must be as frictionless and free-running as possible to prevent waste of the limited energy available. Bought-in components must be to a British Standard Specification wherever possible in the interests of obtaining spares and ensuring uniform quality.

Production planning ensures that the materials and bought-in components arrive at the correct time in the correct quantities to ensure that the production line can run smoothly. Materials and parts failing to arrive can cause loss of production and labour having to be laid off. Materials and parts arriving too soon can clutter up the space round the production line and tie up working capital, leading to cash-flow problems. Many firms adopt a just-in-time (JIT) policy.

Materials and parts must be sourced from suppliers of known and proven reliability. Whilst the purchasing department will always be looking for lower costs, the cheapest is not always the most economical if quality is poor and delivery uncertain and erratic. Nevertheless, the purchasing department must always be on the lookout for new materials and new sources of materials. Such materials must always be rigorously examined and tested before being adopted for production.

The decision must be made whether to manufacture the various components such as frames, wheels, chains, sprockets, etc., in house or to purchase them from specialist manufacturers. Assembly will certainly be carried out in house but finishing will depend upon the size of the company and the finish required. The electroplating of bright parts will nearly always be subcontracted to specialist polishing and plating companies. The frames will have already been enamelled by the framemaker if the frames are bought in ready-made. A large manufacturer of bicycles will make its own frames and finish them in its own paint shop.

Let's consider, for a moment, how *materials requirement planning* (MRP 1) and *manufacturing requirement planning* (MRP 2) can be used to control manufacture. These are both examples of computerized systems of management and control.

4.6.1 Materials requirement planning (MRP 1)

Figure 4.39 shows the relationship between the component parts of the bicycle. The completed assembly is at the *highest level* and is classified as *level 0*. The items making up the constituent parts are classified as the lower levels 1, 2, 3 and 4. Therefore, if the items classified at these lower levels are not being sold as spare parts, they can be classified as *dependent* items. That is, the demand is solely dependent upon the number of complete bicycles assembled.

The completed bicycle is an end in itself and it is not used in, or as part of, a higher level item. It is, in fact, the finished product that the customer purchases from the local cycle shop. The volume of production is based upon sales forecasts for that particular model and is *independent* of the demand for any constituent parts. Thus the demand for a company's products can be classified into two distinct categories:

- *Dependent demand* generated from some higher level item. It is the demand for a component that is dependent upon the number of assemblies being manufactured. As such, it can be accurately calculated and inserted into the *master schedule.*
- *Independent demand* generated by the market for the finished product. The magnitude of the independent demand is determined from firm customer orders or by market forecasting. For MRP 1 purposes, it is independent of other items. It is not used in scheduling any further assemblies.

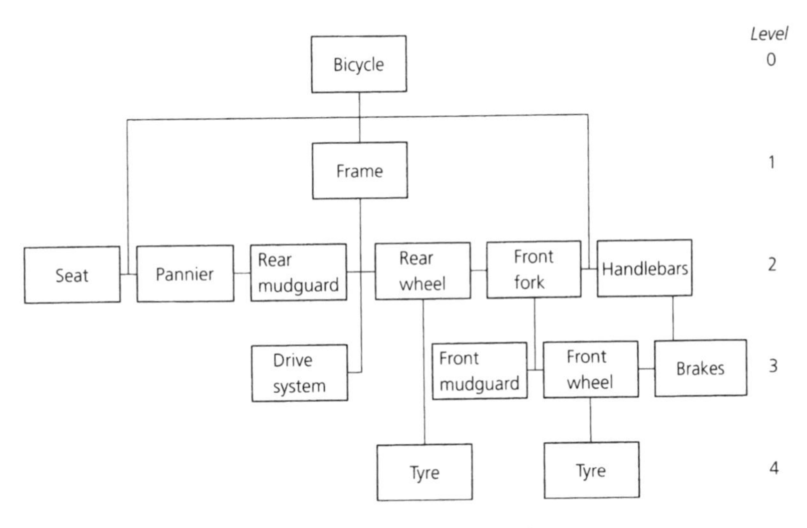

Figure 4.39 Materials requirement planning (MRP 1).

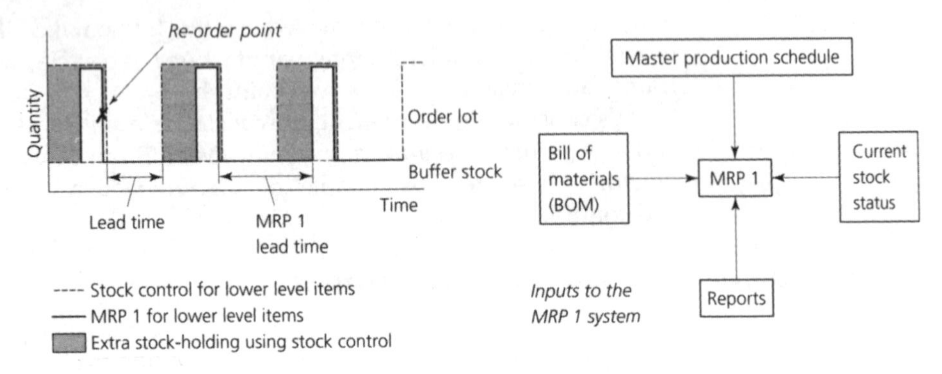

Figure 4.40 Comparison: stock control versus MRP 1.

A stock-control system for all the parts shown in Fig. 4.39 could use either order-point methods or sales forecasts for each item. To do this, each separate item has to be treated as independent. MRP 1 software, however, can calculate the amount of stock of dependent items required to meet the projected demand for the independent higher level item that, in this example, is a bicycle (note how once again manufacturing depends upon computer technology not only for processing, but also for production management).

MRP 1 is a widely used system. Whilst the demand for the independent items (level 0) cannot be accurately predicted – and are therefore unsuitable for stock control methods – lower level items (1, 2, 3, 4,) are dependent on a higher level item and can thus be calculated with accuracy. That is, although it is only possible to forecast next year's sales of complete bicycles approximately, it is possible to predict that exactly as many tyres will be required as there are wheels.

Further, stock control methods assume that usage is at a gradual and continuous rate. In practice, however, this is not the case. The call on parts will occur intermittently, with the quantity depending upon the batch size of the final product. This often results in the holding of excessive stocks. Materials requirement planning assists in the planning of orders so that subsequent deliveries arrive just prior to manufacturing requirements as shown in Fig. 4.40.

Master production schedule

This is the schedule or list of the final products (*independent* or *higher level* items) that have to be produced. It shows these products plotted against a time or period base so that the delivery requirements can be assessed. It is, in fact, the definitive production plan that has been derived from firm orders and market forecasts. In order to minimize the number of components specified, care has to be exercised in deciding how many of the basic level 0 assemblies are to be completed. The total lengths of the time periods (the *planning horizon*) should be sufficient to meet the longest lead-times.

Bill of materials

The bill of materials (BOM) contains all the information regarding the build-up or structure of the final product. This will originate from the design office and will contain such things as the assembly of components and their quantities,

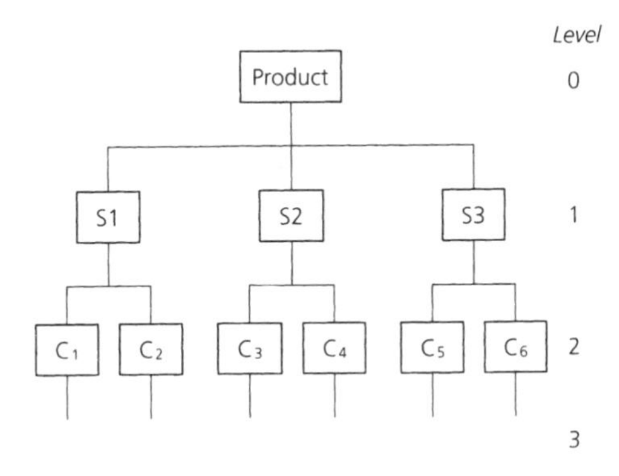

Figure 4.41 Structure for bill of materials: S, subassemblies; C, components.

together with the order of assembly. The structure and presentation of a typical BOM is shown in Fig. 4.41. The number of parts should also be shown at each level.

Current stock status

This is the inventory or latest position of the stock in hand, pending deliveries, planned orders, lead-time, and order batch quantity policy. All components, whether in stores, off-site, or in progress need to be accurately recorded in the inventory record file.

4.6.2 MRP 1 in action

Upon receiving the input data, the system has to compute the size and timing of the orders for the components required to complete the designated number of the final product (the independent or highest level item). It performs this operation by completing a level-by-level calculation of the requirements for each component. Since the latest stock information is available to the system, these calculations will be the net value of the orders to be raised.

$$\text{Requirements} = \text{net requirements} + \text{stock currently available}$$

where

$$\text{Stock currently available} = \text{current stockholding} + \text{expected deliveries}$$

Inherent in the calculations is the factor for lead-time, i.e. ordering and manufacturing times. The system must determine the start date for the various subassemblies required by taking into account the lead-times, i.e. it offsets these lead-times to arrive at the start dates.

A complication that the system has to overcome is the use of some items that may be common to several subassemblies. The requirements of these items that are commonly used must then be collected by the system to provide one single total for each of these items. Figure 4.42 shows a standard planning sheet for one of the outputs of MRP 1. Other reports required in addition to the order release notices are:

- Rescheduling notices indicating changes
- Cancellation notices where changes to the master schedule have been made

	1	2	3	4	5	6	7	8
Gross requirements				140				
Scheduled receipts			50					
On-hand	100		150					
Net requirements				100				
Planned order releases				100				

Figure 4.42 Standard planning sheet for one of the outputs of MRP 1.

- Future planned order release dates
- Performance indicators of such things as costs, stock usage, and comparison of lead-times.

A number of advantages are claimed for MRP 1 software. These include:

- A reduction in stocks held in such things as raw materials, work in progress, and 'bought-in' components
- Better service to the customer (delivery dates are met)
- Reductions in costs and improved cash flow
- Improved productivity and responses to changes in demand.

4.6.3 Manufacturing resource planning (MRP 2)

MRP 1 is a large step forward, compared with manual systems, in the procurement of materials and parts. MRP 1 provides a statement of what is exactly required and when it is required. Unfortunately, it takes no account of the production capacity of a particular manufacturing facility. Most factories have a production output that can only be changed over a period of time either by improved productivity or the injection of capital to purchase more productive equipment.

Clearly, by themselves, the outputs from MRP 1 software have limited value to the *master production schedule* except to detail the material requirements. Without the information regarding production capacity, it is difficult to match MRP 1 to actual production schedules. Remember that *capacity planning* is concerned with matching the production requirements to available resources (human resources and plant). It would be impractical to have a master production schedule that exceeded the capacity of the plant to meet its requirements. This could lead to decisions to extend the plant capacity by increasing the labour force, shift working, or subcontracting, when what is required is more sensible scheduling.

The shortcomings of MRP 1 led to the development of MRP 2 that brought into account the whole of the company's resources, including finance. Thus the initials MRP also stand for *manufacturing resource planning*. This is quite a different, and more difficult, philosophy than the original *materials requirements planning* when one considers the many variables that can affect shop floor production. For example, machines may breakdown, staff may become ill or leave, and rejects may have to be reworked. In addition, MRP 2 takes into account the financial side of the company's business by incorporating the *business plan*. MRP 2 is used to produce reports detailing materials

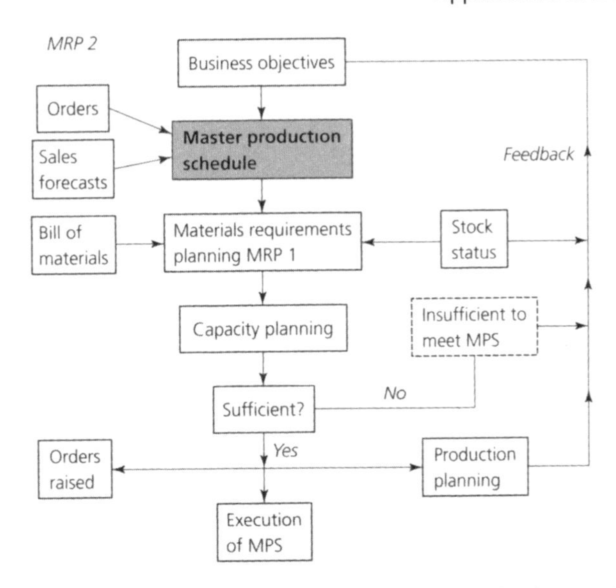

Figure 4.43 Schematic layout of MRP 1 and MRP 2 organized around the master production schedule.

planning together with the detailed capacity plans. This enables control to be effected at both the shop floor and the procurement levels. Note that MRP 2 does not replace MRP 1, but is complementary to it and *the two systems are used together.*

As in all computer systems, the information produced is only as good as the information fed into it. Therefore the system is dependent upon the feedback of information relating to those things under its control, i.e. such data as the state of the manufacturing process and the position of orders. MRP 2 contains enough information, and the power to organize it, to be able to run the whole factory. Figure 4.43 shows a schematic layout of production incorporating MRP 1 and MRP 2 organized around the master production schedule.

Master production statement (mission statement)

MRP 2 is a management technique for highlighting a company's objectives and breaking them down into detailed areas of responsibility for their implementation. It involves the whole plant and not just the materials provision. The key to MRP 2 lies in the master production schedule (also called the *mission statement*) that is a statement of what the company is planning to manufacture and, because it is the master, then all other schedules need to be derived from it. The master schedule should *not* be a statement of what the company would like to produce, as its own derivation lies in the company objectives. It cannot in itself reduce the company's lead-times, but it will provide the incentive for the plant personnel to take action in that area if required. It is important, as in all planning, to monitor adherence to the master production schedule. MRP 2 will not stop over-ambitious planning, but it will show up the consequences.

4.6.4 Packaging and dispatch

Let's return now to the final key stages in the manufacturing process, packaging and dispatch. Packaging is nowadays both a science and an art. With the advent of global manufacture, goods need careful and scientific packaging to prevent damage in transit as they are sent across the world to the customer. The method of packaging and transport will depend upon the size, weight and type of commodity.

The bicycles we have been considering would be delivered to the cycle shop with the minimum of protective packing. Some would be stored as stock, while a limited number would be unpacked and displayed ready for sale.

Beef may be sent from Argentina in special ships so that it can be chilled. Lamb is sent from New Zealand in ships that can keep it frozen. Perishable fruit and vegetables may need to be carefully packed to prevent damage and transported by air to reduce the time taken so that it arrives fresh at your supermarket. Products such as computers and television sets are packed in strong cardboard boxes cradled in preformed polystyrene packing to prevent damage. Many such boxes may be packed into a container to facilitate handling at the docks to prevent damage and theft. Large objects, such as machines, will be transported in strong wooden boxes, called packing cases, when travelling by sea.

Goods packaged for sale to the domestic customer not only have to be packaged to prevent damage, the packaging must be eye-catching and attractive. Consider the boxed and canned goods on the shelves of your supermarket. The carton is the immediate point of contact between the manufacturer and the customer. It must appeal to the customer, it must provide nutritional information, it must list the ingredients and carry warnings if it contains products than may be allergenic or are in any way harmful and it may contain preparation and cooking suggestions.

4.7 Investigating products

Our final case study takes the form of a product investigation. We are going to consider a product with which you are probably already familiar, a compact disc (CD). When you investigate a product you need to be able to research information from manufacturers and suppliers, handle and examine the product, carry out a simple assessment of its properties (including structure, heaviness, colour and surface finish), and evaluate the need for the technology, materials and components used. Research material can be downloaded from manufacturers' Web sites using an appropriate search engine. There is often an associated 'help line' which you can use to get supplementary information or a contact e-mail address.

Case study 4.17: Investigating a compact disc (CD)

CDs can provide 65 minutes of high quality, recorded audio or 650 Mbytes of computer data that is roughly equivalent of 250 00 pages of A4 text. It is hardly surprising that the compact disc has now become firmly established in both the

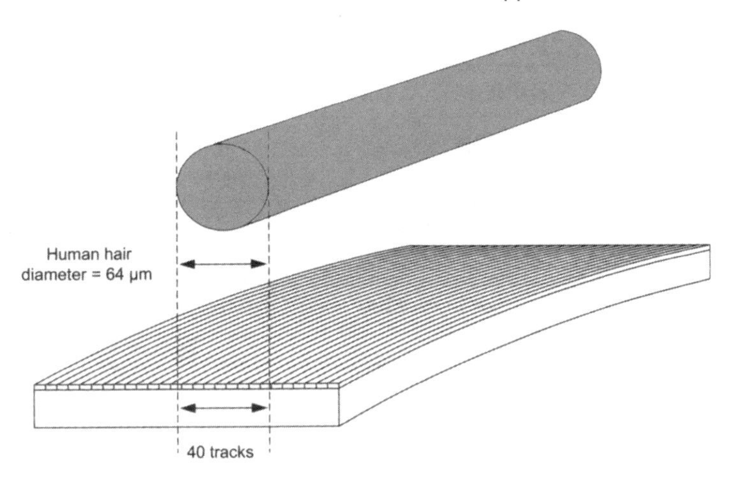

Figure 4.44 A human hair compared with the tracks on a CD.

computing and hi-fi sectors. Figure 4.44 shows the comparison between the size of the 'tracks' on a CD with the average size of a human hair.

As with most 'new' products, the technology used in CDs (and in compact disc players and recorders) relies on several other technologies working together. In the case of CDs, the most important of these are:

- Digital audio technology (being able to represent audio signals using a sequence of digital codes).
- Optical technology (being able to produce a precisely focused beam from a laser light source).
- Control system technology (being able to control the motor speed, as well as the position of the optical unit).
- Manufacturing technology (being able to manufacture the equipment used to play the compact disc and produce the compact media itself both reliably and cost-effectively).

Before explaining how CD technology works, it is worth looking at the previous storage technology in which vinyl discs were used for recording analogue signals in the form of 'long-play' (LP) and 'extended-play' (EP) records. Despite its limitations, vinyl disc recording technology survived for nearly four decades (from around 1950 to 1990). However, the problems associated with recording analogue signal variations in a groove pressed into the relatively soft surface of vinyl plastic eventually led to the downfall of LP recording and its replacement by the digital CD in the late 1980s.

A comparison between the performance specifications of a CD system with those of an LP record player reveals a number of important differences as shown in Table 4.3.

Conventional CD-ROMS, like audio CDs, are made up of three basic layers. The main part of the disc consists of an injection moulded polycarbonate *substrate* which incorporates a spiral track of *pits* and *lands* that are used to encode the data that are stored on the disc. Over the *substrate* is a thin aluminium (or gold) reflective layer, that in turn is protected by an outer protective lacquer coating.

Table 4.3 Comparison of CD and LP record player specifications

Specification	Compact disc player	LP record player
Recording technique	Digital	Analogue
Material	Glass	Vinyl
Dynamic range	90 dB typical	70 dB typical
Frequency response	20 Hz to 20 kHz ± 0.5 dB	20 Hz to 20 kHz ± 3 dB
Signal-to-noise ratio	90 dB typical	60 dB typical
Harmonic distortion	Less than 0.01%	Less than 2%
Channel separation	More than 90 dB	Less than 40 dB
Wow and flutter	Better than 1 part in 10^5	Better than 0.1%
Diameter	120 mm	305 mm
Rotational velocity	196 to 568 rpm	33.3 rpm
Playing time (per side)	65 minutes typical	20 minutes typical

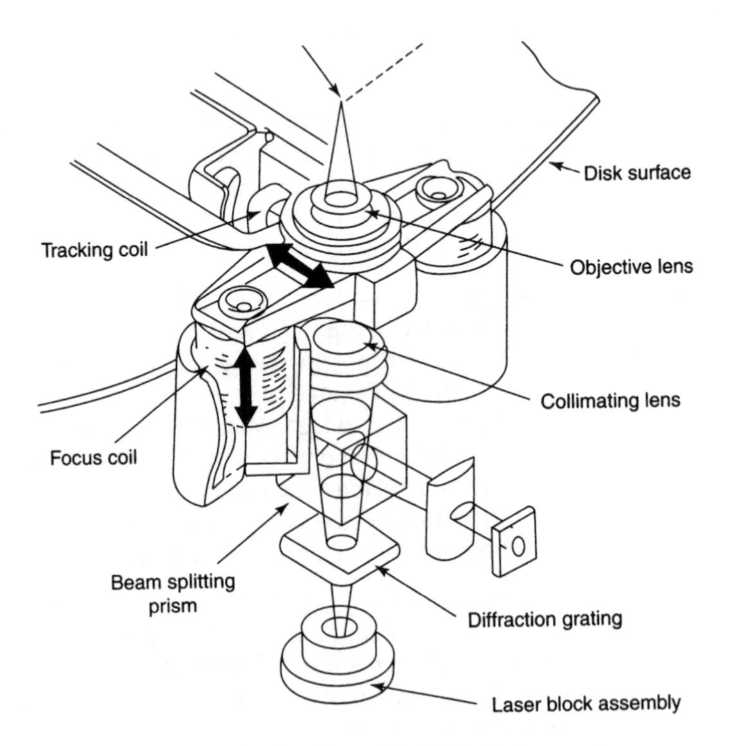

Figure 4.45 Optical assembly fitted to a CD-ROM drive.

Information is retrieved from the CD by focusing a low-power (0.5 mW) infrared (780 nm) laser light beam on to the spiral track of pits and lands in the disc's substrate. The height difference between the pits and the adjacent lands creates a phase shift causing destructive interference in the reflected beam. The effect of the destructive interference and the light scattering is that the intensity of the light returned to a photodiode detector is modulated by the digital data stored on the disc. This modulated signal is then processed, used for tracking, focus and speed control, and then decoded and translated into useable data. The optical system complete with lenses and focus coils, is shown in Fig. 4.45.

Conventional CDs and CD-ROMS only support playback (reading) of the data stored on them. In recent years new technology has appeared that supports both playback (reading) and recording (writing). This technology has resulted in two types of CD: the CD-R (*recordable*) and the CD-E (*erasable*).

Test your knowledge 4.5

1. Name six domestic appliances that depend upon modern electronic technology.
2. Explain, briefly, how the adoption of modern technology in manufacturing has influenced safety and efficiency.
3. Discuss the changes that have occurred in the size and type of the workforce that has occurred as a result of the adoption of modern technology in manufacturing.
4. Discuss:
   (a) The effect on global pollution of the adoption of modern technology in manufacturing.
   (b) The need for sustainability in terms of raw materials and energy in manufacturing.
5. List and discuss the key stages in the manufacturing of a product of your own choosing.
6. Discuss how ICT can assist you in investigating a product of your own choosing.

Key notes 4.3

- The use of CNC machines, PLCs and industrial robots has greatly increased the safety and efficiency of manufacturing by being able to work faster and by removing human errors due to inattention and fatigue. Safety has been improved by removing human operatives from potentially hazardous industrial environments.
- Automation based upon modern technology has resulted in a massive reduction in the size of the manufacturing workforce as more goods can now be produced by fewer people.
- Manufacturing has become a global business as ICT enables firms to communicate quickly and easily with their branches and suppliers throughout the world.
- Automation also means that industries no longer need to be located near traditional pools of skilled labour but can be sited anywhere in the world to take advantage of areas of low-cost labour and materials.
- Pollution is an ever-present problem and manufacturers must strive continually to use less material and energy and to ensure that wherever possible their raw materials and energy come from renewable and sustainable sources. They must also ensure that when no longer required their products can be recycled.
- The key stages of manufacture are: marketing, design, production planning, material sourcing and supply, production processing, assembly and finishing, packaging and despatch.
- MRP 1 refers to material requirement planning.
- MRP 2 refers to manufacturing requirement planning. These are computer software packages used in the management of manufacture.

Assessment activities

1. State the meaning of each of the following abbreviations:
   (a) IP
   (b) HTML
   (c) URL
   (d) ISP
2. (a) Explain why sites on the WWW must all have a unique address.

(b) Name two Web browsers and explain how they are used to locate and view a Web site.
(c) Explain what is meant by a 'search engine'.

3. List the main advantages of preparing engineering drawings using a CAD system.
4. State the meaning of each of the following abbreviations:
(a) CPU
(b) RAM
(c) ROM
(d) LSI
(e) VLSI
(f) I/O
5. With the aid of a sketch, describe the essential features of a microprocessor system.
6. (a) With the aid of a sketch describe a typical DBMS.
(b) With the aid of a sketch describe the structure of a simple database.
7. Name three 'smart' fabrics and explain what they would be used for.
8. Explain what is meant by a 'memory effect' material. Name one example and application.
9. Explain the essential differences between a CNC machine tool and a PLC controlled system.
10. Explain why it is important for manufacturers to make use of sustainable sources of materials and energy, and why it is important to avoid pollution.
(a) State what is meant by the term CD and state what a CD can be used for.
(b) Describe how data are stored on a CD and how those data can be read.
11. (a) List the key stages in manufacturing a product of your choice.
(b) Explain how the cost structure of a manufactured product is built up.
12. Investigate two different makes of the same product and tabulate their advantages and limitations. Use this analysis to arrive at a considered opinion as to which product offers the best value giving your reasons for your choice.

Index

HAVERING COLLEGE OF F & H E
188231